Lecture Note
Mathematics

W9-CLX-008

Edited by A. Dold and B. Eckmann

Subseries: Fondazione C.I.M.E., Firenze
Adviser: Roberto Conti

1092

Complete Intersections

Lectures given at the 1st 1983 Session of the Centro Internationale Matematico Estivo (C.I.M.E.) held at Acireale (Catania), Italy, June 13–21, 1983

Edited by S. Greco and R. Strano

Springer-Verlag
Berlin Heidelberg New York Tokyo 1984

Editors

Silvio Greco
Dipartimento di Matematica, Politecnico di Torino
10129 Torino, Italy

Rosario Strano
Seminario Matematico, Università di Catania
Viale A. Doria 6, 95125 Catania, Italy

AMS Subject Classification (1980): 13H10, 14M10; 14F05, 14H45, 14L30, 14M05

ISBN 3-540-13884-6 Springer-Verlag Berlin Heidelberg New York Tokyo
ISBN 0-387-13884-6 Springer-Verlag New York Heidelberg Berlin Tokyo

Library of Congress Cataloging in Publication Data. Main entry under title: Complete intersections, lectures given at the 1st 1983 session of the Centro Internationale Matematico Estivo (C.I.M.E.) held at Acireale (Catania), Italy, June 13–21, 1983. (Lecture notes in mathematics; 1092) 1. Intersection theory–Addresses, essays, lectures. 2. Algebraic varieties–Addresses, essays, lectures. 3. Local rings–Addresses, essays, lectures. I. Greco, S. (Silvio), 1941-. II. Strano, R. (Rosario), 1944-. III. Centro internazionale matematico estivo. IV. Series: Lecture notes in mathematics (Springer-Verlag); 1092.
QA3.L28 no. 1092 510 s [516.3'53] 84-23570 [QA 564]
ISBN 0-387-13884-6

This work is subject to copyright. All rights are reserved, whether the whole or part of the material is concerned, specifically those of translation, reprinting, re-use of illustrations, broadcasting, reproduction by photocopying machine or similar means, and storage in data banks. Under § 54 of the German Copyright Law where copies are made for other than private use, a fee is payable to "Verwertungsgesellschaft Wort", Munich.

© by Springer-Verlag Berlin Heidelberg 1984
Printed in Germany

Printing and binding: Beltz Offsetdruck, Hemsbach / Bergstr.
2146/3140-543210

Math Sci
Sep

INTRODUCTION

This volume contains the proceedings of the CIME session on Complete Intersections held in Acireale (Catania, Italy) during the period 13-21 June, 1983.

The aim of the session was to present some ideas and techniques from Commutative Algebra, Algebraic Geometry and Analytic Geometry in connection with some problems on Complete Intersections.

The main courses were delivered by O. Forster, R. Lazarsfeld, L. Robbiano and G. Valla. The material developed in the lectures by Forster, Robbiano and Valla has been reshaped by the lecturers for these proceedings; the subject of Lazarsfeld's course is available elsewhere, hence the paper by Lazarsfeld included here contains some further developments and related topics, along with references for the lectures.

The volume contains also a number of original papers, chosen among the ones submitted for the proceedings.

Some of the results were announced during the meeting in special lectures delivered by C. Ciliberto, R. Froberg, S.Kleiman, D. Laksov, P. Valabrega, K. Watanabe.

We wish to thank all the contributors and participants, and the many referees for their collaboration. Our thanks must go also to the CIME for giving us the opportunity to have a meeting on this topic.

Silvio Greco
Rosario Strano

C.I.M.E. Session on "Complete Intersections"

List of Participants

E. AMBROGIO, Strada Comunale Mongreno 361, 10132 Torino

D. AREZZO, Via Sturla 2/5, 16131 Genova

V. BARUCCI, Piazza Sabazio 31, 00199 Roma

G. BECCARI, Politecnico di Torino, Dipartimento di Matematica,
Corso Duca degli Abruzzi 24, 10129 Torino

M. BRUNDU, Istituto Matematico Università, Via L.B. Alberti 4, 16132 Genova

D. CALISTI, Via Imbert 15 (Ognina), 95100 Catania

G. CAMPANELLA, Istituto Matematico Università, Città Universitaria, 00185 Roma

G. CARRA', Via P. Carrera 2, 95123 Catania

M.P. CAVALIERE, Via Pisa 56 A/6, 16146 Genova

L. CHIANTINI, Dipartimento di Matematica, Politecnico di Torino,
Corso Duca degli Abruzzi 24, 10129 Torino

N. CHIARLI, Dipartimento di Matematica, Politecnico di Torino,
Corso Duca degli Abruzzi 24, 10129 Torino

C. CILIBERTO, P.tta Arenella 7/2, 80128 Napoli

M.G. CINQUEGRANI, Via Plaia 220, 95100 Catania

A. COLLINO, Dipartimento di Matematica, Università, Via Principe Amedeo 8,
10123 Torino

A. CONTE, Dipartimento di Matematica,Università, Via Principe Amedeo 8, 10123 Torino

C. CUMINO, Via Pettini 35, 10126 Torino

M. DALE, Universitetet i Bergen, Matematisk Instituut, Allegt. 55, 5014 Bergen

M. D'APRILE, Dipartimento di Matematica, Università della Calabria,
87036 Arcavacata di Rende (Cosenza)

E. DAVIS, Math. Dept., SUNY at Albany, Albany, NY 12203, USA

A. DEL CENTINA, Istituto Matematico Università, Viale Morgagni 67/A, 50134 Firenze

P. DE VITO, Piazza Vanvitelli 15, 80129 Napoli

S. ELIAHOU, Université de Genève, Faculté des Sciences, Section de Mathématiques,
2-4 rue du Lièvre, Case postale 124, 1211 Genève 24, Suisse

M. FIORENTINI, Istituto Matematico, Università di Ferrara, 44100 Ferrara

M. FONTANA, Istituto Matematico Università, Città Universitaria, 00185 Roma

O. FORSTER, Math. Institut, Ludwig-Maxmilians-Universitat, D-8000 Munchen,
Theresienstrasse 39

R. FROBERG, Matematiska Institutionen, Stockholms Universitet, Box 6701, S-11385 Stockholm, Sweden

S. GABELLI, Via Cavalese 25, 00135 Roma

R. GATTAZZO, Istituto di Matematica Applicata, Via Belzoni 7, 35131 Padova

A. GIMIGLIANO, Viale della Repubblica 85, 50019 Sesto Fiorentino (Firenze)

S. GIUFFRIDA, Via Messina 348, 95100 Catania

S. GRECO, Dipartimento di Matematica, Politecnico di Torino, 10129 Torino

C. GREITHER, Mathematisches Institut der Universitat, Theresienstr. 39, 8000 Munchen 2

C. HUNEKE, Math. Dept., Purdue University, W. Lafayette, Ind. 47907, USA

A. JABALLAH, Department of Mathematics, University of Munster, Einsteinstrasse 62, D-4400 Munster, Germany West

S. KLEIMAN, Department of Mathematics, M.I.T., Cambridge, Mass. USA

D. LAKSOV, Univ. of Stockholm, Department of Mathematics, Hagagt. 23, 113 85 Stockholm, Sweden

A. LANTERI, Dipartimento di Matematica, Via C. Saldini 50, 20133 Milano

R. LAZARSFELD, Dept. of Math., Harvard Univ., Science Center, One Oxford Street, Cambridge, Mass. 02138, USA

G. LYUBEZNIK, 6402 23rd Ave., Brooklyn, N.Y. 11204

E.M. LI MARZI, Dipartimento di Matematica, Via C. Battisti 50, 98100 Messina

R. MAGGIONI, Via Leopardi 1, 95030 S. Agata Li Battiati (Catania)

M. MANARESI, Dipartimento di Matematica, Piazza di Porta S. Donato 5, 40127 Bologna

M.G. MARINARI, Via Colombara 4/18, 16016 Cogoleto (Genova)

M.C. MARINO, Via G. Natoli is. 92 n.117, 98100 Messina

P. MAROSCIA, Via Montasio 45, 00141 Roma

C. MARTINENGO, Corso Mazzini 16-8, 17100 Savona

C. MASSAZA, Corso Peschiera 148, 10138 Torino

G. NIESI, Istituto Matematico Università, Via L.B. Alberti 4, 16132 Genova

F. ODETTI, Istituto Matematico Università, Via L.B. Alberti 4, 16132 Genova

P. OLIVERIO, Scuola Normale Superiore, Piazza dei Cavalieri 7, 56100 Pisa

A. ONETO, Via Privata Magnolia 3/1, 16036 Recco (Genova)

M. ORLANDO, Via Carducci 29, 95100 Catania

G. PARIGI, Viale Toscanini 50, 50019 Sesto Fiorentino (Firenze)

G. PAXIA, Facoltà di Ingegneria, Corso Italia 55, 95129 Catania

G. RACITI, Via Vittorio Emanuele 124, 95025 Aci S. Antonio (Catania)

A. RAGUSA, Via Nuovaluce 69, 95030 Catania

L. RAMELLA, Istituto Matematico Università, Via L.B. Alberti 4, 16132 Genova

G. RESTUCCIA, Via Nuova Panoramica dello Stretto, Linea Verde Pal. 23, 98100 Messina

L. ROBBIANO, Istituto Matematico Università, Via L.B. Alberti 4, 16132 Genova

N. RODINO', Via di Vacciano 87, 50015 Grassina (Firenze)

M. ROGGERO, Istituto Matematico Università, Via L.B. Alberti 4, 16132 Genova

G. ROMEO, Via Consolare Pompea 8, 98015 Granzirri (Messina)

P. SALMON, Istituto Matematico Università, Via L.B. Alberti 4, 16132 Genova

N. SANKARAN, Dept. of Math., Panjab University, Chandigarh-14, India

M. SEPPALA, Univ. of Helsinki, Department of Mathematics, Hallituskatu 15,
SF-00100 Helsinki 10, Finland

I. SERGIO, Via Muscatello 28, 95100 Catania

R. STRANO, Seminario Matematico, Università, Viale A. Doria 6, 95125 Catania

N. SUZUKI, 1-2421-21 Sayamagaoka, Tokorozawa 359, Japan

A. SZPIRGLAS, Dept. Informatique, Av. J.B. Clement, 93430 Villetaneuse, France

G. TAMONE, Via P. Negrotto Cambiaso 46-29, 16159 Genova-Rivarolo

G. TEDESCHI, Dipartimento di Matematica, Politecnico di Torino,
Corso Duca degli Abruzzi 24, 10129 Torino

C. TURRINI, Dipartimento di Matematica, Via C. Saldini 50, 20133 Milano

P. VALABREGA, Dipartimento di Matematica, Politecnico di Torino,
Corso Duca degli Abruzzi 24, 10129 Torino

G. VALLA, Istituto Matematico Università, Via L.B. Alberti 4, 16132 Genova

G. VECCHIO, Seminario Matematico dell'Università, Viale A. Doria 6, 95125 Catania

L. VERDI, Istituto Matematico Università, Viale Morgagni 67/A, 50134 Firenze

A. VISTOLI, Via Saffi 18/2, 40131 Bologna

K. WATANABE, Dipartimento di Matematica, Politecnico di Torino,
Corso Duca degli Abruzzi 24, 10129 Torino

P.M.H. WILSON, Dept. of Pure Mathematics, University of Cambridge, 16 Mill Lane,
Cambridge CB2 1SB

TABLE OF CONTENTS

Otto FORSTER - Complete intersections in affine algebraic varieties and Stein spaces 1

Robert LAZARSFELD - Some applications of the Theory of Positive Vector Bundles . 29

Lorenzo ROBBIANO - Factorial and almost factorial schemes in weighted projective spaces 62

Giuseppe VALLA - On set-theoretic complete intersections 85

Haruhisa NAKAJIMA and Kei-ichi WATANABE - The classification of quotient singularities which are complete intersections 102

Ralf FRÖBERG and Dan LAKSOV - Compressed algebras 121

Luca CHIANTINI and Paolo VALABREGA - Some properties of subcanonical curves . . 152

Steven L. KLEIMAN - About the conormal scheme 161

Ciro CILIBERTO and Robert LAZARSFELD - On the uniqueness of certain linear series on some classes of curves 198

Gennady LYUBEZNIK - On the local cohomology module $H^i_{\mathfrak{a}}(R)$ for ideals $\mathfrak{a}$ generated by monomials in an R-sequence 214

Remo GATTAZZO - In characteristic p=2 the Veronese variety $V^m \subset \mathbb{P}^{m(m+3)/2}$ and each of its generic projection is set-theoretic complete intersection 221

Shalom ELIAHOU - Idéaux de définition des courbes monomiales 229

Andrea DEL CENTINA and Alessandro GIMIGLIANO - Curves on rational and elliptic normal cones which are set theoretically complete intersection 241

Edward D. DAVIS and Paolo MAROSCIA - Complete intersections in $\mathbb{P}^2$: Cayley-Bacharach characterizations 253

Maksymilian BORATYNSKI - Poincaré forms, Gorenstein algebras and set theoretic complete intersections. 270

Complete intersections in affine algebraic varieties and Stein spaces

by

Otto Forster

Introduction. Let $(X,\mathcal{O}_X)$ be an affine algebraic variety (or an affine scheme, or a Stein space) and $Y \subset X$ a Zariski-closed (resp. analytic) subspace. We want to describe Y set-theoretically (or ideal-theoretically) by global functions, i.e. find elements $f_1,\dots,f_N \in \Gamma(X,\mathcal{O}_X)$ such that

$$Y = \{x \in X: f_1(x) = \dots = f_N(x) = 0\} ,$$

resp. such that $f_1,\dots,f_N$ generate the ideal of Y (which is a stronger condition). The problem we consider here is how small the number N can be chosen. If in particular N can be chosen equal to the codimension of Y, then Y is called a set-theoretic (resp. ideal-theoretic) complete intersection.

In these lectures we discuss some results with respect to this problem in the algebraic and analytic case. In considering these cases simultaneously, it is interesting to note the analogies and differences of the methods and results. For this purpose we adopt also a more geometric point of view for the algebraic case. We hope that some proofs become more intuitive in this way.

1. Estimation of the number of equations necessary to describe an algebraic (resp. analytic) set

We begin with the following classical result on the set-theoretic description of an algebraic set.

1.1. Theorem (Kronecker 1882). Let $(X,\mathcal{O}_X)$ be an n-dimensional affine algebraic space (or the affine scheme of an n-dimensional noetherian ring) and $Y \subset X$ an algebraic subset. Then there exist functions $f_1,\dots,f_{n+1} \in \Gamma(X,\mathcal{O}_X)$ such that

$$Y = V(f_1,\dots,f_{n+1}) := \{x \in X: f_1(x) = \dots = f_{n+1}(x) = 0\} .$$

Proof (due to Van der Waerden 1941). We prove by induction the following statement

(A.k) There exist $f_1,\dots,f_k \in \Gamma(X,\mathcal{O}_X)$ such that

$$V(f_1,\dots,f_k) = Y \cup Z_k ,$$

where Z_k is an algebraic subset of X with

$$\operatorname{codim} Z_k \geq k.$$

The statement (A.0) is trivial, whereas (A.n+1) gives the theorem. So it remains to prove the induction step

(A.k) → (A.k+1). Let

$$Z_k = Z_k^1 \cup \dots \cup Z_k^s$$

be the decomposition of Z_k into its irreducible components. We may suppose that none of the Z_k^i is contained in Y. Choose a point $p_i \in Z_k^i \setminus Y$ for $i = 1,\dots,s$. Now it is easy to construct a function $f_{k+1} \in \Gamma(X,\mathcal{O}_X)$ with

$$f_{k+1}|Y = 0 \quad \text{and } f_{k+1}(p_i) \neq 0 \text{ for } i = 1,\dots,s.$$

Then $V(f_1,\dots,f_{k+1}) = Y \cup Z_{k+1}$ with

$$\operatorname{codim} Z_{k+1} > \operatorname{codim} Z_k \geq k, \qquad \text{q.e.d.}$$

We want to give an example which shows that in general n equations do not suffice.

1.2. Example. Let $\overline{X}$ be an elliptic curve over $\mathbb{C}$, considered as a torus

$$\overline{X} = \mathbb{C}/\Gamma \ , \quad \Gamma \subset \mathbb{C} \text{ lattice.}$$

Let $p \in \overline{X}$ be an arbitrary point. Then

$$X := \overline{X} \setminus \{p\}$$

is a 1-dimensional affine algebraic variety. Let $Y := \{q\}$ with some $q \in X$. Let $P,Q \in \mathbb{C}$ be representatives of p and q respectively.
Claim. If $P-Q \notin \mathbb{Q}\cdot\Gamma$, then there exists no function $f \in \Gamma(X,\mathcal{O}_{X_{alg}})$ such that

$$Y = \{q\} = V(f).$$

Proof. Such a function f can be considered as a meromorphic function on $\overline{X}$, with poles only in p and zeros only in q. Let $k > 0$ be the vanishing order of f at q. Then k is also the order of the pole of f in p. Thus $k\cdot q - k\cdot p$ would be a principal divisor on $\overline{X}$. By the theorem of Abel, this implies

$$kQ - kP \in \Gamma .$$

But this contradicts our assumption $P - Q \notin \mathbb{Q}\cdot\Gamma$. Hence f cannot exist.

Remark. If we work in the analytic category, i.e. consider X as an open Riemann surface, then there exists a holomorphic function $f \in \Gamma(X,\mathcal{O}_{X_{an}})$ which vanishes precisely in q of order one. This is a special case of the theorem of Weierstraß for open Riemann surfaces, proved by Behnke/Stein 1948, that every divisor on an open Riemann surface is the divisor of a meromorphic function (see e.g.[ ]).

Open Riemann surfaces are special cases of Stein spaces, which are the analogue of affine algebraic varieties in complex analysis. A complex space $(X,\mathcal{O}_X)$ is called a Stein space, if the following conditions are satisfied:

i) X is holomorphically separable, i.e. given two points $x \neq y$ on X, there exists a holomorphic function $f \in \Gamma(X,\mathcal{O}_X)$ such that $f(x) \neq f(y)$.

ii) X is holomorphically convex, i.e. given a sequence $x_1, x_2, \ldots$ of points on X without point of accumulation, there exists $f \in \Gamma(X,\mathcal{O}_X)$ with $\limsup_{k\to\infty} |f(x_k)| = \infty$.

For the general theory of Stein spaces we refer to[ ].

In an n-dimensional Stein space, n equations always suffice to describe an analytic subset:

1.3. Theorem (Forster/Ramspott [12]). Let X be an n-dimensional Stein space and $Y \subset X$ a (closed) analytic subset. Then there exist n holomorphic functions $f_1,\ldots,f_n \in \Gamma(X,\mathcal{O}_X)$ such that

$$Y = V(f_1,\ldots,f_n).$$

Proof. We prove the theorem by induction on n. In order to do so, we have to prove a more precise version, namely, given a cherent ideal sheaf $\mathfrak{I} \subset \mathcal{O}_X$ with $V(\mathfrak{I}) = Y$, we can find functions $f_1,\ldots,f_n \in \Gamma(X,\mathfrak{I})$ such that $Y = V(f_1,\ldots,f_n)$.

n = 1. This is a little generalization of the Weierstraß theorem for open Riemann surfaces. It follows from the fact that for 1-dimensional

Stein spaces (which may have singularities) one has $H^1(X,\mathcal{O}^*_X) = H^1(X,\mathbb{Z}) = 0$.

n-1 → n. First one can find a function $f \in \Gamma(X,\mathfrak{I})$ such that

$$V(f) = Y \cup Z, \quad \text{where} \quad \dim Z \leq n-1.$$

Let $\mathcal{J} \subset \mathcal{O}_Z$ be the image of $\mathfrak{I}$ under the restriction morphism $\mathcal{O}_X \to \mathcal{O}_Z$. Then $\mathcal{J}$ is a coherent ideal sheaf with $V_Z(\mathcal{J}) = Z \cap Y$, and we can apply the induction hypothesis to find $g_1,\dots,g_{n-1} \in \Gamma(Z,\mathcal{J})$ such that

$$Z \cap Y = V_Z(g_1,\dots,g_{n-1}).$$

Since X is Stein, the morphism $\Gamma(X,\mathfrak{I}) \to \Gamma(Z,\mathcal{J})$ is surjective. Let $f_1,\dots,f_{n-1} \in \Gamma(X,\mathfrak{I})$ be functions that are mapped onto $g_1,\dots,g_{n-1}$, then

$$Y = V_X(f_1,\dots,f_{n-1},f), \qquad \text{q.e.d.}$$

As we have seen, in the algebraic case n equations do not suffice in general. However, if one can factor out an affine line from the affine algebraic variety, n equations will suffice.

1.4. Theorem (Storch [33], Eisenbud/Evans[7]). Let X be an affine algebraic space of the form $X = X_1 \times \mathbb{A}^1$, where X_1 is an affine algebraic space of dimension n-1 (or more generally X = Spec R[T] , where R is an (n-1)-dimensional noetherian ring). Then for every algebraic subset $Y \subset X$ there exist n functions $f_1,\dots,f_n \in \Gamma(X,\mathcal{O}_X)$ such that

$$Y = V(f_1,\dots,f_n).$$

In order to carry out the proof, we need a sharper version:
Let there be given an ideal $\mathfrak{a} \subset \Gamma(X,\mathcal{O}_X)$ such that $V(\mathfrak{a}) = Y$. Then the functions $f_1,\dots,f_n$ can be chosen in $\mathfrak{a}$.
However, by the Hilbert Nullstellensatz the rough version of the theorem implies the sharper version.

Proof by induction on n. We may suppose X to be reduced.

n = 1. Then X_1 is a finite set of points, so X is a finite union of affine lines and the assertion is trivial.

Induction step n-1 → n. We have

$$(X,\mathcal{O}_X) = R[T], \quad \text{where} \quad R = \Gamma(X_1,\mathcal{O}_{X_1}).$$

Let S be the set of non-zero divisors of R and

$K = Q(R) = S^{-1}$

the total quotient ring of R. We have

$K = K_1 \times \ldots \times K_r$

Where every K_j is a field. Let $\tilde{\mathfrak{a}} = \mathfrak{a}K[T]$. Since K[T] is a principal ideal ring, there is an $f \in R[T]$ such that $\tilde{\mathfrak{a}} = K[T]f$. Let $\mathfrak{b} = R[T]f$. Then there exists a certain $s \in S$ such that

(*) $\quad \mathfrak{a} \supset \mathfrak{b} \supset s\mathfrak{a}.$

Let $X_2 := V_{X_1}(s)$. Then (*) implies

$Y \subset V(f) \subset Y \cup (X_2 \times \mathbb{A}^1).$

We have $\dim X_2 \leq n-2$. Applying the induction hypothesis to $X_2 \times \mathbb{A}^1$, the algebraic subset $(X_2 \times \mathbb{A}^1) \cap Y$ and the ideal $\mathfrak{a}_2 := \mathrm{Im}(\mathfrak{a} \rightarrow (R/s)[T])$, we get the theorem.

1.5. Corollary. In affine n-space $\mathbb{A}^n$, every algebraic subset is the set of zeros of n polynomials.

Remark. Also in projective n-space $\mathbb{P}^n$ every algebraic set can be described (set-theoretically) by n homogeneous polynomials. This can be proved by methods similar to the affine case, cf. Eisenbud/Evans [7]. For $n = 3$ this had been already proved by Kneser [21].

To conclude this section, we formulate the following

Problem. Find a smooth n-dimensional affine algebraic variety X and a hypersurface $Y \subset X$ that cannot be described set-theoretically by less then $n + 1$ functions.

Example 1.2 is the case $n = 1$. In higher dimensions the problem appears to be much more difficult.

2. Estimation of the number of elements necessary to generate a module over a noetherian ring

Let R be a noetherian ring and M a finitely generated R-module. We want to estimate the minimal number of generators of M over R by a local-global principle. For this purpose, we associate to R the maximal ideal space

$X = \mathrm{Specm}(R),$

indowed with the Zariski topology. Localization of R gives us a sheaf of rings $\mathcal{O}_X$ on X such that

$$R = \Gamma(X, \mathcal{O}_X).$$

To the R-module M there is associated a coherent $\mathcal{O}_X$-module sheaf $\mathcal{M}$ such that

$$M = \Gamma(X, \mathcal{M}).$$

We use the well-known fact: A system of elements $f_1, \ldots, f_m \in M$ generates M over R iff the germs $f_{1x}, \ldots, f_{mx} \in \mathcal{M}_x$ generate $\mathcal{M}_x$ over $\mathcal{O}_{X,x} = R_x$ for every $x \in X = \mathrm{Specm}(R)$.

Let us introduce some further notations:
For $x \in X$ we denote by $\mathfrak{m}_x \subset \mathcal{O}_{X,x}$ the maximal ideal of the local ring $\mathcal{O}_{X,x}$ and by $k(x) := \mathcal{O}_{X,x}/\mathfrak{m}_x$ its residue field. Further let

$$L_x(M) := \mathcal{M}_x/\mathfrak{m}_x\mathcal{M}_x .$$

This is a vector space over k(x). By the Lemma of Nakayama

$$d_x(M) := \dim_{k(x)} L_x(M)$$

is equal to the minimal number of generators of $\mathcal{M}_x$ over $\mathcal{O}_{X,x}$. More precisely:

$\phi_1, \ldots, \phi_m \in \mathcal{M}_x$ generate $\mathcal{M}_x$ over $\mathcal{O}_{X,x}$ iff $\phi_1(x), \ldots, \phi_m(x) \in L_x(M)$ generate $L_x(M)$ over k(x).

Here we denote by $\phi_j(x)$ the image of ϕ_j under the morphism $\mathcal{M}_x \to L_x(M)$. For $f \in M$ we will denote by $f(x) \in L_x(M)$ the image of f under $M \to \mathcal{M}_x \to L_x(M)$.

The module M over R induces a certain stratification of $X = \mathrm{Specm}(R)$, which will be essential for us.

Definition. For $k \in \mathbb{N}$ let

$$X_k(M) := \{x \in X : d_x(M) \geqq k\} .$$

It is easy to prove that $X_k(M)$ is a Zariski-closed subset of X. We have

$$X = X_o(M) \supset X_1(M) \supset \ldots \supset X_r(M) \supset X_{r+1}(M) = \emptyset,$$

where $r := \sup\{\dim_{k(x)} L_x(M) : x \in X\}$. (Since M is finitely generated, $r < \infty$.)

Let us consider some examples:

a) Suppose M is a projective module of rank r over R. Then the associated sheaf $\mathcal{M}$ is locally free of rank r (and defines by definition a vector bundle of rank r over X). We have

$$X = X_o(M) = \ldots = X_r(M) \supset X_{r+1}(M) = \emptyset.$$

b) Let R be a regular noetherian ring and $I \subset R$ a locally complete intersection ideal of height r. If r=1, the ideal I is a projective R-module (example a), so suppose $r \geq 2$. Let

$$X = \mathrm{Specm}(R), \quad Y = V_X(I) = \mathrm{Specm}(R/I)$$

and let $\mathcal{J} \subset \mathcal{O}_X$ be the ideal sheaf associated to I. For $x \in X \setminus Y$, we have $\mathcal{J}_x = \mathcal{O}_{X,x}$, hence $d_x(I) = 1$. For $y \in Y$, the minimal number of generators of $\mathcal{J}_y$ equals r, hence $d_y(I) = r$. This implies

$$X = X_o(I) = X_1(I) \supset X_2(I) = \ldots = X_r(I) \supset X_{r+1}(I) = \emptyset.$$

We visualize the situation by the following picture.

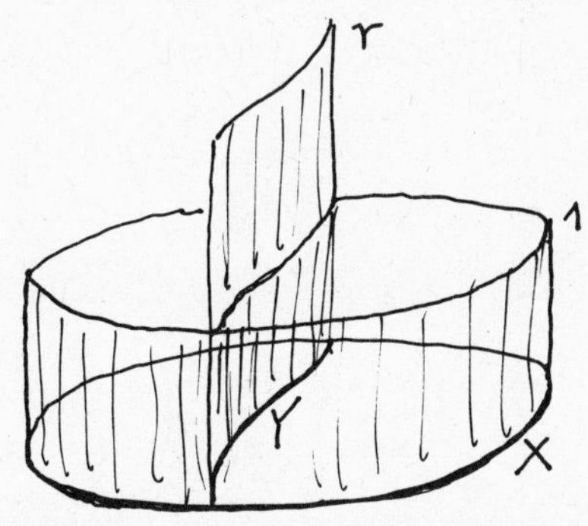

We remark that the topological space X = Specm(R) has a certain combinatorical dimension (finite or infinite). This dimension is less or equal to the dimension of Spec(R), which is the Krull dimension of R. In particular, if $R = k[T_1,\ldots,T_n]$ is a polynomial ring in n indeterminates over a field, dim Specm(R) = dim Spec(R) = n. For a local ring R we have always dim Specm(R)=0.

<u>2.1. Theorem</u> (Forster [9] , Swan [36]). Let R be a noetherian ring and M a finitely generated R-module. Set

$$b(M) := \sup\{k + \dim X_k(M) : k \geq 1, X_k(M) \neq \emptyset\}.$$

Then M can be generated by b(M) elements.

(We set $b(M) = 0$, if $X_1(M) = \emptyset$.)

Proof by induction on b(M). We may suppose $b(M) < \infty$, since otherwise there is nothing to prove.

If $b(M) = 0$, we have $\mathcal{M}_x = 0$ for all $x \in X = \mathrm{Specm}(R)$. This implies $M = 0$, hence M is generated by 0 elements.

Induction step. Let us abbreviate $X_k(M)$ by X_k. We denote by X_k^j the (finitely many) irreducible components of X_k. Let J be the set of all pairs (k,j) such that

$$k \geq 1 \quad \text{and} \quad k + \dim X_k^j = b(M).$$

Then $X_k^j \not\subset X_{k+1}$, since otherwise we would have $(k+1) + \dim X_{k+1} > b(M)$, contradicting the definition of b(M). Choose a point

$$x_{kj} \in X_k^j \setminus X_{k+1}.$$

We have $\dim L_{x_{kj}}(M) = k > 0$ and it is easy to construct an element $f \in M$ such that

$$f(x_{kj}) \neq 0 \quad \text{for all } (k,j) \in J.$$

We consider the quotient module $N := M/Rf$. By the choice of f it follows that

$$\dim L_{x_{kj}}(N) = k - 1 \quad \text{for all } (k,j) \in J,$$

i.e. $x_{kj} \notin X_k(N)$. This implies $k + \dim X_k(N) < b(M)$. By induction hypothesis, N can be generared by $b(M) - 1$ elements, hence M can be generated by b(M) elements.

2.2. Corollary. Let M be a finitely generated projective module of rank r over a noetherian ring R and $n := \dim \mathrm{Specm}(R)$. Then M can be generated by $n + r$ elements.

2.3. Corollary. Let R be a regular noetherian ring and I be a locally complete intersection ideal of height r. Set

$$n := \dim \mathrm{Specm}(R), \quad k := \dim \mathrm{Specm}(R/I).$$

Then I can be generated by $b(I) = \max(n+1, k+r)$ elements

Remark. Let $N = \dim \operatorname{Spec}(R)$ be the Krull dimension of R. Then $n \leq N$ and $k \leq N-r$, so I can always be generated by N+1 elements.

We now consider the problem whether the given estimate is best possible. The answer is yes, if we make no further restrictions on M and R. In order to construct counter-examples, we need some tools from topology.

Topological vector bundles on CW-complexes

Let X be an n-dimensional CW-complex and E be a real vector bundle of rank r over X. We consider the ring $R := \mathcal{C}(X)$ of all (real-valued) continuous functions on X and the vector space $\Gamma_{cont}(X,E)$ of all continuous sections of E. In a natural way, M is an R-module. Suppose M is generated by m elements over R. Then we have a module epimorphism $R^m \to M \to 0$. This corresponds to a vector bundle epimorphism

$$\theta^m \xrightarrow{\beta} E \longrightarrow 0,$$

where θ^m denotes the trivial vector bundle of rank m over X. The kernel of β is a vector bundle F of rank m-r over X. The sequence

$$0 \longrightarrow F \longrightarrow \theta^m \longrightarrow E \longrightarrow 0$$

splits (use a partition of unity), so we get $\theta^m \cong E \oplus F$. We have proved:

If the module $\Gamma_{cont}(X,E)$ can be generated by m elements over the ring $\mathcal{C}(X)$, then there exists a vector bundle F of rank m-r, such that $E \oplus F \cong \theta^m$.

It is easy to see that also the converse implication holds.

Stiefel-Whitney classes

To every real vector bundle E of rank r there are associated Stiefel-Whitney classes

$$\gamma_i(E) \in H^i(X, \mathbb{Z}_2), \qquad (\mathbb{Z}_2 := \mathbb{Z}/2\mathbb{Z}).$$

We have $\gamma_0(E) = 1$ and $\gamma_i(E) = 0$ for $i > r$. It is convenient to consider the total Stiefel-Whitney class

$$\gamma(E) = 1 + \gamma_1(E) + \ldots + \gamma_r(E) \in H^*(X, \mathbb{Z}_2) = \bigoplus_{i \geq 0} H^i(X, \mathbb{Z}_2)$$

in the (commutative) cOhomology ring $H^*(X, \mathbb{Z}_2)$. We will need the following properties of the Stiefel-Whitney classes (for more information see e.g. Husemoller [20]):

a) $\gamma(\theta^m) = 1$ for the trivial vector bundle θ^m.

b) If E,F are two vector bundles, then

$$\gamma(E \oplus F) = \gamma(E)\gamma(F).$$

In particular, if $E \oplus F \cong \theta^m$, then

$$\gamma(F) = \gamma(F)^{-1}.$$

We remark that every element of the form $1+\xi_1+\ldots+\xi_r$, $\xi_i \in H^i(X, \mathbb{Z}_2)$, is invertible in $H^*(X, \mathbb{Z}_2)$.

Example. Consider the real projective n-space $X := \mathbb{P}^n(\mathbb{R})$ as a topological space and let E be the line bundle on X corresponding to a hyperplane section. The cohomology ring of X is

$$H^*(X, \mathbb{Z}_2) \cong \mathbb{Z}_2[t]/(t^{n+1}),$$

i.e. $H^i(X, \mathbb{Z}_2) \cong \mathbb{Z}_2$ for $i = 0,\ldots,r$, and $\xi^i := t^i \mathrm{mod}(t^{n+1})$ is the non-zero element of $H^i(X, \mathbb{Z}_2)$. It is well-known that

$$\gamma(E) = 1 + \xi .$$

Suppose now that F is a vector bundle of rank m-1 such that $E \oplus F \cong \theta^m$. Then

$$\gamma(F) = (1+\xi)^{-1} = 1 + \xi +\ldots+ \xi^n.$$

From this follows $n \geqq \operatorname{rank} F = m-1$. Therefore we have proved:

The module $\Gamma_{cont}(X,E)$ of sections of E cannot be generated by less than n+1 elements over $\mathcal{C}(X)$.

However this example gives not yet an answer to our original problem, since the ring $\mathcal{C}(X)$ is not noetherian. But one can modify this example to construct an n-dimensional noetherian ring A and a projective A-module M of rank 1 such that the minimal number of generators is n+1. For this purpose we represent projective n-space as the quotient of the n-sphere

$$S^n := \{x \in \mathbb{R}^{n+1} : x_o^2 + \ldots + x_n^2 = 1\}$$

by identifying antipodal points:

$$\mathbb{P}^n(\mathbb{R}) = S^n/\sim .$$

Let $\mathbb{R}[x_o,\ldots,x_n]^{ev}$ be the ring of all even polynomials in $x_o,\ldots,x_n$, i.e. polynomials f satisfying $f(x) = f(-x)$. We define

$$A := \mathbb{R}[x_o,\ldots,x_n]^{ev}/(x_o^2+\ldots+x_n^2-1).$$

It is clear that the elements of A can be considered as continuous functions on $\mathbb{P}^n(\mathbb{R})$. The ring A is noetherian and has dimension n. The hyperplane section $x_o=0$ corresponds to the ideal $M \subset A$ generated by the classes of

$$x_o^2,\ x_o x_1,\ \ldots\ ,\ x_o x_n\ .$$

M is a projective A-module of rank 1. If M is generated by m elements, we have an epimorphism $A^m \to M \to 0$. This leads to an epimorphism of vector bundles on $\mathbb{P}^n(\mathbb{R})$

$$\theta^m \longrightarrow E \longrightarrow 0\ ,$$

where E is the line bundle corresponding to a hyperplane section. As we have proved above, $m \geq n+1$. Thus we have got the desired example.

This example illustrates also a general theorem of Lønsted [ ],which says the following: Let X be a finite n-dimensional CW-complex. Then there exists an n-dimensional noetherian ring A and a natural bijective correspondence between the isomorphism classes of real vector bundles over X and finitely generated projective modules over A.

3. Estimation of the number of global generators of a coherent sheaf on a Stein space

There exists a simple analogue of Theorem 2.1 on Stein spaces, cf.[10]. But in the analytic case the estimates can be made much better.

Let $\mathcal{M}$ be a coherent sheaf on a complex space $(X,\mathcal{O}_X)$. For $k \in \mathbb{N}$ we set

$$X_k(\mathcal{M}) := \{x \in X: \dim_{\mathbb{C}} L_x(\mathcal{M}) \geq k\}\ ,$$

where $L_x(\mathcal{M}) := \mathcal{M}_x/\mathfrak{m}_x\mathcal{M}_x$. The $X_k(\mathcal{M})$ are analytic subsets of X.

3.1. Theorem. Let X be a Stein space and $\mathcal{M}$ a coherent analytic sheaf on X. Set

$$\tilde{b}(\mathcal{M}) := \sup\ \{k + [\tfrac{1}{2}\dim X_k(\mathcal{M})] : k \geq 1,\ X_k(\mathcal{M}) \neq \emptyset\}\ .$$

Then the module $\Gamma(X,\mathcal{M})$ of global sections of $\mathcal{M}$ can be generated by $\tilde{b}(\mathcal{M})$ elements over the ring $\Gamma(X,\mathcal{O}_X)$.

(For $a \in \mathbb{R}$ the symbol $[a]$ denotes the greatest integer $p \leq a$.)

Before we come to the proof, we give some corollaries.

<u>3.2. Corollary.</u> Let X be an n-dimensional Stein space and E a holomorphic vector bundle of rank r on X. Then the module $\Gamma(X,E)$ of holomorphic sections can be generated by $r + [n/2]$ elements over $\Gamma(X,\mathcal{O}_X)$.

<u>In particular:</u> Over a 1-dimensional Stein space every holomorphic vector bundle is trivial.

<u>3.3. Corollary.</u> Let X be a pure n-dimensional Stein space, $n \geq 3$, and $Y \subset X$ a curve (not necessarily reduced), which is a locally complete intersection. Then Y is a global ideal-theoretic complete intersection.

<u>Proof.</u> Let $\mathcal{J} \subset \mathcal{O}_X$ be the ideal sheaf of Y. We have

$$X_1(\mathcal{J}) = X, \quad X_2(\mathcal{J}) = \ldots = X_{n-1}(\mathcal{J}) = Y, \quad X_n(\mathcal{J}) = \emptyset.$$

Therefore

$$\tilde{b}(\mathcal{J}) = \max\,\{1+[n/2],\ n-1+[1/2] = n-1\}$$

since $n \geq 3$. By Theorem 3.1, $\Gamma(X,\mathcal{J})$ can be generated by n-1 elements, so Y is a complete intersection.

<u>Remark.</u> The Corollary 3.3 is not valid in 2-dimensional Stein manifolds. For example in $X = \mathbb{C}^* \times \mathbb{C}^*$ there exist divisors which are not principal. Of course, in $\mathbb{C}^2$ every curve is a complete intersection.

<u>Proof of Theorem 3.1.</u> We have to consider only the case $\tilde{b}(\mathcal{M}) < \infty$. From this follows $\dim \operatorname{Supp}(\mathcal{M}) < \infty$. So we may suppose $\dim X < \infty$. Since our hypothesis implies that the minimal number of generators of $\mathcal{M}_x$ over $\mathcal{O}_{X,x}$ is bounded for $x \in X$, it is relatively easy to see that there exist finitely many elements $f_1,\ldots,f_N \in \Gamma(X,\mathcal{M})$ generating this module over $\Gamma(X,\mathcal{O}_X)$.

Let $m \in \mathbb{N}$. In order to find a system of generators $g_1,\ldots,g_m \in \Gamma(X,\mathcal{M})$ consisting of m elements we make the following ansatz: Take a holomorphic $m\times N$-matrix

$$A = (a_{ij}) \in M(m\times N,\Gamma(X,\mathcal{O}_X))$$

and define

$$g_i = \sum_{j=1}^{N} a_{ij} f_j \in \Gamma(X,\mathcal{M}) \quad \text{for } i = 1,\dots,m,$$

or in matrix notation

$$g = Af , \qquad \text{where} \quad f = \begin{pmatrix} f_1 \\ \vdots \\ f_N \end{pmatrix}, \quad g = \begin{pmatrix} g_1 \\ \vdots \\ g_m \end{pmatrix}.$$

We now study the problem what conditions the matrix A has to satisfy so that $g = (g_1,\dots,g_m)$ becomes again a system of generators. For this purpose we define for $x \in X$ the set of matrices

$$(*) \qquad E(x) := \{S \in M(m\times N,\mathbb{C}): \operatorname{rank}(Sf(x)) = d_x(\mathcal{M})\} .$$

Here $f_j(x)$ is the image of f_j in

$$L_x(\mathcal{M}) = \mathcal{M}_x/\mathfrak{m}_x\mathcal{M}_x \text{ and } d_x(\mathcal{M}) = \dim L_x(\mathcal{M}).$$

Note that, since $f_1,\dots,f_N$ generate $\Gamma(X,\mathcal{M})$, the elements $f_1(x),\dots,$ $f_N(x)$ generate $L_x(\mathcal{M})$, hence $\operatorname{rank}(f(x)) = d_x(\mathcal{M})$ for every $x \in X$.

Claim. $g = Af$ is a system of generators of $\Gamma(X,\mathcal{M})$ iff

$$A(x) \in E(x) \quad \text{for every } x \in X.$$

The necessity is clear. Suppose conversely, that $A(x) \in E(x)$ for all $x \in X$. Since $g(x) = A(x)f(x)$, we get $\operatorname{rank}(g(x)) = d_x(\mathcal{M})$, hence by the Lemma of Nakayama the germs $g_{1x},\dots,g_{mx}$ generate $\mathcal{M}_x$ over $\mathcal{O}_{X,x}$. Since X is Stein, this implies that $g_1,\dots,g_m$ generate $\Gamma(X,\mathcal{M})$ over $\Gamma(X,\mathcal{O}_X)$.

We can reformulate the condition as follows. Define the following subset of the trivial bundle $X \times M(m\times N,\mathbb{C})$:

$$E(\mathcal{M},f,m) := \{(x,S) \in X \times M(m\times N,\mathbb{C}): S \in E(x)\} ,$$

where $E(x)$ is defined by (*). It is easy to see that $E(\mathcal{M},f,m)$ is an open subset of $X \times M(m\times N,\mathbb{C})$. We have a natural projection

$$p: E(\mathcal{M},f,m) \longrightarrow X$$

and $p^{-1}(x) = \{x\} \times E(x) \cong E(x)$. We call $E(\mathcal{M},f,m)$ the endromis bundle of $\mathcal{M}$ with respect to the system of generators $f = (f_1,\dots,f_N)$ and the natural number m. Note however, that in general this is not a locally trivial bundle. We have proved:

3.2. Proposition. The module $\Gamma(X,\mathcal{M})$ can be generated by m elements over $\Gamma(X,\mathcal{O}_X)$ iff the endromis bundle $E(\mathcal{M},f,m) \to X$ admits a holomorphic section.

The essential tool is now an Oka principle for endromis bundles, which allows to reduce the problem to a topological problem.

3.3. Theorem (Forster/Ramspott [14]). The endromis bundle $E(\mathcal{M},f,m) \to X$ admits a holomorphic section iff it admits a continuous section.

We cannot give a proof here, but refer to[13],[14]. It is a generalization of the Oka principle proved by Grauert [17].

It is now necessary to study some topological properties of the endromis bundle.

3.4. Proposition. For $x \in X_k(\mathcal{M}) \setminus X_{k+1}(\mathcal{M})$, the topological space $E(x)$ is homeomorphic to $W_{km} \times \mathbb{R}^t$, where W_{km} is the Stiefel manifold of orthonormal k-frames in $\mathbb{C}^m$.

Proof. By definition we have for $x \in X_k(\mathcal{M}) \setminus X_{k+1}(\mathcal{M})$

$$E(x) = \{S \in M(m\times N,\mathbb{C}) : \operatorname{rank}(SF) = k\},$$

where F is a certain fixed $N\times k$-matrix of rank k. After a change of coordinates we may assume $F = \binom{1_k}{0}$, where 1_k is the unit $k\times k$-matrix and 0 denotes the zero $(N-k)\times k$-matrix. If we decompose $S = (S_1,S_2)$ with $S_1 \in M(m\times k,\mathbb{C})$, $S_2 \in M(m\times(N-k),\mathbb{C})$, then $SF = S_1$. Therefore $E(x)$ is homeomorphic to $W'_{km} \times M(m\times(N-k),\mathbb{C})$, where W'_{km} is the space of all $m\times k$-matrices of rank k. But W'_{km} is up to a factor $\mathbb{R}^s$ homeomorphic to the Stiefel manifold W_{km}.

More precisely one can prove:

3.5. Proposition. $E(\mathcal{M},f,m) | X_k(\mathcal{M}) \setminus X_{k+1}(\mathcal{M})$ is a locally trivial fibre bundle with fibre homeomorphic to $W_{km} \times \mathbb{R}^t$.

To be able to apply topological obstruction theory to the endromis bundle, we have to know some homotopy groups of the Stiefel manifolds.

3.6. Proposition. $\pi_q(W_{km}) = 0$ for all $q \leq 2(m-k)$.

Proof by induction on k.

k=1. The Stiefel manifold W_{1m} is nothing else than the (2m-1)-sphere S^{2m-1}, hence $\pi_q(W_{1m}) = 0$ for $q \leq 2(m-1)$.

k-1 → k. By associating to a k-frame its first vector, we get a fibering

$$W_{k-1,m-1} \rightarrowtail W_{km} \twoheadrightarrow S^{2m-1},$$

hence an exact homotopy sequence

$$\ldots \longrightarrow \pi_{q+1}(S^{2m-1}) \longrightarrow \pi_q(W_{k-1,m-1}) \longrightarrow \pi_q(W_{km}) \longrightarrow \pi_q(S^{2m-1}).$$

For $q < 2m-1$ we have therefore isomorphisms $\pi_q(W_{k-1,m-1}) \cong \pi_q(W_{km})$. By induction hypothesis the assertion follows.

We will apply the following theorem of obstruction theory for fibre bundles (cf. Steenrod [32]):

3.7. Theorem. Let X be a CW-complex, Y a subcomplex and $E \to X$ a locally trivial fibre bundle with typical fibre F and connected structure group. Let $s: Y \to E$ be a section of E over Y. If

$$H^{q-1}(X,Y;\pi_q(F)) = 0 \text{ for all } q \geq 1,$$

then there exists a global section $\bar{s}: X \to E$ with $\bar{s}|Y = s$.

This theorem can in particular be applied to complex spaces with countable topology since these spaces can be triangulated (Giesecke [15], Łojasiewicz [23]). Note that every connected component of a Stein space has countable topolgy (Grauert [16]).

Theorem 3.1 will now be a consequence of the following proposition.

3.8. Proposition. If $m \geq \tilde{b}(\mathcal{M})$, then the endromis bundle $E(\mathcal{M},f,m) \to X$ admits a continuous section.

Proof. Let $r = \sup_{x \in X} \dim L_x(\mathcal{M})$ and write X_k for $X_k(\mathcal{M})$. We have

$$X = X_o \supset X_1 \supset \ldots \supset X_r \supset X_{r+1} = \emptyset.$$

We construct a section $s_k: X_k \to E(\mathcal{M},f,m)$ by descending induction on k.

k = r. $E(\mathcal{M},f,m)|X_r$ is a locally trivial fibre bundle with fibre homotopically equivalent to W_{rm}. The obstructions to finding a section lie in

$$H^{q+1}(X_r,\pi_q(W_{rm})), \quad q \geq 1.$$

By Proposition 3.6 we have only to consider the case $q \geq 2(m-r) + 1$. Since $m \geq \tilde{b}(\mathcal{M})$, we have in particular

$$r + [\tfrac{1}{2} \dim X_r] \leq m,$$

hence $\dim X_r \leq 2(m-1) + 1$. But for an arbitrary Stein space Z and an arbitrary abelian group G we have

$$H^{q+1}(Z,G) = 0 \quad \text{for all } q \geq \dim Z$$

(Theorem of Andreotti-Frankel [1], Hamm [19]). Thus $H^{q+1}(X_r,\pi_q(W_{rm}))$ $= 0$ for all $q \geq 1$ and the section $s_r\colon X_r \to E(\mathcal{M},f,m)$ can be constructed.

<u>k+1 → k.</u> From Proposition 3.6 and the theorem of Andreotti-Frankel-Hamm we conclude again that

$$H^{q+1}(X_k,X_{k+1};\pi_q(W_{km})) = 0 \quad \text{for all } q \geq 1.$$

This will allow us to extend the section $s_{k+1}\colon X_{k+1} \to E(\mathcal{M},f,m)$ over X_k. However, we cannot apply Theorem 3.7 directly, since $E(\mathcal{M},f,m)$ is not a locally trivial fibre bundle. So we proceed as follows: We first extend the section s_{k+1} to a section $\tilde{s}$ over a small neighborhood T of X_{k+1} in X_k . We can choose T in such a way that X_{k+1} is a deformation retract of T. Over $X_k \setminus X_{k+1}$ the endromis bundle is locally trivial and we can apply Theorem 3.7 to extend the section $\tilde{s}|T\setminus X_{k+1}$ over all of $X_k \setminus X_{k+1}$. This is possible, since the relative cohomology of the pair $(X_k \setminus X_{k+1}, T\setminus X_{k+1})$ is the same as of the pair (X_k,X_{k+1}).

The technique of endromis bundles also allows to prove the following theorem.

<u>3.9. Theorem</u> [14]. Let $Y \subset \mathbb{C}^n$ be a pure m-dimensional locally complete intersection with $m \leq \frac{2}{3}(n-1)$. Then Y is a (global, ideal-theoretic) complete intersection if and only if the conormal bundle of Y is trivial.

<u>Proof.</u> Let $\mathfrak{J} \subset \mathcal{O}_{\mathbb{C}^n}$ be the ideal sheaf of Y. The conormal bundle of Y is given by $\mathfrak{J}/\mathfrak{J}^2$, which is a locally free sheaf of rank $r = n-m$ over $\mathcal{O}_Y = \mathcal{O}_{\mathbb{C}^n}/\mathfrak{J}$. If Y is a complete intersection, $\Gamma(\mathbb{C}^n,\mathfrak{J})$ is generated by r elements over $\Gamma(\mathbb{C}^n,\mathcal{O}_{\mathbb{C}^n})$. Then also $\Gamma(Y,\mathfrak{J}/\mathfrak{J}^2)$ is generated by r elements over $\Gamma(Y,\mathcal{O}_Y)$, hence $\mathfrak{J}/\mathfrak{J}^2$ is free, i.e. the conormal bundle of Y is trivial.

Conversely, suppose that the conormal bundle is trivial. Then there exist functions $f_1,\dots,f_r \in \Gamma(\mathbb{C}^n,\mathfrak{J})$, whose classes modulo $\mathfrak{J}^2$ generate $\Gamma(Y,\mathfrak{J}/\mathfrak{J}^2)$. Therefore the germs $f_{1x},\dots,f_{rx}$ generate $\mathfrak{J}_x$ for all x in some neighborhood of Y.

Consider now the endromis bundle $E = E(\mathfrak{J},g,r) \to \mathbb{C}^n$ for some system of generators $g = (g_1,\dots,g_N)$ of $\Gamma(\mathbb{C}^n,\mathfrak{J})$. The functions $f_1,\dots,f_r$ give rise to a section s of E over some neighborhood of Y. We have to extend this section continuously over $\mathbb{C}^n$. As in the proof of Proposition 3.8, the obstructions to this extension lie in $H^{q+1}(\mathbb{C}^n,Y;\pi_q(W_{1r}))$. The Hypothesis $m \leq \frac{2}{3}(n-1)$ implies $2r-1 > m = \dim Y$, hence by the theorem of Andreotti-Frankel-Hamm the groups

$$H^{q+1}(\mathbb{C}^n,Y;\pi_q(W_{1r})) \cong H^q(Y,\pi_q(S^{2r-1}))$$

vanish, q.e.d.

4. Theorems of Mohan Kumar

In the algebraic case one cannot apply the strong tools of algebraic topology as in the theory of Stein spaces. One has to use other methods. We expose here some results of Mohan Kumar [25], [26].

We begin with a simple proposition.

4.1. Proposition. Let R be a noetherian ring and $I \subset R$ an ideal. If I/I^2 can be generated by m elements over R/I, then I can be generated by m+1 elements over R.

Proof. Let X = Spec(R) be the affine scheme associated to R and Y = $V(I) \subset X$ the subspace defined by I. We denote by $\mathfrak{J} \subset \mathcal{O}_X$ the ideal sheaf associated to I.
Let $f_1,\dots,f_m \in I$ be elements generating I mod I^2. By the Lemma of Nakayama the germs $f_{1x},\dots,f_{mx}$ generate the ideal $\mathfrak{J}_x \subset \mathcal{O}_{X,x}$ for all $x \in Y$ and by coherence this is true even for all x in a certain neighborhood of Y. Therefore

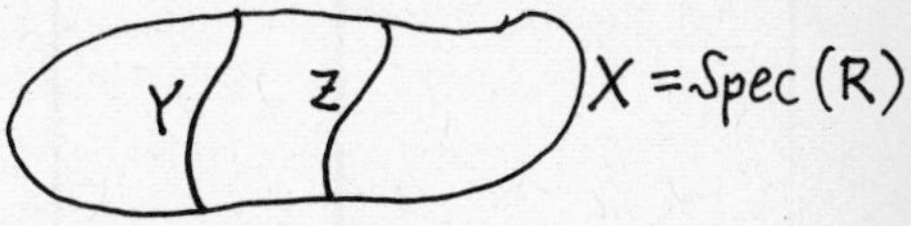

$$V(f_1,\dots,f_m) = Y \cup Z,$$

Y Z X = Spec(R)

where Z X is a closed subset disjoint from Y. In prticular we have that $f_{1x},\dots,f_{mx}$ generate $_x$ for all $x \in X \setminus Z$. Now there exists a function $f_{m+1} \in \Gamma(X,\mathcal{O}_X) = R$ such that $f_{m+1}|Z = 1$ and $f_{m+1}|Y = 0$ (i.e. $f_{m+1} \in I$). Then $f_1,\dots,f_{m+1}$ generate I over R (since this is true locally).

4.2. Theorem (Mohan Kumar [25]). Let Y be a smooth pure m-dimensional algebraic subvariety in affine n-space $\mathbb{A}^n$ (over an algebraically closed field). Suppose $2m+1 < n$. Then Y is a complete intersection (in the ideal theoretic sense) if and only if the normal bundle of Y is trivial.

Remark. This theorem is only a special case of the next theorem. We will prove it here, since the method of proof is interesting for itself.

We have to recall some notions of algebraic K-theory.

Definition. Two vector bundles E,F over an algebraic variety X are called stably isomorphic, if there exist trivial bundles θ^k, θ^l over X such that $E \oplus \theta^k \cong F \oplus \theta^l$. A vector bundle E is called stably trivial, if it is stably isomorphic to a trivial bundle.

One has the following

Cancellation Theorem. Let E and F be stably isomorphic vector bundles of the same (constant) rank r over an n-dimensional affine algebraic variety X. If $r \geq n+1$, then E and F are isomorphic.

More generally, this Cancellation Theorem holds for projective modules over n-dimensional noetherian rings, see e.g. Bass [3].

Remark. The same Cancellation Theorem holds also in the topological category for real vector bundles:

If E,F are two stably isomorphic real vector bundles over an n-dimensional CW-complex X and if $r \geq n+1$, then E and F are topologically isomorphic.
For complex vector bundles, Cancellation is already possible for $r \geq n/2$. The Oka principle for holomorphic vector bundles on Stein spaces then implies the following:

Let E,F be two holomorphic vector bundles of rank r over an n-dimensonal Stein space X. Suppose $r \geq n/2$. If E and F are stably isomorphic, they are analytically isomorphic.

Proof of Theorem 4.2. Since $n > 2 \dim Y + 1$, we can choose coordinates in $\mathbb{A}^n$ such that, denoting by $p: \mathbb{A}^n \to \mathbb{A}^{n-1}$ the projection to the first n-1 coordinates, p maps Y isomorphically onto a smooth algebraic subvariety $Y' \subset \mathbb{A}^{n-1}$.
Over Y we have the exact sequence

$$0 \longrightarrow T_Y \longrightarrow T_{\mathbb{A}^n}|Y \longrightarrow N_{Y/\mathbb{A}^n} \longrightarrow 0,$$

where T stands for the tangent bundle and N for the normal bundle. Since Y is affine, the sequence splits. Thus we have

$$T_Y \oplus N_{Y/\mathbb{A}^n} = \theta^n,$$

where $\theta^n = T_{\mathbb{A}^n}|Y$ is the trivial n-bundle over Y.

Now we suppose that the normal bundle of Y is trivial. This implies that the tangent bundle T_Y is stably trivial. Because $Y \cong Y'$, also the tangent bundle $T_{Y'}$ is stably trivial. From the isomorphism $T_{Y'} \oplus N_{Y'/\mathbb{A}^{n-1}} = \theta^{n-1}$ we conclude then that $N_{Y'/\mathbb{A}^{n-1}}$ is stably trivial. But rank $N_{Y'/\mathbb{A}^{n-1}} = (n-1) - m > m = \dim Y'$, so by the Cancellation Theorem $N_{Y'/\mathbb{A}^{n-1}}$ is in fact trivial. This means that $I_{Y'}/I_{Y'}^2$ is a free module of rank r-1 over $\Gamma(Y',\mathcal{O}_{Y'}) = K[T_1,\dots,T_{n-1}]/I_{Y'}$,where $r = n-m$. Choose polynomials $f_1,\dots,f_{r-1} \in K[T_1,\dots,T_{n-1}]$ which generate $I_{Y'}$ mod $I_{Y'}^2$. We have

$$V_{\mathbb{A}^{n-1}}(f_1,\dots,f_{r-1}) = Y' \cup Z',$$

where $Z' \subset \mathbb{A}^{n-1}$ is an algebraic subset disjoint from Y' (cf. the proof of Proposition 4.1). We can consider the f_j also as elements of $K[T_1,\dots,T_n]$ and have

$$V_{\mathbb{A}^n}(f_1,\dots,f_{r-1}) = (Y'\times \mathbb{A}^1) \cup (Z'\times \mathbb{A}^1).$$

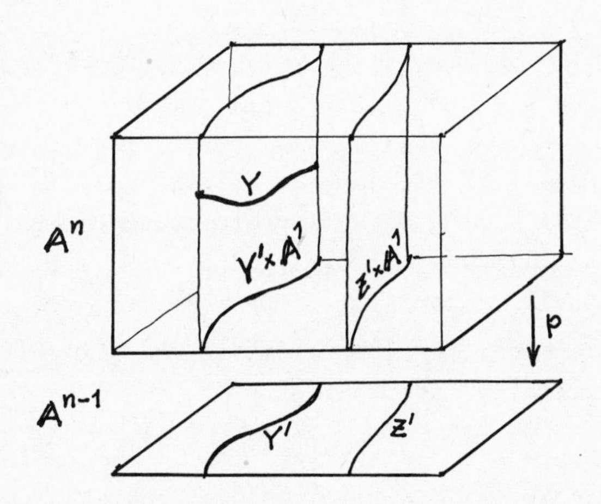

Since $p|Y \to Y'$ is an isomorphism, Y is a graph over Y', hence there exists an element $\varphi \in \Gamma(Y',\mathcal{O}_{Y'})$ such that the ideal of Y in

$$\Gamma(Y'\times \mathbb{A}^1,\mathcal{O}_{Y'\times\mathbb{A}^1}) = \Gamma(Y',\mathcal{O}_{Y'})[T_n]$$

is generated by $\Phi := T_n - \varphi$. Now choose an element $f_r \in K[T_1,\ldots,T_n]$ such that

$$f_r|Y'\times \mathbb{A}^1 = \Phi \quad \text{and} \quad f_r|Z'\times \mathbb{A}^1 = 1.$$

Then $f_1,\ldots,f_r$ generate the ideal of Y in $K[T_1,\ldots,T_n]$, hence Y is a complete intersection.

Problem. Let $Y \subset \mathbb{A}^n$ be a smooth subvariety (or a locally complete intersection) with trivial normal bundle. Can one conclude that Y is a complete intersection without the dimension restriction $2 \dim Y + 1 < n$ of Mohan Kumar's theorem?

In the case codim $Y \leq 2$ this is always true (cf. Sec. 5). The simplest case that remains open are surfaces in $\mathbb{A}^5$.

4.3. Theorem (Mohan Kumar [26]). Let $I \subset K[T_1,\ldots,T_n]$ be an ideal such that I/I^2 is generated by s elements (K arbitrary field). If $s > m+1$, where $m = \dim V(I)$, then also I can be generated by s elements.

Remark. If we take I to be the ideal of a locally complete intersection with trivial normal bundle, we get a generalization of Theorem 4.2 to locally complete intersections.

Proof. We set $Y = V(I) \subset \mathbb{A}^n$.
We first reduce the general case to the case codim $Y \geq 2$. If Y contains components of dimension n-1, then one can write $I = h\cdot J$, where codim $V(J) \geq 2$ and $h \in K[T_1,\ldots,T_n]$ is a generator of the intersection of all primary components of dimension n-1 of I. The ideals I and J have the same number of generators.
So we may suppose $m = \dim Y \leq n-2$. After a change of coordinates we may suppose that I contains a monic polynomial g with respect to T_n. If $p: \mathbb{A}^n \to \mathbb{A}^{n-1}$ denotes the projection to the first n-1 coordinates, then $p|V(g) \to \mathbb{A}^{n-1}$ is proper, in particular $p|Y \to \mathbb{A}^{n-1}$ is proper. Therefore

$$Y' := p(Y) \subset \mathbb{A}^{n-1}$$

is an algebraic subset of dimension $m < n-1$.
By hypothesis, I/I^2 can be generated by s elements. Let $f_1,\ldots,f_s \in I \subset K[T_1,\ldots,T_n]$ be representatives of a system of generators of I/I^2. By adding suitable elements of I^2, we may suppose that

a) f_1 is monic with respect to T_n.

(If this is not the case, add a sufficiently high power of g.)

b) $V(f_1,\dots,f_s) \cap (Y' \times \mathbb{A}^1) = Y$.

(This is possible since $\dim(Y' \times \mathbb{A}^1) = m+1 < s$, by an argument similar to the proof of Theorem 1.1.)

Write $V(f_1,\dots,f_s) = Y \cup Z$, $Z \cap Y = \emptyset$. By condition a), $p|Z \to \mathbb{A}^{n-1}$ is proper, hence $Z' := p(Z)$ is an algebraic subset of $\mathbb{A}^{n-1}$ and by condition b) we have $Y' \cap Z' = \emptyset$. There exist affine open subsets $U',V' \subset \mathbb{A}^{n-1}$ such that

$$Y' \subset U' \subset \mathbb{A}^{n-1} \setminus Z',$$
$$Z' \subset V' \subset \mathbb{A}^{n-1} \setminus Y'$$

and $U' \cup V' = \mathbb{A}^{n-1}$. Let $U := p^{-1}(U')$, $V := p^{-1}(V')$.

Denote by $\mathfrak{J} \subset \mathcal{O}_{\mathbb{A}^n}$ the ideal sheaf associated to I. Since $U \cap Z = \emptyset$, we get an exact sequence

$$(1) \qquad \mathcal{O}^s_{\mathbb{A}^n} \xrightarrow{(f_1,\dots,f_s)} \mathfrak{J} \longrightarrow 0 \quad \text{over } U.$$

Since $V \cap Y = \emptyset$, we have $\mathfrak{J}|V = \mathcal{O}_{\mathbb{A}^n}|V$, hence an exact sequence

$$(2) \qquad \mathcal{O}^s_{\mathbb{A}^n} \xrightarrow{(1,0,\dots,0)} \mathfrak{J} \longrightarrow 0 \quad \text{over } V.$$

We want to patch together these two sequences over $U \cap V$, which is affine algebraic. To do this, we remark that $(f_1,\dots,f_s)|U \cap V$ generates the unit ideal of the ring

$$\Gamma(U\cap V,\mathcal{O}_{\mathbb{A}^n}) = \Gamma(U'\cap V',\mathcal{O}_{\mathbb{A}^{n-1}})[T_n] =: A[T_n]$$

and f_1 is monic with respect to T_n. Therefore by a theorem of Quillen-Suslin [28],[34] $(f_1,\dots,f_s)|U\cap V$ can be completed to an invertible $s\times s$-matrix

$$F \in GL(s,A[T_n]),$$

whose first row is $(f_1,\dots,f_s)|U\cap V$. This matrix defines an isomorphism $F: \mathcal{O}^s_{\mathbb{A}^n} \longrightarrow \mathcal{O}^s_{\mathbb{A}^n}$ over $U \cap V$ and we get a commutative diagram

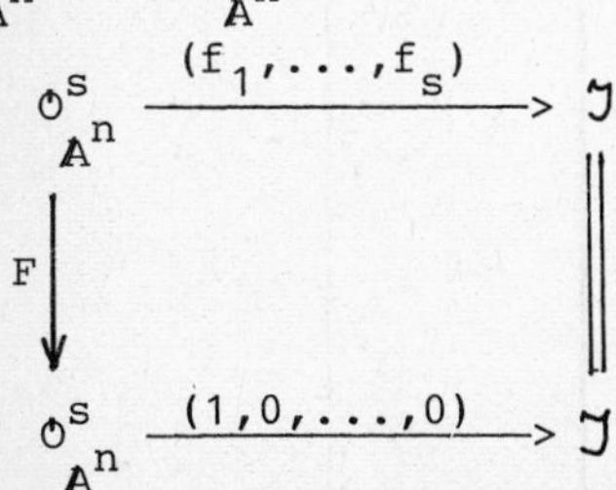

over $U \cap V$.

Let $\mathcal{M}$ be the locally free module sheaf over $\mathbb{A}^n$ obtained by glueing $\mathcal{O}^s_{\mathbb{A}^n}|U$ and $\mathcal{O}^s_{\mathbb{A}^n}|V$ over $U\cap V$ by means of the isomorphism F. The sequences (1) and (2) now patch together to a single exact sequence

$$\mathcal{M} \longrightarrow \mathcal{J} \longrightarrow 0 \quad \text{over } \mathbb{A}^n.$$

Again by Quillen-Suslin's solution of the Serre problem, the sheaf $\mathcal{M}$ is free, i.e. globally isomorphic to $\mathcal{O}^s_{\mathbb{A}^n}$. This means that $I = \Gamma(\mathbb{A}^n, \mathcal{J})$ can be generated by s elements, q.e.d.

<u>4.4. Corollary.</u> Let $Y \subset \mathbb{A}^n$ be a locally complete intersection (not necessarily of pure dimension). Then the ideal I of Y can be generated by n elements.

(For smooth Y this was a conjecture of Forster [9].)

<u>Proof.</u> As in the proof of Theorem 4.3, we may suppose that $m := \dim Y \leq n-2$. Now we can apply Theorem 4.3, if we have proved that I/I^2 can be generated by n elements. But this follows from Theorem 2.1.

Mohan Kumar [26] has also proved by similar techniques a conjecture of Eisenbud-Evans [6], which is a generalization of Corollary 4.4. (It was also proved by Sathaye [29] under some restrictions.) This can be formulated as follows:

Let A be a noetherian ring of finite Krull dimension, $R := A[T]$ and M a finitely generated R-module. Let $X := \mathrm{Spec}(R)$ and define in analogy to Sec. 2

$$X_k(M) = \{x \in X: \dim_{k(x)}(\mathcal{M}_x/\mathfrak{m}_x\mathcal{M}_x) \geq k\} \quad ,$$

where $\mathcal{M}$ is the module sheaf on X associated to M. Let

$$b_k^*(M) := \begin{cases} 0 & \text{if } X_k(M) = \emptyset, \\ k + \dim X_k(M) & \text{if } 0 \leq \dim X_k(M) < \dim X, \\ k + \dim X_k(M) - 1, & \text{if } \dim X_k(M) = \dim X \end{cases}$$

Then M can be generated by

$$b^*(M) := \sup\{b_k^*(M): k \geq 1\}$$

elements.

5. Set-theoretic complete intersections

By the theorem of Mohan Kumar, an m-dimensional locally complete intersection Y with trivial normal bundle in affine n-space is a complete intersection, if $2m + 2 \leq n$. Boratyński [4] has proved that without any restriction on the dimension, Y is at least a set-theoretic complete intersection.

5.1. Theorem (Boratyński). Let $Y \subset \mathbb{A}^n$ (resp. $Y \subset \mathbb{C}^n$) be an algebraic (resp. analytic) locally complete intersection with trivial normal bundle. Then Y is a set-theoretic complete intersection.

Proof. Set $X = \mathbb{A}^n$ (resp. $X = \mathbb{C}^n$). Let $r = \operatorname{codim} Y$ and $f_1,\dots,f_r \in \Gamma(X, \mathcal{J}_Y)$ functions such that the classes $f_j \bmod \mathcal{J}_Y^2$ form a global basis of the conormal bundle $\nu_{Y/X} = \mathcal{J}_Y/\mathcal{J}_Y^2$. Then the zero set of $f_1,\dots,f_r$ can be written as

$$V(f_1,\dots,f_r) = Y \cup Y',$$

where $Y' \subset X$ is an algebraic (analytic) set dijoint from Y. The sets Y, Y' are contained in disjoint hypersurfaces,

$$Y \subset H, \quad Y' \subset H', \quad H \cap H' = \emptyset.$$

The set $U := X \setminus (H \cup H')$ is affine algebraic (resp. Stein). Since $f_1,\dots,f_r$ have no common zeros on U, they generate the unit ideal in the ring $\Gamma(U,\mathcal{O}_X)$. By a theorem of Suslin [35] there exists a matrix $F \in GL(r,\Gamma(U,\mathcal{O}_X))$, whose first row is $(f_1,\dots,f_{r-1},f_r^{(r-1)!})$. Consider the ideal $\mathcal{J}_Z \subset \mathcal{O}_X$ generated by $(f_1,\dots,f_{r-1},f_r^{(r-1)!})$ over $X \setminus H'$ and equal to $\mathcal{O}_X$ over $X \setminus H$. Then $V(\mathcal{J}_Z) = V(\mathcal{J}_Y) = Y$. The vectors

$$\varphi := (f_1,\dots,f_{r-1},f_r^{(r-1)!}), \quad \psi := (1,0,\dots,0)$$

define epimorphisms

$$\varphi\colon \mathcal{O}_X^r \longrightarrow \mathcal{J}_Z \longrightarrow 0 \quad \text{over } X \setminus H',$$

$$\psi\colon \mathcal{O}_X^r \longrightarrow \mathcal{J}_Z \longrightarrow 0 \quad \text{over } X \setminus H.$$

The matrix F defines an isomorphism $F\colon \mathcal{O}_X^r \to \mathcal{O}_X^r$ over $X \setminus (H \cup H')$ such that $\varphi = \psi \circ F$. Therefore, denoting by $\mathcal{M}$ the locally free sheaf on X obtained by glueing $\mathcal{O}_X^r|X\setminus H'$ and $\mathcal{O}_X^r|X\setminus H$ over $X \setminus (H \cup H')$ by means of F, we get an epimorphism

$$\mathcal{M} \longrightarrow \mathcal{J}_Z \longrightarrow 0 \quad \text{over } X.$$

Since $X = \mathbb{A}^n$ (resp. $X = \mathbb{C}^n$), the sheaf $\mathcal{M}$ is globally free of rank r, i.e. Z is a complete intersection. Hence Y is a set-theoretic complete intersection.

Remark. If the codimension $r = 2$, then Y is an ideal-theoretic complete intersection, since in this case $\mathfrak{J}_Z = \mathfrak{J}_Y$.

The Ferrand construction

If the conormal bundle of a locally complete intersection is not trivial, one can try to change the structure Of the subvariety by adding nilpotent elements in order to make the conormal bundle trivial. Such a device has been invented by Ferrand [8] and Szpiro [37] to prove that locally complete intersection curves in $\mathbb{A}^3$ are set-theoretic complete intersections.

Let Y be a locally complete intersection in a complex manifold X (resp. smooth algebraic variety) with conormal bundle $\nu_{Y/X} = \mathfrak{J}_Y/\mathfrak{J}_Y^2$. Suppose there is given a line bundle L on Y and an epimorphism $\beta: \nu_{Y/X} \to L$. Then we can define a new ideal sheaf $\mathfrak{J}_Z \subset \mathcal{O}_X$ with $\mathfrak{J}_Y^2 \subset \mathfrak{J}_Z \subset \mathfrak{J}_Y$ by the exact sequence

$$(1) \qquad 0 \longrightarrow \mathfrak{J}_Z/\mathfrak{J}_Y^2 \longrightarrow \mathfrak{J}_Y/\mathfrak{J}_Y^2 \xrightarrow{\beta} L \longrightarrow 0 .$$

Then $Z = (|Y|, \mathcal{O}_X/\mathfrak{J}_Z)$ is again a locally complete intersection. This can be seen as follows: For $y \in Y$ let $f_1,\dots,f_r \in \mathfrak{J}_{Y,y}$, $r = \operatorname{codim}_y Y$, be a minimal system of generators. Then the classes $[f_j] := f_j \bmod \mathfrak{J}_{Y,y}^2$ form a basis of $(\mathfrak{J}_Y/\mathfrak{J}_Y^2)_y$ over $\mathcal{O}_{Y,y}$. We can choose the f_j in such a way that $[f_1],\dots,[f_{r-1}]$ generate the kernel of β. Then

$$\mathfrak{J}_{Z,y} = (f_1,\dots,f_{r-1}) + \mathfrak{J}_{Y,y}^2 = (f_1,\dots,f_{r-1},f_r^2) .$$

Conormal bundle of Z

Since Z is again a locally complete intersection, $\nu_{Z/X} = \mathfrak{J}_Z/\mathfrak{J}_Z^2$ is a locally free sheaf over $\mathcal{O}_Z = \mathcal{O}_X/\mathfrak{J}_Z$. We consider its analytic restriction to Y,

$$\nu_{Z/X}|Y = (\mathfrak{J}_Z/\mathfrak{J}_Z^2) \otimes (\mathcal{O}_X/\mathfrak{J}_Y) = \mathfrak{J}_Z/\mathfrak{J}_Y\mathfrak{J}_Z ,$$

which fits into an exact sequence

$$(2) \qquad 0 \longrightarrow \mathfrak{J}_Y^2/\mathfrak{J}_Y\mathfrak{J}_Z \longrightarrow \mathfrak{J}_Z/\mathfrak{J}_Y\mathfrak{J}_Z \longrightarrow \mathfrak{J}_Z/\mathfrak{J}_Y^2 \longrightarrow 0.$$

Using the isomorphisms $L \cong \mathfrak{J}_Y/\mathfrak{J}_Z$ and $L^2 \cong \mathfrak{J}_Y^2/\mathfrak{J}_Y\mathfrak{J}_Z$, we can combine the exact sequences (1) and (2) to obtain the exact sequence

$$0 \longrightarrow L^2 \longrightarrow \nu_{Z/X}|Y \longrightarrow \nu_{Y/X} \longrightarrow L \longrightarrow 0 .$$

From this follows in patricular

$$\det(\nu_{Z/X}|Y) \cong \det(\nu_{Y/X}) \otimes L .$$

5.2. Theorem. Let $Y \subset \mathbb{A}^n$ be a curve, which is a locally complete intersection. Then Y is a set-theoretic complete intersection.

This theorem is due to Szpiro for n = 3 (cf. [37]) and to Mohan Kumar [26] for n > 3.

Proof. For $n < 3$ the theorem is trivial, so suppose $n \geq 3$.

Let $\nu_{Y/X}$ be the conormal bundle of Y in $X := \mathbb{A}^n$ and set $L := \det(\nu_{Y/X})^*$. The bundle $E := \nu^*_{Y/X} \otimes L$ has rank $n-1 \geq 2$. Since Y is affine algebraic and 1-dimensional, E admits a section without zeros. This section corresponds to an epimorphism $\beta: \nu_{Y/X} \to L$. Applying the Ferrand construction, we get a new structure Z on $|Y|$ such that

$$\det(\nu_{Z/X}|Y) \cong \det(\nu_{Y/X}) \otimes L \cong \mathcal{O}_Y .$$

A vector bundle on a 1-dimensional affine algebraic space is already determined by its determinant. Therefore $\nu_{Z/X}|Y$ is trivial, hence also $\nu_{Z/X}$ is trivial. The assertion follows by applying Theorem 5.1.

Remark. Cowsik-Nori [5] have proved, that in affine n-space over a field of characteristic $p > 0$ every curve is a set-theoretic complete intersection. But the proof cannot be carried over to characteristic zero.

5.3. Theorem ([27],[2],[30]). Let $Y \subset \mathbb{C}^n$ be an analytic subspace whichis a locally complete intersection of (pure) dimension $m \leq 3$. Then Y is a set-theoretic complete intersection.

Proof. We proceed as in the proof of Theorem 5.2. That in the analytic case Y may have dimension up to 3, is due to the following facts on vector bundles over Stein spaces.

5.4. Proposition. Let E be a holomorphic vector bundle of rank r over an m-dimensional Stein space Y. If $r > m/2$, then E admits a holomorphic section without zeros.

Proof. Let E_0 be the bundle with fibre $\mathbb{C}^r \setminus 0$ obtained by deleting the zero section from E. By the Oka principle, it suffices to con struct a continuous section of E_0. The obstructions lie in

$$H^{q+1}(Y,\pi_q(S^{2r-1})).$$

Since $2r - 1 \geq \dim Y$, these groups vanish by the theorem of Andreotti-Frankel-Hamm.

5.5. Proposition. Let E be a holomorphic vector bundle of rank r over a Stein space Y of dimension $m \leq 3$. If the line bundle det(E) is trivial,then E is trivial itself.

Proof. By multiple application of Proposition 5.4 one gets

$$E \cong L \oplus \theta^{r-1},$$

where L is a line bundle and θ^{r-1} the trivial vector bundle of rank r-1 over X. But then $\det(E) \cong L$. If this is trivial, E must be trivial.

We refer to [2],[30],[31] for more results on (ideal-theoretic and set-theoretic) complete intersections in Stein spaces.

Problem (Murthy). Is every locally complete intersection in $\mathbb{C}^n$ (resp. $\mathbb{A}^n$) a set-theoretic complete intersection?

In order to make substantial progress in the problem of set-theoretic complete intersections it seems necessary to devise new techniques (besides the Ferrand construction) to change the structure of subvarieties and influence their conormal bundle.

References

[1] A. Andreotti, T. Frankel: The Lefschetz theorem on hyperplane sections. Ann. of Math. 69 (1959) 713-717.

[2] C. Bănică, O. Forster: Complete intersections in Stein manifolds. Manuscr. Math. 37 (1982) 343-356.

[3] H. Bass: Algebraic K-theory. Benjamin 1968.

[4] M. Boratyński: A note on set theoretic complete intersections. J. of Algebra 54 (1978) 1-5.

[5] R.C. Cowsik, M.V. Nori: Curves in characteristic p are set theoretic complete intersections. Inv. Math. 45 (1978) 111-114.

[6] D. Eisenbud, E.G. Evans: Three conjectures about modules over polynomial rings. Conf. on Commutative Algebra. Springer Lecture Notes in Math. 311 (1973) 78-89.

[7] D. Eisenbud, E.G. Evans: Every algebraic set in n-space is the intersection of n hypersurfaces. Inv. Math. 19 (1973) 278-305.

[8] D. Ferrand: Courbes gauches et fibrés de rang deux. C.R. Acad. Sci. Paris 281 (1975) 345-347.

[9] O. Forster: Über die Anzahl der Erzeugenden eines Ideals in einem Noetherschen Ring. Math. Zeits. 84 (1964) 80-87.

[10] O. Forster: Zur Theorie der Steinschen Algebren und Moduln. Math. Zeits. 97 (1967) 376-405.

[11] O. Forster: Lectures on Riemann surfaces. Springer 1981.

[12] O. Forster, K.J. Ramspott: Über die Darstellung analytischer Mengen. Sb. Bayer. Akad. Wiss., Math.-Nat. Kl., Jg. 1963, 89-99.

[13] O. Forster, K.J. Ramspott: Okasche Paare von Garben nicht-abelscher Gruppen. Inv. Math. 1 (1966) 260-286.

[14] O. Forster, K.J. Ramspott: Analytische Modulgarben und Endromisbündel. Inv. Math. 2 (1966) 145-170.

[15] B. Giesecke: Simpliziale Zerlegung abzählbarer analytischer Räume. Math. Zeits. 83 (1964) 177-213.

[16] H. Grauert: Charakterisierung der holomorph-vollständigen Räume. Mth. Ann. 129 (1955) 233-259.

[17] H. Grauert: Analytische Faserunden über holomorph-vollständigen Räumen. Math. Ann. 135 (1958) 263-273.

[18] H. Grauert, R. Remmert: Theory of Stein spaces. Springer 1979.

[19] H. Hamm: Zum Homotopietyp Steinscher Räume. Journal f.d.r.u.a. Math. (Crelle) 338 (1983) 121-135.

[20] D. Husemoller: Fibre bundles. 2nd ed. Springer 1975.

[21] M. Kneser: Über die Darstellung algebraischer Raumkurven als Durchschnitte von Flächen. Arch. Math. 11 (1960) 157-158.

[22] E. Kunz: Einführung in die kommutative Algebra und algebraische Geometrie. Vieweg 1979.

[23] S. Łojasiewicz: Triangulation of semi-analytic sets. Ann. Scuola Sup. Pisa (3) 18 (1964) 449-474.

[24] K. Lønsted: Vector bundles over finite CW-complexes are algebraic. Proc. AMS 38 (1973) 27-31.

[25] N. Mohan Kumar: Complete intersections. J. Kyoto Univ. 17 (1977) 533-538.

[26] N. Mohan Kumar: On two conjectures about polynomial rings. Inv. Math. 46 (1978) 225-236.

[27] P. Murthy: Affine varieties as complete intersections. Int. Symp. Algebraic Geometry Kyoto (1977) 231-236.

[28] D. Quillen: Projective modules over polynomial rings. Inv. Math. 36 (1976) 167-171.

[29] A. Sathaye: On the Forster-Eisenbud-Evans conjecture. Inv. Math. 46 (1978) 211-224.

[30] M. Schneider: Vollständige, fast-vollständige und mengentheoretisch vollständige Durchschnitte in Steinschen Mannigfaltigkeiten. Math. Ann. 260 (1982) 151-174.

[31] M. Schneider: On the number of equations needed to describe a variety. Conference on Several Complex Variables, Madison 1982. To appear in Proc. Symp. Pure Math. AMS 1983.

[32] N. Steenrod: The topology of fibre bundles. Princeton Univ. Press 1951.

[33] U. Storch: Bemerkung zu einem Satz von M. Kneser. Arch. Math. 23 (1972) 403-404.

[34] A.A. Suslin: Projective modules over a polynomial ring are free (Russian). Dokl. Acad. Nauk SSSR 229 (1976) 1063-1066.

[35] A.A. Suslin: On stably free modules (Russian). Mat. Sbornik 102 (1977) 537-550.

[36] R.G. Swan: The number of generators of a module. Math. Zeits. 102 (1967) 318-322.

[37] L. Szpiro: Lectures on equations defining a space curve. Tata Inst. of Fund. Research, Bombay. Springer 1979.

O. Forster

Mathematisches Institut der LMU

Theresienstr. 39

D-8000 München 2

West Germany

Some Applications of the Theory of Positive Vector Bundles

by

Robert Lazarsfeld*

Introduction 30

§1. Ample Line Bundles and Ample Vector Bundles 32

§2. Degeneracy Loci and a Theorem of Ghione 40

§3. A Theorem of Barth-Larsen Type on the Homotopy Groups of Branched Coverings of Projective Space 47

§4. A Problem of Remmert and Van de Ven 55

References 59

*Partially supported by an N.S.F. Grant.

Introduction

A considerable body of work has developed over the last few years loosely centered about the notion of positivity in algebraic geometry. On the one hand, numerous results have appeared on what might be called the geometry of projective space, the theme being the often remarkable special properties enjoyed by low codimensional subvarieties of, and mappings to, projective space (cf. [1], [27], [9], [11], [15], [31], [43], [32]). These results depend on the positivity of projective space itself, as manifested for example in various theorems of Bertini type. In another direction, the general theory of positive vector bundles has recently been extended and, more interestingly, applied in various geometric situations (cf. [42], [12], [13], [31], [7]). Bridging these two groups of results, in a class all by itself, one has Mori's far-reaching proof of the Frankel-Hartshorne conjecture ([35], [6]).

Our lectures at the C.I.M.E. conference were largely concerned with the geometry of projective space, and especially with the work of F. L. Zak on linear normality ([43], [32]). In addition, we discussed a recent theorem of Z. Ran [39] related to Hartshorne's conjecture [27] on complete intersections. Most of this material has been surveyed elsewhere (cf. [23], [11], [32]), and we do not propose to duplicate the existing literature here.

The present paper will rather constitute the notes to a course that we _might_ have given at the Acireale conference, focusing on positive vector bundles and their applications. We start (§1) with an elementary overview of the general theory, emphasizing the similarities and differences between the cases of line bundles and vector bundles of higher rank. The remaining sections are devoted to expositions of several previously unpublished proofs and results. In §2 we give a simple topological proof of a theorem guaranteeing that under suitable positivity and dimensional hypotheses a map of vector bundles must drop rank. We then sketch how this may be applied, along the lines of [12], to give a quick proof of (a slight generalization of) a recent theorem of Ghione [16] concerning the existence of special divisors associated to a vector bundle on an algebraic curve. In §3 we use a theorem of Goresky-MacPherson [17] to prove a homotopy Lefschetz-type result for the zero-loci of sections of certain positive vector bundles (Thm. 3.5). We deduce from this the homotopy analogue of the Barth-type theorem for branched coverings of projective space given in [31]. Finally, in §4 we show how Mori's arguments in [35] lead to the proof of an old conjecture of Remmert and Van de Ven, to the effect that if X is a smooth projective variety of dimension ≥ 1 which is the target of a surjective mapping $f : \mathbb{P}^n \to X$, then X is isomorphic to $\mathbb{P}^n$. There are many interesting questions related to this circle of ideas, and the reader will find open problems - some well-known - scattered throughout the paper.

It may seem at this point that the subject of these notes bears little relationship to the theme of the conference, complete intersections. In reality, however, there is an intimate connection. Suppose, for example, that $X \subseteq \mathbb{P}^n$ is the complete intersection of hypersurfaces $F_1,\ldots,F_e$ of positive degrees $d_1,\ldots,d_e$. If we think of F_i as being the zero-locus of a section s_i of the line bundle $\mathcal{O}_{\mathbb{P}^n}(d_i)$, then it is natural to view X as the zero-locus of the section $s = (s_1,\ldots,s_e)$ of the rank e vector bundle $E = \mathcal{O}_{\mathbb{P}^n}(d_1) \oplus \cdots \oplus \mathcal{O}_{\mathbb{P}^n}(d_e)$. But E is the very prototype of a positive, or _ample_, vector bundle, and in fact most of the basic results about complete intersections (e.g. Lefschetz-type results) are special cases of general results for positive vector bundles (e.g. (1.8) and (3.5) below). In this sense, the theory of ample vector bundles is a natural generalization of the study of complete intersections in projective space.

We work throughout with algebraic varieties over the complex numbers, although the results of §2 remain valid over an arbitrary algebraically closed ground field. If E is a vector bundle on a variety X, we denote by $\mathbb{P}(E)$ the projective bundle of one-dimensional subspaces of E. We shall follow Hartshorne's definition [24] of an ample vector bundle. The reader should be aware that there is a great deal of conflicting terminology in the literature; in particular, ample vector bundles are called "cohomologically positive" in [19] (where "ample" is used in another sense).

I'd like to take this opportunity to express my gratitude to the many Italian mathematicians - and especially to G. Ceresa, L. Chiantini, N. Chiarli, C. Ciliberto, A. Collino, S. Greco, P. Maroscia, and E. Sernesi - who made my stay in Italy a valuable and enjoyable one.

§1. Ample Line Bundles and Ample Vector Bundles.

Our purpose in this section is to give an elementary survey of the general theory of ample vector bundles.

We start by reviewing the basic facts about positivity in the line bundle case. Let X be an irreducible projective variety, and let L be a line bundle on X. Recall that L is very ample if there is a projective embedding

$$X \subseteq \mathbb{P}^N$$

such that L is the restriction to X of the hyperplane line bundle on $\mathbb{P}^N$:

$$L \cong \mathcal{O}_{\mathbb{P}^N}(1)|X.$$

This is perhaps the most appealing notion of positivity from an intuitive point of view, but unfortunately it is technically rather difficult to work with. For example, even when X is a smooth curve, it can be subtle to determine whether or not a given line bundle is very ample - the canonical bundle is a simple case in point.

It is found to be much more convenient to deal instead with a somewhat weaker notion. Specifically, recall that L is ample if $L^{\otimes k}$ is very ample for some $k > 0$. What this definition may lack in intuitive content is made up in the simplicity it yields. For example, if X is a smooth curve, then L is ample if and only if its degree is positive. Ample line bundles behave well functorially: if $f: X \to Y$ is finite (eg. an embedding), and L is an ample line bundle on Y, then f^*L is an ample line bundle on X. When X is smooth, amplitude is equivalent to Kodaira's differential geometric notion of positivity (cf [21]).

There are essentially four basic theorems on ample line bundles. First, one has Serre's cohomological criterion:

(1.1). A line bundle L on X is ample if and only if for every sheaf $\mathfrak{F}$ on X there exists a positive integer $k(\mathfrak{F})$ such that

$$H^i(X, \mathfrak{F} \otimes L^{\otimes k}) = 0$$

for all $i > 0$ and $k \geq k(\mathfrak{F})$.

When X is smooth, one has the more precise:

(1.2). *Kodaira Vanishing Theorem:* If L is ample, then

$$H^i(X,L^*) = 0$$

for $i \leq \dim X - 1$.

The basic topological fact is given by the

(1.3). *Lefschetz Hyperplane Theorem: Assume that* X *is smooth, and that* L *is an ample line bundle on* X. *Let* $s \in \Gamma(X,L)$ *be a section of* L, *and let* $Z = Z(s)$ *be the zero-locus of* s. *Then*

$$\pi_i(X,Z) = 0 \quad \textit{for} \quad i \leq \dim X - 1.$$

A variant of (1.3), which holds for arbitrary irreducible X, states that $X - Z$ has the homotopy type of a CW complex of (real) dimension $\leq \dim X$. When $X - Z$ is smooth, this is a well known fact about affine varieties; the result in general was recently established by Goresky-MacPherson [17] and by Hamm [22]. These authors also show that if X is a local complete intersection, then (1.3) itself remains true.

Finally, one has the theorem of Nakai *et al.* which characterizes ample line bundles numerically:

(1.4) *A line bundle* L *on* X *is ample if and only if for every irreducible subvariety* $Y \subseteq X$ *the Chern number*

$$\int_Y c_1(I)^k$$

is strictly positive, where $k = \dim(Y)$.

We refer to [26, Ch. 1] or [21, Ch. 1] for fuller accounts of the theory of ample line bundles.

In the 1960's, a number of authors - notably Grauert [18], Griffiths [19], and

Hartshorne [24] - undertook to generalize the notion of ampleness to vector bundles of higher rank. One of the goals was to prove analogues for vector bundles of the basic theorems (1.1) - (1.4), and this led initially to a number of competing notions of positivity. (Indeed, the literature of the period is marked by a certain terminological chaos.) With the passage of time, however, it has become clear that the weakest of these definitions is also the most useful. The idea is simply to reduce the definition of amplitude for vector bundles to the case of line bundles.

Suppose, then, that X is an irreducible projective variety, and that E is a vector bundle on X of rank e. Following [24] one defines E to be _ample_ if the Serre line bundle $\mathcal{O}_{\mathbb{P}(E^*)}(1)$ on the projective bundle $\mathbb{P}(E^*)$ is ample.* The first indication that this definition is the correct one is that it leads to various desirable formal properties (cf. [24]):

(1.5) (i) _A quotient of an ample vector bundle is ample._

(ii) _A direct sum of vector bundles is ample if and only if each summand is._

(iii) E _is ample if and only if the symmetric power_ $S^k(E)$ _is for some (or all) positive integer(s)_ k. _If_ E _and_ F _are ample, then so is_ $E \otimes F$.

(iv) _If_ $f: X \to Y$ _is a finite map, and if_ E _is an ample vector bundle on_ Y, _then_ f^*E _is an ample vector bundle on_ X. _If_ f _is in addition flat, then the converse holds._

The basic results (1.1) - (1.3) have good analogues for ample vector bundles:

(1.6) _A vector bundle_ E _on_ X _is ample if and only if for every coherent sheaf_ $\mathcal{F}$ _on_ X _there exists a positive integer_

*The presence of $\mathbb{P}(E^*)$ here, rather than $\mathbb{P}(E)$, may be explained by the observation that if $E = L$ is a line bundle, so that $\mathbb{P}(E) \cong \mathbb{P}(E^*) \cong X$, then $\mathcal{O}_{\mathbb{P}(E^*)}(1) = L$, whereas $\mathcal{O}_{\mathbb{P}(E)}(1) = L^*$.

$k(\mathfrak{F})$ such that

$$H^i(X, \mathfrak{F} \otimes S^k(E)) = 0$$

for $i > 0$ *and* $k \geq k(\mathfrak{F})$ ([19], [24]).

The analogue of Kodaira's vanishing theorem (1.2) is due to Le Poitier [33]:

(1.7) *If* X *is smooth, and* E *is ample, then*

$$H^i(X, E^*) = 0$$

for $i \leq \dim X - \operatorname{rk} E$.

The strongest general Lefschetz-type result was proved by Sommese in [42]:

(1.8) *Assume that* X *is smooth, and that* E *is an ample vector bundle on* X *of rank* e. *Let* $s \in \Gamma(X, E)$ *be a section of* E, *and let* $Z = Z(s)$ *be the zero-locus of* s. *Then*

$$H^i(X, Z; \mathbb{Z}) = 0$$

for $i \leq \dim X - e$.

Because Sommese's argument deserves to be better known that it is, we give the

Proof. We will show, for arbitrarily singular irreducible X, that

$$H_i(X - Z; \mathbb{Z}) = 0 \tag{1.9}$$

for $i \geq \dim X + e$. When X is smooth, the theorem as stated follows by Lefschetz duality.

Consider then the projective bundle

$$\pi : \mathbb{P}(E^*) \to X.$$

The Serre line bundle $\mathcal{O}_{\mathbb{P}(E^*)}(1)$ is a quotient of π^*E, and so the given section s determines a section $s^* \in \Gamma(\mathbb{P}(E^*), \mathcal{O}_{\mathbb{P}(E^*)}(1))$:

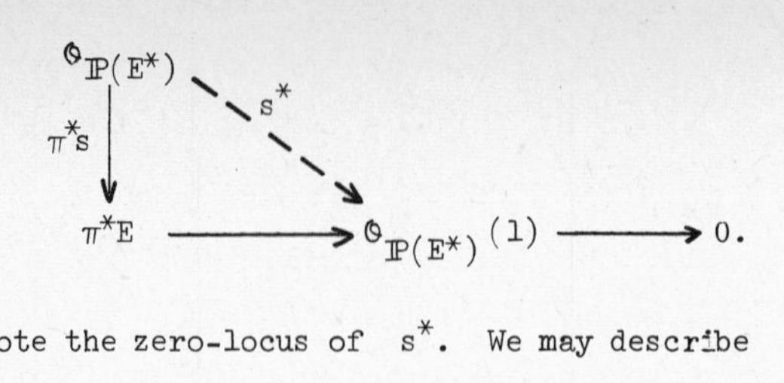

Let $Z^* \subseteq \mathbb{P}(E^*)$ denote the zero-locus of s^*. We may describe Z^* very concretely as follows. Thinking of $\mathbb{P}(E^*)$ as the bundle of hyperplanes in E:

$$\mathbb{P}(E^*) = \{(x,\Lambda) | \Lambda \subseteq E(x) \text{ a cod } 1 \text{ subspace}\},$$

a moment's thought shows that

$$Z^* = \{(x,\Lambda) | s(x) \in \Lambda\}.$$

In particular, the bundle map π restricts to a morphism

$$p : \mathbb{P}(E^*) - Z^* \longrightarrow X - Z,$$

and in fact, p <u>is a</u> $\mathbb{C}^{e-1}$ - <u>bundle</u> (but not, in general, a vector bundle, i.e. p may not section).

On the other hand, since E is ample, Z^* is an ample divisor on $\mathbb{P}(E^*)$. Therefore $\mathbb{P}(E^*) - Z^*$ is an affine variety, of dimension $\dim X + e - 1$, and hence has the homotopy type of a CW complex of (real) dimension $\leq \dim X + e - 1$. In particular, $H_i(\mathbb{P}(E^*) - Z^*; \mathbb{Z}) = 0$ for $i \geq \dim X + e$. But since p is an affine space bundle, this implies that $H_i(X - Z; \mathbb{Z}) = 0$ for $i \geq \dim X + e$, as desired. □

<u>Problem</u>. In the situation of (1.8), is it true that $\pi_i(X,Z) = 0$ for $i \leq \dim X - e$?

Under a stronger positivity condition on E, and a transversality assumption on the section s, Griffiths [19] has proven such a homotopy statement. Another result of this nature is given in §3. (Thm. 3.5).

Observe that there is no genericity or transversality hypothesis on the section s in (1.8). Hence the result gives topological obstructions to expressing a variety set-theoretically as the zero-locus of a section of an ample vector bundle.

For example, consider the Segre variety

$$S = \mathbb{P}^1 \times \mathbb{P}^2 \subseteq \mathbb{P}^5$$

Since $b_2(S) = 2$ while $b_2(\mathbb{P}^5) = 1$, we conclude that there cannot exist an ample vector bundle E of rank 2 on $\mathbb{P}^5$ with a section vanishing precisely on E. In particular, (1.8) gives an elementary proof of the well known fact that S is not a set-theoretic complete intersection. (Compare the lectures of Forster and Valla in this volume.)

Another result along the lines of (1.8) has been established by Ein [7], who proves a Noether-type theorem on the Picard group of the zero-locus of a generic section of certain ample rank $n-2$ bundles on $\mathbb{P}^n$; he also treats determinantal surfaces. In the same paper, Ein uses the vanishing theorem (1.7) of Le Poitier to give a simple proof of a theorem of Evans and Griffith on the cohomology of vector bundles of small rank on projective space.

Turning to the numerical properties of ample vector bundles, one finds that there are two questions to ask if one hopes to generalize (1.4). First:

A. What are the numerically positive polynomials for ample vector bundles?

(Recall that a homogeneous polynomial $P \in Q[c_1, \ldots, c_e]$ of weighted degree n is <u>numerically positive</u> if for every irreducible projective variety X of dimension n, and for every ample vector bundle E of rank e on X, the Chern number

$$\int_X P(c_1(E), \ldots c_e(E))$$

is strictly positive. For example, if $e = rk(E) = 1$, the positive polynomials are just $\alpha\, c_1^n$ $(\alpha > 0)$.) And secondly:

B. Is there a numberical criterion for ampleness analogous to the theorem of Nakai *et al.* for line bundles?

Question (B) was the first to be answered. A theorem of Hartshorne [25] states that a vector bundle E on a smooth curve X is ample if and only if every quotient of E has positive degree. For X of dimension ≥ 2, however, simple examples [8] show that there cannot be a numerical criterion, at least in the form suggested by Griffiths [20].

As for question (A), the numerically positive polynomials may be described succinctly - if unrevealingly - as follows. Let $\Lambda(n,e)$ denote the set of partitions of n into a sum of non-negative integers $\leq e$. Given $\lambda \in \Lambda(n,e)$, λ being the partition $\lambda_1 \geq \cdots \geq \lambda_n \geq 0$, one forms the so-called Schur polynomial P_λ, defined as the $n \times n$ determinant

$$P_\lambda = \begin{vmatrix} c_{\lambda_1} & c_{\lambda_1+1} & \cdots & & \\ c_{\lambda_2-1} & c_{\lambda_2} & \cdots & & \\ & & \cdots & & \\ & \cdots & & c_{\lambda_n-1} & c_{\lambda_n} \end{vmatrix}$$

where one makes the convention that $c_o = 1$ and $c_i = 0$ for $i \notin [0,e]$. The P_λ's from a basis for the homogeneous polynomials of weighted degree n, and the result is:

(1.10) $P \in Q[c_1,\ldots,c_e]$ <u>is numerically positive for ample vector bundles if and only if</u> $P \neq 0$, <u>and</u> P <u>is a non-negative linear combination of the</u> P_λ $(\lambda \in \Lambda(n,e))$.

We refer to [13] for the proof, and for a discussion of earlier work on question (A).

The determinantal definition of the P_λ is evidently rather awkward to deal with. There is a more conceptual approach, which makes (1.10) seem quite natural. For simplicity, we explain this only for bundles generated by their global sections.

Suppose, then, that E is an ample vector bundle of rank e on X^n which is given as a quotient of a trivial bundle of rank m. Then there is a classifying map

$$\varphi : X \longrightarrow G,$$

where $G = G(m-e,m)$ is the Grassmannian of codimension e subspaces of an m-dimensional vector space; the bundle E is recovered as the pull-back of the universal quotient bundle Q on G. Consider now a codimension n cycle $z \in H^{2n}(G)$. Then $\varphi^*(z)$ is a top dimensional cohomology class on X, and one may ask when $\int_X \varphi^*(z) > 0$. It is not hard to show that the ampleness of E implies that this degree is positive whenever z is represented by an <u>effective</u> algebraic cycle. Conversely, if z is not effective, then there exists an X and E so that $\int \varphi^*(z) < 0$. But the cone of effective cycles on the Grassmannian G is well-understood: it is generated by the codimension n Schubert cycles $\{\Omega_\lambda\}_{\lambda \in \Lambda(n,e)}$ (cf. [21]). And it turns out that the cycle Ω_λ represents the cohomology class $P_\lambda(c_1(Q),\ldots,c_e(Q))$. Thus $\varphi^*(\Omega_\lambda) = P_\lambda(c_1(E),\ldots,c_e(E))$, which proves (1.10) for bundles generated by their global sections. In general, one thinks of the classes $P_\lambda(c_1(E),\ldots,c_e(E))$ as representing "virtual" Schubert cycles; up to now, the explicit formula for the P_λ has not proved to be of any particular significance in itself.

<u>Problem</u>. Find a wider class of vector bundles for which the P_λ are numerically positive.

It seems certain that one could weaken the hypothesis of ampleness and yet retain the positivity of the Schur polynomials. For applications, such a strengthening of (1.10) should prove useful. What seems difficult, however, is to find a suitable class of bundles with which to deal. It might well be that this problem is most sensibly attacked only with some particular application in mind.

<u>Problem</u>. Determine whether the following conjecture of Hartshorne [26, III. 4.5] is true or false:

(*) Let M be a smooth variety, and let $X, Y \subseteq M$ be smooth projective subvarieties with ample normal bundles. If $\dim X + \dim Y \geq \dim M$, then X meets Y.

A number of conjectures have appeared suggesting global consequences of ampleness of normal bundles (eg. [26, III. 4.4], [9]). Simple counter-examples dispose of many of these (cf. [14]), although they tend to be true when the ambient space is a rational homogeneous manifold. What's fascinating about Hartshorne's conjecture (*) is that several approaches to the construction of counter-examples seem systematically to fail. Hence it seems likely that the resolution of the conjecture one way or the other could involve some interesting new ideas.

§2. Degeneracy Loci, and a Theorem of Ghione.

A theorem on the non-emptiness of degeneracy loci.

Let X be an irreducible projective variety of dimension n, and let

$$u : E \longrightarrow F$$

be a homomorphism of vector bundles of ranks e and f respectively. A number of interesting geometric problems can be formulated in terms of the degeneracy loci associated to such a map, i.e. the sets

$$D_k(u) \underset{\text{def}}{=} \{x \in X \mid rk(u(x)) \leq k\}$$

Recall that the set $D_k(u)$ is Zariski-closed, and its postulated codimension in X is $(e-k)(f-k)$; if non-empty, its actual codimension is $\leq (e-k)(f-k)$.

It may happen, of course, that $D_k(u)$ is empty even when its expected dimension is non-negative. Our purpose here is to give a simple proof that this cannot occur under suitable positivity hypotheses:

Theorem 2.1. *Assume that the vector bundle*

$$Hom(E,F) = E^* \otimes F$$

is ample. If $n \geq (e-k)(f-k)$, *then* $D_k(u)$ *is non-empty.*

The proof below arose in the course of the author's work on [12], where a more elaborate argument was given to show that in fact $D_k(u)$ is connected if $n > (e-k)(f-k)$.

We shall actually prove a slight strengthening of (2.1). Specifically

(2.2). <u>Assume that</u> Hom(E,F) <u>is ample</u>. <u>Fix an integer</u> ℓ, <u>and let</u>

$$Y \subseteq D_\ell(u)$$

<u>be an irreducible projective variety of dimension</u> $m \geq (e+f) - 2\ell + 1$. <u>Then</u>

$$D_{\ell-1}(u) \cap Y \neq \emptyset.$$

Note that $(e+f) - 2\ell + 1$ is the expected codimension of $D_{\ell-1}(u)$ in $D_\ell(u)$. Theorem 2.1 follows by applying (2.2) successively to each of the varieties in the chain.

$$X = D_r(u) \supseteq D_{r-1}(u) \supseteq \cdots \supseteq D_k(u),$$

where $r = \min(e,f)$. The idea of the proof is to exploit the observation that if the assertion were false, then the kernel and image of u would be vector bundles on Y. This approach has been taken up again in [10].

<u>Proof of (2.2)</u>. We assume that $Y \subseteq D_\ell(u)$ is a projective variety of dimension m which does not meet $D_{\ell-1}(u)$; we will show that $m \leq (e+f) - 2\ell$. Evidently we may suppose that $\ell \leq \min(e,f)$, and for simplicity of notation we write E and F for the restrictions of these bundles to Y.

Let $N = \ker(u|Y)$ and $K = \mathrm{im}(u|Y)$. Since u has rank ℓ everywhere on Y, N and K are vector bundles of ranks $e - \ell$ and ℓ respectively. Consider the projective bundle $\pi : \mathbb{P}(E) \longrightarrow Y$. On $\mathbb{P}(E)$ one has the diagram:

$$\begin{array}{ccccccc} & & \mathcal{O}_{\mathbb{P}(E)}(-1) & & & & \\ & & \downarrow & \searrow^{s} & & & \\ 0 \longrightarrow & \pi^*N \longrightarrow & \pi^*E & \xrightarrow[\pi^*u]{} & \pi^*K & \longrightarrow & 0, \end{array}$$

which defines a section $s \in \Gamma(\mathbb{P}(E), \pi^*K \otimes \mathcal{O}_{\mathbb{P}(E)}(1))$ as shown. Note that the zero-locus $Z(s)$ of s is exactly the subvariety $\mathbb{P}(N) \subseteq \mathbb{P}(E)$. The idea is to apply the Lefschetz theorem (1.9) to study $\mathbb{P}(E) - \mathbb{P}(N)$.

To this end, let t denote the composition

$$\mathcal{O}_{\mathbb{P}(E)}(-1) \xrightarrow{s} \pi^*K \hookrightarrow \pi^*F.$$

Then evidently

$$Z(t) = Z(s).$$

On the other hand, we shall show below that

(2.3) If $E^* \otimes F$ is ample on Y then

$$\pi^*F \otimes \mathcal{O}_{\mathbb{P}(E)}(1)$$

is an ample vector bundle on $\mathbb{P}(E)$.

Thus $\mathbb{P}(N)$ is the zero-locus of the section t of the ample vector bundle $\pi^*F \otimes \mathcal{O}_{\mathbb{P}(E)}(1)$. Hence by (1.9):

$$H_i(\mathbb{P}(E) - \mathbb{P}(N)) = 0 \quad \text{if} \quad i \geq (m + e - 1) + f.$$

But there is a natural map

$$\begin{array}{ccc} \mathbb{P}(E) - \mathbb{P}(N) & \xrightarrow{p} & \mathbb{P}(K) \\ & {\scriptstyle\pi}\searrow \quad \swarrow & \\ & Y & \end{array} ;$$

fibre by fibre, p is just the linear projection centered at $\mathbb{P}(N(y)) \subseteq \mathbb{P}(E(y))$. In particular, p is a $\mathbb{C}^{e-\ell}$ - bundle map, and hence

$$H_i(\mathbb{P}(E) - \mathbb{P}(N)) = H_i(\mathbb{P}(K))$$

for all i. Therefore $H_i(\mathbb{P}(K)) = 0$ for $i \geq m + e + f - 1$. But $\mathbb{P}(K)$ is a <u>compact</u> variety, of dimension $m + \ell - 1$, and so $H_{2(m+\ell-1)}(\mathbb{P}(K)) \neq 0$. We conclude that

$$2(m + \ell - 1) < m + e + f - 1,$$

i.e.

$$m \leq e + f - 2\ell,$$

as desired.

It remains to check (2.3), for which we use an argument suggested by W. Fulton. Consider the projectivization $\mathbb{P} = \mathbb{P}(\pi^* F^* \otimes \mathcal{O}_{\mathbb{P}(E)}(-1)) \longrightarrow \mathbb{P}(E)$. We need to show that $\mathcal{O}_{\mathbb{P}}(1)$ is an ample line bundle. But $\mathbb{P}$ is isomorphic to the fibre product $\mathbb{P}(E) \times_X \mathbb{P}(F^*)$, and $\mathcal{O}_{\mathbb{P}}(1)$ is the restriction of the Serre line bundle $\mathcal{O}_{\mathbb{P}(E\otimes F^*)}(1)$ under the Segre embedding

$$\mathbb{P}(E) \times_X \mathbb{P}(F^*) \subseteq \mathbb{P}(E \otimes F^*).$$

Hence $\mathcal{O}_{\mathbb{P}(E\otimes F^*)}(1)$ is ample since $E^* \otimes F$ is. □

Ghione's generalization of the Kempf-Kleiman-Laksov existence theorem.

One of the most famous examples of determinantal loci are the varieties of special divisors on a smooth projective algebraic curve C of genus g. Specifically, let $J = \mathrm{Pic}^0(C)$ be the Jacobian of C, and fix once and for all a base point $P_0 \in C$. One is interested in the set

$$W_d^r(C) = \{x \in J \,|\, h^0(L_x(dP_0)) \geq r + 1\},$$

where L_x is the line bundle of degree 0 on C corresponding to the point $x \in J$. Thus $W_d^r(C)$ parametrizes linear equivalence classes of divisors of degree d moving in a linear system of (projective) dimension $\geq r$.

Let us recall how these varieties of special divisors are realized as determinantal loci. Choose some integer $n \geq \max(d, 2g)$, and $n - d$ points $p_1, \ldots, p_{n-d} \in C$ (say distinct, to fix ideas). Then for each $x \in J$, evaluation at the p_i yields a homomorphism

$$u(x) : H^0(C, L_x(nP_0)) \longrightarrow \bigoplus_{i=1}^{n-d} H^0(C, L_x(nP_0) \otimes \mathcal{O}_{p_i}).$$

As x varies over J, the vector spaces $H^0(C, L_x(nP_0))$ and $\oplus\, H^0(C, L_x(nP_0) \otimes \mathcal{O}_{p_i})$ fit together to form vector bundles E and F on J, of ranks $n+1-g$ and

$n-d$ respectively. Furthermore, the maps $u(x)$ globalize to a vector bundle homomorphism

$$u: E \longrightarrow F.$$

Since $\ker u(x) = H^0(C, L_x(nP_0 - \Sigma P_i))$, we see that up to translation

$$W^r_d(C) = D_{n-g-r}(u).$$

(Cf. [28], [29] or [12] for details.) It follows in particular that if $W^r_d(C)$ is non-empty, then

$$\dim W^r_d(C) \geq \rho^r_d(C) \underset{\text{def}}{=} g - (r+1)(g-d+r).$$

The celebrated existence theorem of Kempf [28] and Kleiman-Laksov [29] asserts that in fact $W^r_d(C) \neq \emptyset$ provided that $\rho^r_d(C) \geq 0$.

The traditional approach to the Kempf-Kleiman-Laksov theorem is to compute via Porteous' formula the (postulated) fundamental class of $W^r_d(C)$ (or of a closely related variety). This turns out to be non-zero when $\rho \geq 0$, and the theorem follows. This <u>quantitative</u> approach, as we may call it, has the advantage that a formula for $[W^r_d(C)]$, which is useful in enumerative questions, emerges as a by-product. However there is an alternative <u>qualitative</u> approach based on positivity considerations. Specifically, it was shown in [12, §2] that

(2.4) $E^* \otimes F$ <u>is an ample vector bundle on</u> J.

Thus in fact the existence theorem follows from the elementary result (2.1), and this is one of the quickest proofs available.

Ghione [16] has recently proved an interesting generalization of the Kempf-Kleiman-Laksov theorem. Specifically, fix a vector bundle

M of degree a and rank e

on C. Then set

$$W^r_d(C,M) = \{x \in J \mid h^0(M(dP_0) \otimes L_x) \geq r+1\}.$$

Thus the classical set $W^r_d(C)$ correspond to taking $M = \mathcal{O}_C$. As before, the loci $W^r_d(C,M)$ may be realized determinantally. To do so, following [16], we fix an integer $n \geq 2g - d$ large enough so that $M^*(nP_0)$ is generated by its global sections. Choosing $e = rk(M)$ general sections gives an exact sequence

$$0 \longrightarrow M \longrightarrow \mathcal{O}^e_C(nP_0) \longrightarrow \tau \longrightarrow 0$$

on C, where τ is a torsion sheaf of length $en - a$. Then for each $x \in J$ we have homomorphisms

$$u(x) : H^0(C,\mathcal{O}^e_C((n+d)P_0 \otimes L_x)) \longrightarrow J^0(C,\tau \otimes L_x(dP_0)),$$

which as before fit together to form a vector bundle map

$$(2.5) \qquad u : E \longrightarrow F,$$

where E and F are now vector bundles on J of ranks $e(n+d+1-g)$ and $en-a$ respectively. Then $\ker u(x) = H^0(C,M(dP_0 \otimes L_x)$, so

$$W^r_d(C) = D_{e(n+d+1-g)-(r+1)}(u).$$

In particular, if $W^r_d(C,M) \neq \emptyset$, then

$$\dim W^r_d(C,M) \geq \rho^r_d(C,M) \underset{\text{def}}{=} g - (r+1)(e(g-d+1) + r + 1 - a).$$

Ghione's generalization of the Kempf-Kleiman-Laksov theorem is:

<u>Theorem 2.6</u>. ([16]). <u>If</u> $\rho^r_d(C,M) \geq 0$, <u>then</u> $W^r_d(C,M)$ <u>is non-empty</u>.*

Ghione takes the quantitative approach to Theorem 2.6, and obtains also a formula for $[W^r_d(C,M)]$ valid when $\dim W^r_d(C,M) = \rho^r_d(C,M)$. For Theorem 2.6 the qualitative approach is very much quicker, and essentially involves nothing beyond what was proved in [12].

<u>Proof of (2.6)</u>. Replacing M by $M(dP_0)$, we may as well assume that $d = 0$. It suffices to show that $E^* \otimes F$ is ample, E and F being the vector bundles defined informally in (2.5). Let us start by defining these bundles more precisely.

---

*Ghione assumes that M is general in a suitable sense. However the proof below shows that this is not necessary.

Denote by f and π the projections of $J \times C$ onto J and C respectively. Let $\mathcal{L}$ be the Poincaré line bundle on $J \times C$, normalized so that $\mathcal{L}|J \times \{P_0\} = \mathcal{O}_J$. We take

$$E = f_*(\mathcal{O}_C^e(nP_0)) \otimes \mathcal{L})$$

and

$$F = f_*(\pi^*\tau \otimes \mathcal{L}).$$

The map u arises by taking direct images from the exact sequence

$$0 \longrightarrow \pi^*M \otimes \mathcal{L} \longrightarrow \pi^*(\mathcal{O}_C^e(nP_0)) \otimes \mathcal{L} \longrightarrow \pi^*\tau \otimes \mathcal{L} \longrightarrow 0.$$

Since $E = \bigoplus_{i=1}^{e} E_1$, where $E_1 = f_*(\pi^*(\mathcal{O}_C(nP_0) \otimes \mathcal{L})$, it is enough to show that $E_1^* \otimes F$ is ample. On the other hand, τ - like any torsion sheaf on C - has a filtration whose successive quotients are torsion sheaves of length one, and hence isomorphic to $\mathcal{O}_{P_i}$ for suitable points $P_i \in C$. Therefore F has a filtration whose successive quotients are line bundles of the form

$$f_*(\pi^*\mathcal{O}_{P_i} \otimes \mathcal{L}) \underset{def}{=} \mathcal{L}_{P_i}.$$

Recalling that an extension of ample vector bundles is ample, we are reduced to proving the amplitude of $E_1^* \otimes \mathcal{L}_{P_i}$. But this is the assertion of Lemma 2.2 of [12]. (The proof in brief: observing that $\mathcal{L}_{P_i}$ is a deformation of $\mathcal{L}_{P_0} = \mathcal{O}_J$, one shows that it suffices to prove that E_1^* is ample. But $\mathbb{P}(E_1) = C_n$, the nth symmetric product of C, and $\mathcal{O}_{\mathbb{P}(E_1)}(1) = \mathcal{O}_{C_n}(C_{n-1})$, C_{n-1} being embedded in C_n via $D \longrightarrow D + P_0$. And it is elementary - eg. by Nakai's criterion - that C_{n-1} is an ample divisor on C_n.) □

Note that by §1 of [12] we conclude also that

(2.7) <u>In the situation of Theorem 2.6, if</u> $\rho_d^r(C,M) > 0$, <u>then</u> $W_d^r(C,M)$ <u>is connected</u>.

Problem. Work out concretely the varieties $W^r_d(C,M)$ for various vector bundles M on curves of low genus.

The question is whether the geometry of C is reflected in the geometry of $W^r_d(C,M)$ as it is in the geometry of $W^r_d(C)$. (cf. [38, Chapt. 1]).

Problem. Are there theorems of Martens-Mumford type ([34], [37]) for $W^r_d(C,M)$, say when M is stable?

The examples of Raynaud [40] show that the cohomological properties of stable vector bundles can be quite subtle.

§3. A Theorem of Barth-Larsen Type on the Homotopy Groups of Branched Coverings of Projective Space.

A celebrated theorem of Barth and Larsen ([1], [2], [4], [30]) asserts that if $X \subseteq \mathbb{P}^{n+e}$ is a smooth variety of dimension n and codimension e, then the maps $\pi_i(X) \longrightarrow \pi_i(\mathbb{P}^{n+e})$ induced by inclusion are bijective for $i \leq n-e$, and surjective if $i = n-e+1$ (cf. also [11], §9). Our goal in this section is to prove an analogue for branched coverings of projective space:

Theorem 3.1. *Let X be an irreducible, non-singular, projective variety of dimension n, and let $f: X \longrightarrow \mathbb{P}^n$ be a finite mapping of degree d. Fix $x \in X$. Then the induced homomorphisms*

$$f_* : \pi_i(X,x) \longrightarrow \pi_i(\mathbb{P}^n, f(x))$$

are bijective for $i \leq n+1-d$, and surjective if $i = n+2-d$.

Corollary 3.2. *In the setting of the theorem, the maps*

$$f_* : H_i(X;\mathbb{Z}) \longrightarrow H_i(\mathbb{P}^n : \mathbb{Z})$$

and

$$f^* : H^i(\mathbb{P}^n;\mathbb{Z}) \longrightarrow H^i(X;\mathbb{Z})$$

are isomorphisms if $i \leq n+1-d$. *When* $i = n+2-d$, f_* *is surjective and* f^* is *injective.* □

It follows for example that if $d \leq n$ then X is simply connected, while if $d \leq n-1$ then $\mathrm{Pic}(X) \cong \mathrm{Pic}(\mathbb{P}^n)$. The theorem was announced in [31], where the analogous result for complex cohomology was proved. The material in this section was part of the author's Ph.D. thesis (unpublished).

It is shown in [31, §1] that canonically associated to a branched covering $f : X \longrightarrow \mathbb{P}^n$ satisfying the hypotheses of (3.1), there exists a vector bundle

$$E \longrightarrow \mathbb{P}^n$$

of rank $d-1$ having the property that f factors through an embedding of X in the total space of E. The bundle E may be defined as the dual of the kernel of the trace $\mathrm{Tr}_{X/\mathbb{P}^n} : f_* \mathcal{O}_X \longrightarrow \mathcal{O}_{\mathbb{P}^n}$. The crucial fact for our purposes is that the bundle associated to a branched covering of projective space satisfies the strong positivity property:

(3.3) $E(-1)$ *is generated by its global sections, ie.* E *arises as a quotient of a direct sum* $\oplus\, \mathcal{O}_{\mathbb{P}^n}(1)$ *of copies of the hyperplane line bundle.*

Proof. ([31], §1). According to a theorem of Mumford [36, Lect. 14], it suffices to show that E is (-1) - regular, i.e. that $H^i(\mathbb{P}^n, E(-i-1)) = 0$ for $i > 0$. This is equivalent by duality to the assertion that

$$(*) \qquad H^{n-i}(\mathbb{P}^n, E^*(i-n)) = 0 \quad \text{for} \quad i > 0.$$

It follows from the definition of E that

$$f_*\mathcal{O}_X = \mathcal{O}_{\mathbb{P}^n} \oplus E^*$$

and hence

$$\mathbb{C} = H^0(X, \mathcal{O}_X) = H^0(\mathbb{P}^n, \mathcal{O}_{\mathbb{P}^n}) \oplus H^0(\mathbb{P}^n, E^*) = \mathbb{C} \oplus H^0(\mathbb{P}^n, E^*).$$

Thus $H^0(\mathbb{P}^n, E^*) = 0$ which proves (*) for $i = n$. When $1 \leq i \leq n-1$ we have similarly

$$H^{n-i}(\mathbb{P}^n, E^*(i-n)) = H^{n-i}(\mathbb{P}^n, f_*\mathcal{O}_X(i-n))$$
$$= H^{n-i}(X, f^*\mathcal{O}_{\mathbb{P}^n}(i-n)).$$

But $f^*\mathcal{O}_{\mathbb{P}^n}(i-n)$ is the dual of an ample line bundle, whence $H^{n-i}(X, f^*\mathcal{O}_{\mathbb{P}^n}(i-n)) = 0$ by the Kodaira vanishing theorem. □

Theorem 3.1 is therefore a consequence of

Theorem 3.4. Let $E \longrightarrow \mathbb{P}^n$ *be a vector bundle of rank* e *satisfying the positivity condition* (3.3). *Suppose that* X *is a compact, connected, local complete intersection variety of pure dimension* n *embedded in the total space of* E:

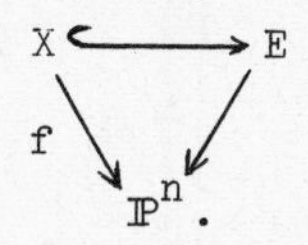

Define f *as shown, and fix* $x \in X$. *Then*

$$f_* : \pi_i(X,x) \longrightarrow \pi_i(\mathbb{P}^n, f(x))$$

is bijective for $i \leq n-e$, *and surjective if* $i = n-e+1$ (*i.e.* $\pi_i(E,X,x) = 0$ *for* $i \leq n-e+1$).

Note that f, being affine and proper, is finite. When E is the direct sum of e copies of the hyperplane line bundle, the theorem is equivalent to the Barth-Larsen theorem for embeddings $X \subseteq \mathbb{P}^{n+e}$ (cf [31, Rmk. 2.4]). We leave it to the reader to formulate the corresponding results for integral homology and cohomology implied by (3.4). Note that the latter in turn imply that if X is smooth, and if $e \leq n-2$, then $f^* : \mathrm{Pic}(\mathbb{P}^n) \longrightarrow \mathrm{Pic}(X)$ is an isomorphism.

Turning to the proof of (3.4), the strategy is to derive from Deligne's generalization [11, §9] of the Fulton-Hansen connectedness theorem, an analogoue for the diagonal embedding $X = \Delta_X \longrightarrow X \times X$. This will imply (3.4) in much the same way that [11, (9.2)] can be used, as W. Fulton remarked, to prove the Barth-Larsen theorem. The one additional ingredient we shall need is the following Lefschetz-type result, which is proved below.

Theorem 3.5. *Let* X *be a complete, connected, but possibly reducible local complete intersection variety of pure dimension* n, *and let* A *be an ample line bundle on* X *which is generated by its global sections. Suppose that* E *is a vector bundle of rank* e *on* X *having the property that* $E \otimes A^*$ *is generated by its global sections. Let* $s \in \Gamma(X,E)$ *be a section of* E, *and let* $Z = Z(s) \subseteq X$ *be the zero locus of* s. *Then, fixing* $x \in Z$, *one has*

$$\pi_i(X,Z,x) = 0$$

for $i \leq n-e$.

Proof of Theorem 3.4. Put $Y = (f \times f)^{-1}(\Delta_{\mathbb{P}^n})$, so that the diagonal embedding $\delta : X \longrightarrow X \times X$ factors through an embedding of X in Y. The set-up we shall deal with is summarized in the diagram (3.6) below. Each of the three squares is cartesian, and we henceforth make free use of the natural identifications indicated in that diagram. The inclusion $E \hookrightarrow E \oplus E = E \times_{\mathbb{P}^n} E$ is the evident diagonal homomorphism over $\mathbb{P}^n = \Delta_{\mathbb{P}^n}$.

(3.6)

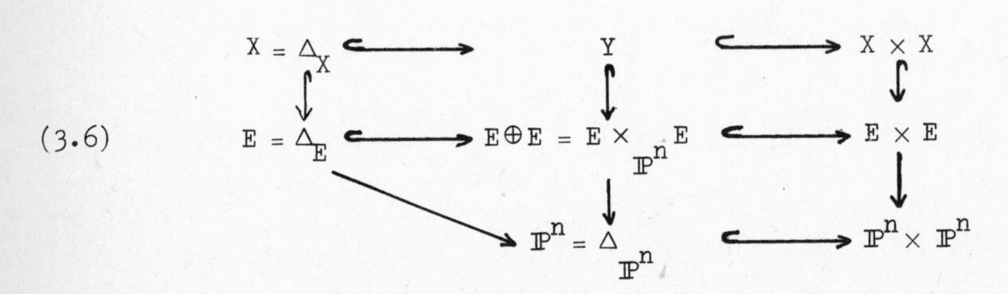

Note that Y is a complete, connected, local complete intersection of pure dimension n. Indeed, Y is locally cut out in $X \times X$ by n equations, and maps finitely to $\mathbb{P}^n$. It follows that Y has pure dimension n, and hence is a local complete intersection variety. The connectedness of Y follows, for instance, from Deligne's theorem [11, Thm. 9.2].

Consider first the inclusion $X = \Delta_X \hookrightarrow Y$. We assert that

(3.7) $$\pi_i(Y,X) = 0$$

for $i \leq n-e$.* Letting h denote the composition $Y \longrightarrow E \oplus E \longrightarrow \mathbb{P}^n$, the point to observe is that X is defined in Y as the zero-locus of a section of h^*E. In fact, the embedding of Y in the total space of $E \oplus E$ determines a tautological section of $h^*(E \oplus E)$, i.e. two sections $s_1\ s_2 \in \Gamma(Y, h^*E)$, and $X = \text{Zeroes}(s_1 - s_2) \subseteq Y$. But the positivity assumption (3.3) on E implies that $h^*E(-1)$ is generated by its global sections, and hence since h is finite (3.7) is a consequence of Theorem 3.5.

On the other hand, Deligne's theorem [11, (9.2)] applies to the inclusion $Y \longrightarrow X \times X$. In the case at hand, the theorem in question states that

$$\pi_i(X \times X, Y) = 0$$

for $i \neq 2$, $i \leq n$, and that if $n \geq 2$ there is an exact sequence

(*) $$\pi_2(Y) \longrightarrow \pi_2(X \times X) \longrightarrow \mathbb{Z} \longrightarrow \pi_1(Y) \longrightarrow \pi_1(X \times X) \longrightarrow 0$$

Moreover the map to $\mathbb{Z}$ in (*) may be identified with the difference of the homomorphisms

$$(pr_1 \circ (f \times f))_*,\ (pr_2 \circ (f \times f))_* : \pi_2(X \times X) \longrightarrow \pi_2(\mathbb{P}^n) = \mathbb{Z}.$$

Consider now the composition δ_*:

$$\pi_i(X) \longrightarrow \pi_i(Y) \longrightarrow \pi_i(X \times X).$$

(with δ_* the composite arrow $\pi_i(X) \to \pi_i(X \times X)$)

This is just the diagonal map, so δ_* is in any event injective. But it follows from (3.7) and Deligne's theorem that δ_* is surjective when $i \leq n-e$, $i \neq 2$, which implies $\pi_i(X) = 0$. This proves Theorem 3.4 in the range $i \leq n-e$, $i \neq 2$. If $i = 2 \leq n-e$, then one obtains the commutative diagram

*As we deal exclusively with path-connected spaces, we will henceforth omit base-points.

$$\begin{array}{ccccccccc}
0 & \longrightarrow & \pi_2(X) & \longrightarrow & \pi_2(X \times X) & \longrightarrow & \mathbb{Z} & \longrightarrow & 0 = \pi_1(X) \\
 & & \downarrow f_* & & \downarrow f_* \times f_* & & \wr\wr & & \\
0 & \longrightarrow & \pi_2(\mathbb{P}^n) & \longrightarrow & \pi_2(\mathbb{P}^n \times \mathbb{P}^n) & \longrightarrow & \mathbb{Z} & \longrightarrow & 0 = \pi_1(\mathbb{P}^n)
\end{array}$$

of exact sequences. Hence $\ker f_* = \ker(f_* \times f_*)$, and $\operatorname{coker} f_* = \operatorname{coker}(f_* \times f_*)$. But this forces $\ker f_* = \operatorname{coker} f_* = 0$, i.e. f_* is an isomorphism on π_2. Finally, the surjectivity of $\pi_{n-e+1}(X) \longrightarrow \pi_{n-e+1}(\mathbb{P}^n)$ is non-trivial only if $n - e = 1$, and we leave this case to the reader. (Hint: the diagram above remains exact on the right.) □

Proof of the Lefschetz-type theorem (3.5).

The strategy will be to reduce the result to the following theorem of Goresky and MacPherson, which one may view as a non-compact strengthening of the classical Lefschetz theorem:

(3.8). *Let* Y *be a connected local complete intersection variety of pure dimension* n, *possibly reducible and non-compact, and let*

$$f : Y \longrightarrow \mathbb{P}^m$$

be a finite-to-one morphism. Let $L \subseteq \mathbb{P}^m$ *be a linear space of co-dimension* d, *and denote by* L_ε *an* ε-*neighborhood of* L *with respect to some Riemannian metric on* $\mathbb{P}^n$. *Then for sufficiently small* ε *one has*

$$\pi_i(Y, f^{-1}(L_\varepsilon)) = 0 \quad \underline{\text{for}} \quad i \leq n - d.$$

See [17, §4] for an announcement with indications of proof.

Returning to the situation of (3.5), we start with

Lemma 3.9. *Let* X *be a compact irreducible variety, and* A *an ample line bundle on* X *which is generated by its global sections. Let* $T \longrightarrow X$ *be the direct sum of* t *copies of* A, *and denote by* $X_0 \longrightarrow T$ *the zero section. Suppose that* Y

<u>is a connected, local complete intersection of pure dimension</u> n, <u>and that</u> $g : Y \longrightarrow T$ <u>is a finite (i.e. finite-to-one and proper) map. Then</u>

$$\pi_i(Y, g^{-1}(X_0)) = 0 \quad \underline{\text{for}} \quad i \leq n - t.$$

<u>Proof</u>. The assumption on A means that there is a finite map $\varphi : X \longrightarrow \mathbb{P}^r$ such that $A = \varphi^* \mathcal{O}_{\mathbb{P}^r}(1)$. Let S denote the direct sum of t copies of $\mathcal{O}_{\mathbb{P}^r}(1)$. In a standard manner, one can represent S as a Zariski open subset of $\mathbb{P}^{r+t}$. Specifically, fix disjoint linear spaces $L_0, L \subseteq \mathbb{P}^{r+t}$, of dimensions r and $t - 1$ respectively. Then $S = \mathbb{P}^{r+t} - L$, the bundle map $S \longrightarrow L_0 = \mathbb{P}^r$ being linear projection from L onto L_0. The natural inclusion $L_0 \subseteq S$ is identified with the zero section. Hence we can realize the bundle T on X as the fibre product $X \times_{\mathbb{P}^r} S$. The projection $\Phi : T \longrightarrow S$ is finite, and $X_0 = \Phi^{-1}(L_0)$:

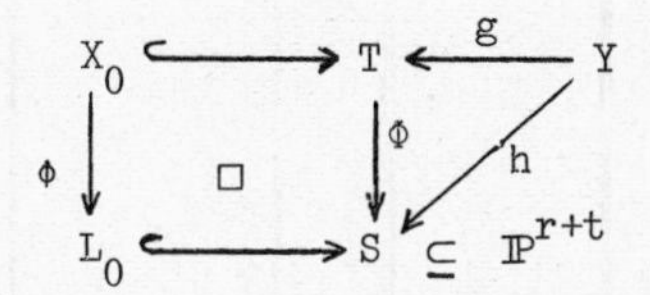

Now let h, as indicated, be the composition $\Phi \circ g : Y \longrightarrow S$, which is finite. We apply the theorem (3.8) of Goresky-MacPherson to h, and to the linear space $L_0 \subseteq S \subseteq \mathbb{P}^{r+t}$. Denoting by L_ε an ε-neighborhood of L_0, we conclude that for sufficiently small ε,

$$\pi_i(Y, h^{-1}(L_\varepsilon)) = 0$$

for $i \leq \dim Y - \operatorname{codim} L_0 = n - t$. But since h is proper, and L_0 is closed, $h^{-1}(L_0)$ is a deformation retract of $h^{-1}(L_\varepsilon)$ when ε is small, and the lemma follows. □

<u>Proof of Theorem 3.5</u>. The hypothesis on E implies that there is a surjective homomorphism $p : T \longrightarrow E$, where T is the direct sum of some number - say t - copies of A. Let $X_s \longrightarrow E$ be the image of the given section $s \in \Gamma(X,E)$, and set $Y = X_s \times_E T$. Denote by $X_0 \longrightarrow T$ and $X \longrightarrow E$ the zero sections:

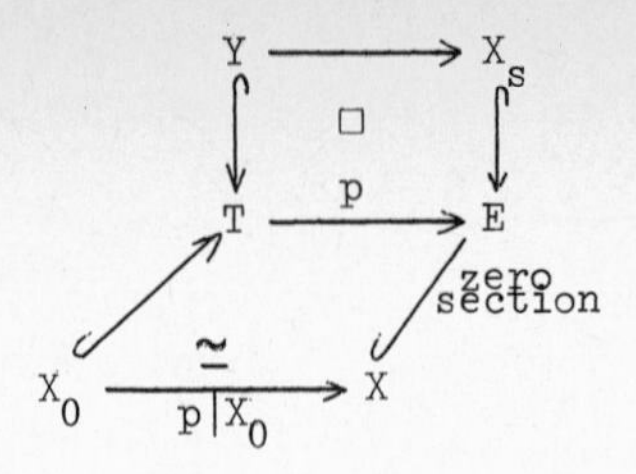

Since p restricts to an isomorphism on zero-sections, we have

$$X_0 \cap Y \xrightarrow{\simeq} X \cap X_s = \left\{ \begin{matrix} \text{zero-locus} \\ Z \quad \text{of} \quad s \end{matrix} \right\}.$$

Bearing in mind that Y is a $\mathbb{C}^{t-e}$-bundle over X_s, on the level of homotopy groups one gets:

$$(*) \qquad \begin{array}{ccc} \pi_i(Y) & \xrightarrow{\simeq} & \pi_i(X_s) \\ \uparrow & & \uparrow \\ \pi_i(X_0 \cap Y) & \xrightarrow{\simeq} & \pi_i(Z) \end{array}$$

But Y is a connected, local complete intersection variety of pure dimension $n+t-e$, and the inclusion $Y \hookrightarrow T$ is a closed embedding, and in particular finite. Hence by Lemma 3.9, $\pi_i(Y, X_0 \cap Y) = 0$ for $i \leq (n+t-e) - t = n-e$, and the theorem follows from (*). □

We conclude this section with a problem on branched coverings of projective space. A well-known, and elementary, theorem states that if $X \subseteq \mathbb{P}^m$ is a smooth variety of degree 3 and dimension n, and if $n \geq 4$, then X is a hypersurface. This was generalized in [31] to branched coverings: if $f : X \longrightarrow \mathbb{P}^n$ has degree three, and if $n \geq 4$, then f factors through an embedding in the total space of a line bundle on $\mathbb{P}^n$:

$$\begin{array}{ccc} X & \hookrightarrow & L \\ & {\scriptstyle f}\searrow \quad \swarrow & \\ & \mathbb{P}^n. & \end{array}$$

One direction in which the classical results on subvarieties generalize is through the "Babylonian" theorems of Barth and Van de Ven [2], [5]. The problem, which was suggested by W. Fulton, is to generalize the result to branched coverings.

Specifically, suppose given for each $n \geq 1$ a branched covering:

$$f_n : X_n \longrightarrow \mathbb{P}^n .$$

Suppose also that $X_n = f_{n+1}^{-1}(\mathbb{P}^n)$ for a suitable hyperplane $\mathbb{P}^n \subseteq \mathbb{P}^{n+1}$. Then describe X_n explicitly. For instance, is X_n a complete intersection in the total space of a direct sum of line bundles on $\mathbb{P}^n$?

We note that the example at the end of [31] suggests that the possibilities for coverings are more varied than those for subvarities.

§4. A Problem of Remmert and Van de Ven.

One of the most elementary results in algebraic geometry is that any projective variety can be mapped onto some projective space. What's less clear, however, is whether projective space is the only smooth variety that plays this role. Our purpose in this section is to show how Mori's results in [35] lead to a proof that this property does in fact characterize projective space:

Theorem 4.1. Let X be a smooth projective variety of dimension ≥ 1, and let

$$f : \mathbb{P}^n \longrightarrow X$$

be a surjective map. Then $X \cong \mathbb{P}^n$.

This was conjectured by Remmert and Van de Ven (cf. [41]). Note that one cannot assert that f is an isomorphism, for there are non-trivial branched coverings $\mathbb{P}^n \longrightarrow \mathbb{P}^n$ (obtained by projections of Veronese embeddings). Observe also that the non-singularity of X is crucial. In fact, if one drops this hypothesis then one can take $X = \mathbb{P}^n/G$, where G is a finite group acting on $\mathbb{P}^n$. We refer the reader to Demazure's paper [6] for a highly readable account of Mori's theorem.

The proof of (4.1) is an elementary application of results proved (but not stated) by Mori in the course of his spectacular proof of the Frankel-Hartshorne conjecture that projective space is the only projective manifold with ample tangent bundle. Specifically, we shall use two results:

(4.2) <u>Let</u> X <u>be a smooth projective variety of dimension</u> n <u>such that the anti-canonical bundle</u> $\Lambda^n(TX)$ <u>is ample. Then for a generic point</u> $P \in X$, <u>there exists a map</u>

$$u = (\mathbb{P}^1, a) \longrightarrow (X,P),^*$$

<u>birational onto its image, with</u> P <u>a smooth point of</u> $u(\mathbb{P}^1)$, <u>and</u>

$$\int_{\mathbb{P}^1} u^* c_1(X) \leq n + 1.$$

This is essentially proved in §2 of [35]. (cf. Thm. 6). Mori's statement does not mention the possibility of finding a rational curve through a general point, but it was observed by Kollar that this is in fact what a small elaboration of Mori's proof yields. Note that the result implies that if X is as in (4.2), then X is uniruled.

The second theorem we need is:

(4.3) <u>Let</u> X <u>be a smooth projective variety of dimension</u> n, <u>and let</u>

$$u : (\mathbb{P}^1, a) \longrightarrow (X,P)$$

<u>be a map, birational onto its image, with</u> $\int u^* c_1(X) \leq n+1$. <u>Suppose that</u> $P(= u(a))$ <u>is a smooth point of</u> $u(\mathbb{P}^1)$, <u>and that the following is satisfied</u>:

(*) <u>For any morphism</u>

$$v : (\mathbb{P}^1, a) \longrightarrow (X,P)$$

<u>arising as a deformation of</u> u <u>through maps taking</u> a <u>to</u> P, <u>the pull-back</u> v^*TX <u>of the tangent bundle of</u> X <u>is ample</u>.

---

*i.e. u is a map $\mathbb{P}^1 \longrightarrow X$, and $a \in \mathbb{P}^1$ is a point with $u(a) = P$.

Then

$$X \cong \mathbb{P}^n .$$

The condition in (*) is that the maps u and v correspond to points in the same connected component of the scheme $\mathrm{Hom}((\mathbb{P}^1, a), (X,P))$ parametrizing maps $\mathbb{P}^1 \longrightarrow X$ taking a to P. (4.3) is the essence of [35], §3. If one knows that TX is ample, then (*) is automatic, and in fact this, plus the amplitude of $\Lambda^n TX$, is the only way in which Mori uses the ampleness of TX (cf. [35], p. 594).

Proof of Theorem 4.1. Note to begin with that X has dimension n, and that f is finite (hence flat). In fact, projective space does not map to any variety other than a point with any fibres of positive dimension. We observe next that $\Lambda^n TX$ is ample. To check this, it suffices by (1.5 (iv)) to show that $f^*\Lambda^n TX$ is ample. But $f^*\Lambda^n TX = \mathcal{O}_{\mathbb{P}^n}(k)$ for some $k \in \mathbb{Z}$, and the inclusion $\Lambda^n T\mathbb{P}^n \longrightarrow \Lambda^n f^* TX$ of sheaves shows that $k \geq n+1$. Thus Mori's theorem (4.2) applies.

Denote by $R \subseteq \mathbb{P}^n$ the ramification divisor of f, and by $B = f(R) \subseteq X$ the branch divisor. By (4.2) there exists a map $u = (\mathbb{P}^1, a) \longrightarrow (X,P)$ as in (4.3) with $P \notin B$. To prove the theorem, it then suffices to show:

(4.4) If $w : (\mathbb{P}^1, a) \longrightarrow (X,P)$ is any non-constant map, with $P \notin B$, then $w^* TX$ is ample.

For once (4.4) is known, (4.3) applies to yeild $X \cong \mathbb{P}^n$.

To prove (4.4), choose a smooth irreducible projective curve C fitting into a commutative diagram

$$\begin{array}{ccc} C & \xrightarrow{\overline{w}} & \mathbb{P}^n \\ {\scriptstyle \overline{f}}\downarrow & & \downarrow{\scriptstyle f} \\ \mathbb{P}^1 & \xrightarrow{w} & X \end{array} ;$$

where $\overline{w}$ and $\overline{f}$ are finite. For example, one may take C to be the normalization of an irreducible component of $\mathbb{P}^1 \times_X \mathbb{P}^n$. Observe that since $P = w(a) \notin B$, the image $\bar{w}(C)$ is not contained in R. Since $\bar{f}$ is flat, it suffices by (1.5 (iv)) to show that $\bar{f}^* w^* TX = \bar{w}^* f^* TX$ is an ample vector bundle on C.

But on $\mathbb{P}^n$ one has the exact sequence

$$(*) \quad 0 \longrightarrow T\mathbb{P}^n \xrightarrow{df} f^*TX \longrightarrow \mathcal{R} \longrightarrow 0$$

of sheaves, where $\mathcal{R}$ is a torsion sheaf supported on the ramification divisor R. Then $\bar{w}^* df : w^* T\mathbb{P}^n \longrightarrow \bar{w}^* f^* TX$ is an isomorphism away from the finite set $\bar{w}^{-1}(R)$, so pulling (*) back by $\bar{w}^*$ expresses $\bar{w}^* f^* TX$ as an extension of the ample vector bundle $\bar{w}^* T\mathbb{P}^n$ by the torsion sheaf $\bar{w}^* \mathcal{R}$. Hence (4.4) is a consequence of

Lemma 4.5. *Let C be a smooth irreducible projective curve, E an ample vector bundle on C, and F a vector bundle on C arising as an extension*

$$0 \longrightarrow E \longrightarrow F \longrightarrow \tau \longrightarrow 0,$$

where τ *is a torsion sheaf. Then F is ample.*

Proof. By Hartshorne's numerical criterion [25], it is equivalent to show that any quotient bundle of F has positive degree. Given such a quotient $F \longrightarrow Q \longrightarrow 0$, we have the exact commutative diagram

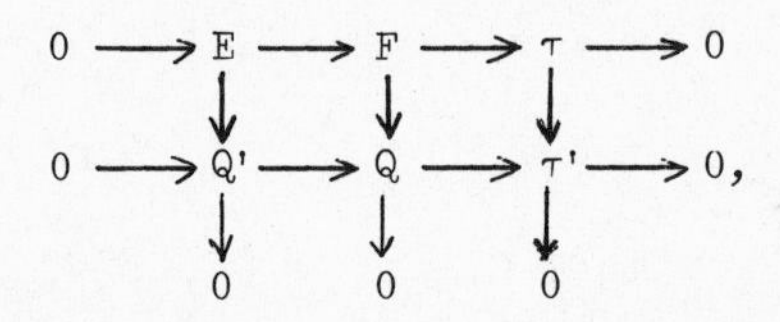

where Q' is the image of the composition $E \longrightarrow F \longrightarrow Q$. Thus Q' is locally free, and τ' is a torsion sheaf on C, and since E is ample, $\deg Q' > 0$. But

$$\deg Q = \deg Q' + \text{length}\ (\tau'),$$

so $\deg Q \geq \deg Q'$. □

This completes the proof of Theorem 4.1.

Problem. Does Theorem 4.1 generalize when $\mathbb{P}^n$ is replaced by a homogeneous space G/P, where G is a semi-simple algebraic group, and $P \subseteq G$ is a maximal parabolic subgroup? For instance, if Q is a quadric of dimension ≥ 3, or a Grassmannian, and if $f = Q \longrightarrow X$ is a non-trivial branched covering, with X smooth, is X a projective space?

References.

1. W. Barth, Transplanting cohomology classes in complex projective space, Am. J. Math. 92 (1970), 951-967.

2. W. Barth, Larsen's theorem on the homotopy groups of projective manifolds of small embedding codimension, Proc. Symp. Pure Math. 29 (1975), 307-311.

3. W. Barth, Submanifolds of low codimensions in projective space, Proc. ICM. Vancouver (1975), 409-413.

4. W. Barth and M. Larsen, On the homotopy groups of complex projective algebraic manifolds, Math. Scand. 30 (1972), 88-94.

5. W. Barth and A. Van de Ven, A decomposability criterion for algebraic 2-bundles on projective spaces, Invent. Math. 25 (1974), 91-106.

6. M. Demazure, Characterisations de l'espace projectif (conjectures de Hartshorne et de Frankel), d'après S. Mori, Sem. Bourbaki, 1979/80, No. 544, Lect. Notes in Math. 842 (1981), 11-19.

7. L. Ein, An analogue of Max Noether's theorem, to appear.

8. W. Fulton, Ample vector bundles, Chern classes, and Numerical criteria, Invent. Math. 32 (1976), 171-178.

9. W. Fulton and J. Hansen, A connectedness theorem for projective varieties, with applications to intersections and singularities of mappings, Ann. Math 110 (1979), 159-166.

10. W. Fulton, J. Harris, and R. Lazarsfeld, Excess linear series on an algebraic curve, to appear.

11. W. Fulton and R. Lazarsfeld, Connectivity and its applications in algebraic geometry, in Algebraic Geometry Proceedings, Lect. Notes in Math 862 (1981), 26-92.

12. W. Fulton and R. Lazarsfeld, On the connectedness of degeneracy loci and special divisors, Acta Math. 146 (1981), 271-283.

13. W. Fulton and R. Lazarsfeld, Positive polynomials for ample vector bundles, Ann. Math. 118 (1983), 35-60.

14. W. Fulton and R. Lazarsfeld, Positivity and excess intersections, in Enumeritive Geometry and Classical Algebraic Geometry, Progress in Math. 24 (1982), 97-105.

15. T. Gaffney and R. Lazarsfeld, On the ramification of branched coverings of $\mathbb{P}^n$, Invent. Math. 59 (1980), 53-58.

16. F. Ghione, Un probleme du type Brill-Noether pour les fibrés vectoriels, in Algebraic Geometry-Open Problems, Lect. Notes in Math 997 (1983), 197-209.

17. M. Goresky and R. MacPherson, Stratified Morse theory, Proc. Symp. Pure Math. 40 (1983), 517-534.

18. H. Grauert, Uber Modifikationen und exceptionelle analytische Mengen, Math Ann. 146 (1962), 331-368.

19. P. Griffiths, Hermitian differential geometry, Chern classes, and positive vector bundles, in D. Spencer and S. Iyanaga (eds.), Global Analysis, Princeton Math. Series No. 29, Tokyo, 1969.

20. P. Griffiths, Some transcendental methods in the study of algebraic cycles, Proceedings of the Maryland Conference on Complex Analysis, Lect. Notes in Math. 185 (1971), 1-46.

21. P. Griffiths and J. Harris, Principles of Algebraic Geometry, Wiley Interscience, New York (1978).

22. H. Hamm, Lefschetz theorems for singular varieties, Proc. Symp. Pure Math. 40 (1) (1983), 547-558.

23. J. Hansen, Connectedness theorems in algebraic geometry, Proceedings of the 18th Scandinavian Congress of Mathematics, Progress in Math. 11 (1980).

24. R. Hartshorne, Ample vector bundles, Publ. Math. I.H.E.S. 29 (1966), 63-94.

25. R. Hartshorne, Ample vector bundles on curves, Nagoya Math. J. 43 (1971).

26. R. Hartshorne, Ample Subvarieties of Algebraic Varieties, Lect. Notes in Math. 156 (1970).

27. R. Hartshorne, Varieties of small codimension in projective space, Bull. A.M.S. 80 (1974), 1017-1032.

28. G. Kempf, Schubert methods with an application to algebraic curves, Publ. Math. Centrum, Amsterdam, 1971.

29. S. Kleiman and D. Laksov, On the existence of special divisiors, Am. J. Math. 94 (1972), 431-436.

30. M. Larsen, On the topology of complex projective manifolds, Invent. Math. 19 (1973), 251-260.

31. R. Lazarsfeld, A Barth-type theorem for branched coverings of projective space, Math. Ann. 249 (1980), 153-162.

32. R. Lazarsfeld and A. Van de Ven, Recent work of F. L. Zak, to appear.

33. J. LePoitier, Annulation de la cohomologie à valeurs dans un fibré vectoriel holomorphe positif de rang quelconque, Math. Ann 218 (75), 35-53.

34. H. Martens, On the variety of special divisors on a curve, J. Reine Angew. Math. 227 (1967, 111-120.

35. S. Mori, Projective manifolds with ample tangent bundles, Ann. Math. 110 (1979) 593-606.

36. D. Mumford, Lectures on Curves on an Algebraic Surface, Ann. Math. Studies 59 (1966).

37. D. Mumford, Prym Varieties I, in Contributions to Analysis, Academic Press (1974), 325-350.

38. D. Mumford, Curves and their Jacobians, Univ. of Mich. Press, Ann Arbor (1975).

39. Z. Ran, On projective varieties of codimension 2, Inv. Math. 73 (1983), 333-336.

40. M. Raynaud, Sections des fibrés vectoriels sur un courbe, Bull. Soc. Math. France, 110 (1982), 103-125.

41. R. Remmert and A. Van de Ven. Über holomorphe Abbildung projektiv-algebraischer Manningfaltigkeiten auf komplexe Räume, Math. Ann. 142 (1961), 453-486.

42. A. Sommese, Submanifolds of abelian varieties, Math. Ann. 233 (1978), 229-256.

43. F. L. Zak, Projections of algebraic varieties, Math. USSR Sbornik, 44 (1983), 535-544.

Department of Mathematics
University of California at Los Angeles
Los Angeles, CA 90024

FACTORIAL AND ALMOST FACTORIAL SCHEMES IN WEIGHTED PROJECTIVE SPACES

by

Lorenzo Robbiano

§ 0 Introduction 62

§ 1 Introduction to weighted projective spaces 64

§ 2 Rings associated to divisors on weighted projective spaces 71

§ 3 Application to factorial and almost factorial rings 76

§ 4 Further remarks and examples 79

§ 0 Introduction

In 1982 I was invited by S. Greco to give four lectures at the CIME meeting on "Complete Intersections". At that time I was interested in the theory of weighted projective spaces (w.p.s.), since some seminars given at the University of Genoa by L. Badescu revealed to me the strong connection between this theory and some problems on complete intersections and factorial rings which I had in mind since the early seventies (see [Ro]). Therefore my first intention was to give lectures on " Complete intersections in w.p.s. " and this became the title of my course. However later on my attention was drawn by the beautiful seminar [Dem] of Demazure, where the theory of w.p.s. is connected with that one of "rational coefficient Weil divisors" on normal projective schemes.

The fact that the methods of [Dem] are very powerful became more and more clear to me along the lines of a serie of seminars given at the University of Genoa by Keiichi Watanabe. He was the first one who used those methods for a better understanding of some deep theorems contained in the paper [Mo 2] of S. Mori, whose main point is the classification of graded 2-dimensional factorial finitely generated k-algebras over an algebraically closed field k.

Therefore my intention of lecturing on complete intersections in w.p.s. evolved to that one of focusing on problems of factoriality and almost factoriality and this explains the reason why the title of the paper does not coincide with that one of the course.

The notion of w.p.s. was introduced by Mori in [Mo 1], developed by Dolgachev in [Do] and Delorme in [Del] and recently used by many authors in a wide range of problems. A fully detailed introduction is going to be given in [B-R], from which I essentially borrow (without proofs) the first section; while the second one is devoted to the description of the rings associated to "rational coefficient Weil divisors" on w.p.s. as it is explained in [Dem]. The main point is Theorem 2.5, which was proved in [Wa] and which states the existence of a very powerful exact sequence relating the class group of the ring associated to a divisor on a projective scheme X and the class group of X itself. This allows us to give a complete description of the Picard groups and the class groups of the w.p.s. (Theorem 2.7).

The third section contains the main results; namely Proposition 3.1, Theorem 3.4 and Theorem 3.5 give a precise description of the family of finitely generated graded factorial k-algebras inside the larger class of finitely generated graded almost factorial k-algebras. In particular Theorem 3.5 describes the "shape" that a rational coefficient Weil divisor must have to give rise to a factorial ring. The last section is devoted to the explicit computation of some interesting examples, which, I hope, should convince the reader that the techniques and the methods explained in the paper not only can give answers to unsolved problems, but also can give new simple form to some classical statements. This is exactly what one should require from a theory to say that it is beautiful and efficient and this was the reason why I felt happy to have the chance of lecturing on it.

For that I must thank S.Greco for the invitation, Lucian Badescu, Igor Dolgachev, Keiichi Watanabe and many others for a large amount of valuable conversations.

<u>In the present paper the symbol k denotes an algebraically closed field of characteristic 0.</u>

§1 Introduction to weighted projective spaces

Let X be an affine variety defined over k, G a finite subgroup of $Aut_k(X)$, $Y = X/G$ the topological quotient i.e. the space of orbits with the induced topology. Let $k[X]$ be the coordinate ring of X and $k(X)$ the field of rational functions on X; then it is clear that G also acts on $k[X]$ and $k(X)$ and we denote by $k[X]^G$ and $k(X)^G$ respectively the subrings of invariants.

Theorem 1.1. *Y is an affine variety with coordinate ring* $k[X]^G$ *and field of rational functions* $k(X)^G$. *The canonical projection* $p: X \longrightarrow Y$ *corresponding to the embedding* $k[X]^G \hookrightarrow k[X]$ *is a finite morphism. Moreover, if G acts freely on X, then p is an étale morphism.*

Pf. See [Se] p. 57 and [Mu 1] p. 65.

Corollary 1.2. *If X is normal, Y is normal.*

Pf. $k[X]^G = k[X] \cap k(X)^G$.

Theorem 1.3. *Let X be a variety (not necessarily affine) and G a finite subgroup of* $Aut_k(X)$. *Assume that every orbit Gx is contained in an affine open set* (e.g. X *is projective*). *Then* $Y = X/G$ *is a variety and if p denotes the canonical projection* $p: X \longrightarrow Y$, *then* $\mathcal{O}_Y = (p_* \mathcal{O}_X)^G$

Pf. See [Mu 1] p. 65.

Let now $\mathbb{G}_m$ denote the multiplicative group variety Spec $k[X, X^{-1}]$ and let A be a finitely generated k-algebra.

Theorem 1.4. *To every* $\mathbb{Z}$*-graduation on A it corresponds an action of* $\mathbb{G}_m$ *on* Spec(A)

Pf. If $A = \bigoplus_{n \in \mathbb{Z}} A_n$, we may consider $A \longrightarrow A[T, T^{-1}]$ given by $a_d \rightsquigarrow a_d T^d$. This is a k-homomorphism and $A[T, T^{-1}] \cong A \otimes_k k[T, T^{-1}]$ hence we get a k-morphism $\mathbb{G}_m \times \mathrm{Spec}(A) \longrightarrow \mathrm{Spec}(A)$ which gives rise to an action of G_m on Spec(A).

Remark This is what we need in the following; however more can be said about this correspondence (see for instance E.G.A. vol II p. 167).

Since we are dealing with a finitely generated graded k-algebra, we may take a minimal set of homogeneous generators $x_0,\dots,x_r$ with degrees $q_0,\dots,q_r$ respectively. Therefore A can be represented as a quotient of $k[T_0,\dots,T_r]$ where $q_i = \deg T_i$ for $i=0,\dots,r$. If $t \in k^* = k-\{0\}$, then it corresponds to a point of $\mathbb{G}_m$ and the maps $A \to A \otimes k[X,X^{-1}] \to A \otimes k[X,X^{-1}]/(X-t) \simeq A$, whose composite is given by $a_d \rightsquigarrow t^d a_d$, describe the corresponding action of $\mathbb{G}_m$ on A (in fact $\mathbb{G}_m$ is commutative; otherwise we should consider $a_d \rightsquigarrow (t^{-1})^d a_d$). If P is a closed point of Spec(A), $P=(a_0,\dots,a_r)$, then the action of $\mathbb{G}_m$ is given by $(t,(a_0,\dots,a_r)) \rightsquigarrow (t^{q_0}a_0,\dots,t^{q_r}a_r)$. Therefore the orbit of P is the curve of $\mathbb{A}^{r+1}$ given parametrically by $(t^{q_0}a_0,\dots,t^{q_r}a_r)$ $t \in k^*$.

__Lemma__ 1.5. $A^{\mathbb{G}_m} = A_0$

__Pf__. Easy exercise.

__Lemma__ 1.6. __Let__ $x_0,\dots,x_r$ __be a minimal set of homogeneous generators of__ A; __put__ $q_i = \deg(x_i)$ __and assume that not all of them are negative__. __Then__ T.F.A.E.

1) $q_i > 0$ $i=0,\dots r$

2) $q_i \geq 0$ $i=0,\dots,r$ __and__ $A_0 = k$

3) __The closures of the orbits in__ $\mathbb{A}^{r+1}$ __only meet at the origin__.

__Pf__. Easy exercise.

This Lemma allows us to give the following

__Definition__. A closed subscheme of $\mathbb{A}^{r+1}$ is called __quasicone__ if it verifies the equivalent conditions of Lemma 1.6. The origin is the __vertex__ of the quasicone.

__Remark__. Quasicones are the closed subschemes of $\mathbb{A}^{r+1}$ which are invariant under an action of $\mathbb{G}_m$ of "positive degree". If $q_i=1$ $i=0,\dots,r$ a quasicone is actually a cone.

Now every graduation on A extends to a graduation on A_f if $f \in A_d$ or, equivalently, the action of $\mathbb{G}_m$ on A, which corresponds to the graduation, extends to the following action of $\mathbb{G}_m$ on A_f

$$A_f \longrightarrow A_f[X,X^{-1}]$$

$$\frac{a_n}{f^s} \rightsquigarrow \frac{a_n}{f^s}.T^{n-sd}$$

and, by 1.5 $A_f^{\mathbb{G}_m} = A_{(f)} = \left\{ \frac{a_n}{f^s} \;/\; n = sd \right\}$. Therefore, if V^+ denotes the punctured spectrum Spec(A) $-\{\mathfrak{m}\}$, then we can say that Proj(A) is the geometric quotient of V^+ by the action of $\mathbb{G}_m$. However $\mathbb{G}_m$ is not a finite group, so there is a need of a better description of Proj(A). Let μ_d denote the group scheme of dth roots of unity; μ_d = Spec(k[X]/(X^d-1)) = Spec(k[X,X^{-1}]/(X^d-1)), hence it is a subgroup-scheme of $\mathbb{G}_m$, and the action of $\mathbb{G}_m$ on Spec(A) restricts to an action of μ_d on Spec(A) in the following way

$$A \longrightarrow A[X,X^{-1}] \longrightarrow A[X,X^{-1}]/(X^d-1)$$

Of course if $d \in A_d$, f is invariant under the action of μ_d, hence (f-1) is an invariant ideal. Therefore we get an action on A/(f-1) which extends to an action on A/(f-1)[U,U^{-1}], where U is an indeterminate of degree 1,

$$A/(f-1)[U,U^{-1}] \longrightarrow A/f-1)[U,U^{-1}][X,X^{-1}]\ /(X^d-1)$$

$$U \rightsquigarrow U\cdot\bar{X}^{-1}$$

$$\bar{a}_n \rightsquigarrow \bar{a}_n\cdot\bar{X}^n$$

Consider now the following homomorphisms

$$A_{(f)} \xrightarrow{\alpha} A/(f-1)$$

$$\frac{a_n}{f^s} \rightsquigarrow \bar{a}_n$$

$$A_f \xrightarrow{\beta} A/(f-1)[U,U^{-1}]$$

$$\frac{a_n}{f^s} \rightsquigarrow \bar{a}_n\cdot U^{n-ds}$$

Theorem 1.7. a) α,β <u>are injective</u>

b) $A_{(f)} = (A/(f-1))^{\mu_d}$

c) $A_f = (A/(f-1)[U,U^{-1}])^{\mu_d}$

d) β <u>is étale</u>

<u>Pf</u>. See [F1] p.37.

<u>Corollary</u> 1.8. <u>Let</u> V <u>be a quasicone and</u> A <u>its coordinate ring</u>. <u>Assume that</u> $V^+ = V-\{0\}$ <u>has one of the following properties</u>: <u>irreducibility</u>, <u>normality</u>, <u>C-M</u> (Cohen-Macaulay). <u>Then</u> Proj(A) <u>has the corresponding property</u>. <u>If</u> V^+ <u>is regular then</u> Proj(A) <u>has only cyclic quotient singularities</u>.

Pf. The properties are stable under the operations of extending by étale morphisms, suppressing indeterminates and taking invariants with respect to the action of finite cyclic groups. Namely, irreducibility is clear, normality follows by Corollary 1.2; as to the property C-M see for instance [Ke] Lemma 8.

Corollary 1.9. If V^+ is regular and $\dim V = 2$, then the curve Proj(A) is regular.

Pf. Proj(A) has dimension 1 and it is normal by Corollary 1.8.

We have seen that every quasicone has a coordinate ring which is a quotient of a ring of polynomials $k[T_0,\dots,T_r]$, graded by $\deg(T_i) = q_i$. This fact leeds naturally to the concept of weighted projective space (w.p.s.).

Let $q_0,\dots,q_r$ be positive integers $Q = (q_0,\dots,q_r)$ $|Q| = \Sigma q_i$, $S(Q) = k[T_0,\dots,T_r]$ graded by $\deg(T_i) = q_i$.

Definition. The weighted projective space of weights Q is $\mathbb{P}(Q) = \operatorname{Proj}(S(Q))$ and U denotes $\mathbb{A}^{r+1} - \{0\}$ i.e. the associated punctured quasicone.

Theorem 1.10. a) $\mathbb{P}(Q)$ is the geometric quotient of U under the action of $\mathbb{G}_m$ "given" by the grading of $S(Q)$.

b) $D_+(T_i) \simeq V_i/\mu_{q_i}$ where $V_i = \operatorname{Spec} k[T_0,\dots,\hat{T}_i,\dots,T_r]$

c) $\mathbb{P}(Q)$ is irreducible, normal, C-M and it has only cyclic quotient singularities.

d) $\mathbb{P}(Q) \simeq \mathbb{P}^r/\mu_Q$ where $\mu_Q = \mu_{q_0} \times \dots \times \mu_{q_r}$

Pf. a) Follows from the remark after Lemma 1.6 (see also [Mu 2]).

b) It is a consequence of Theorem 1.7 b), since

$k[T_0,\dots,T_i,\dots,T_r]/(T_i - 1) \simeq k[T_0,\dots,\hat{T}_i,\dots,T_r]$.

c) It is a consequence of Corollary 1.8 and b).

d) Look at the following commutative diagram

$$\begin{array}{ccc} k[X_0,\dots,X_r] & \longrightarrow & k[X_0,\dots X_r][T,T^{-1}] \\ \wr\wr & & \uparrow \\ k[X_0]\otimes\dots\otimes k[X_r] & \longrightarrow & k[X_0][U_0,U_0^{-1}]\otimes\dots\otimes k[X_r][U_r,U_r^{-1}] \\ (1\otimes\dots\otimes x_i\otimes\dots\otimes 1) & \rightsquigarrow & (1\otimes\dots\otimes x_iU_i\otimes\dots\otimes 1) \end{array}$$

where the top arrow is given by $x_i \rightsquigarrow x_i T$, hence it defines the usual grading, while the vertical arrow is given by $U_i \rightsquigarrow T$. The "degrees" given gy the bottom arrow induce actions of μ_{q_i} and it is clear that $k[X_i]^{\mu_{q_i}} = k[X_i^{q_i}]$.

Therefore $k[X_0,\dots,X_r]^{\mu_Q} = k[X_0^{q_0},\dots, X_r^{q_r}]$.

In the following, if A is a graded algebra over $\mathbb{N}$, we denote as usual by $\mathcal{O}_{\mathrm{Proj}(A)}(n)$ the sheaf $\widetilde{A(n)}$. Therefore $\mathcal{O}_{\mathbb{P}(Q)}(n) = \widetilde{S(Q)(n)}$.

<u>Remark</u> The canonical morphism $p\colon \mathbb{P}^r \longrightarrow \mathbb{P}(Q)$ of Theorem 1.10 d) corresponds, as we have seen, to the equivariant homomorphism

$$k[T_0,\dots,T_r] \simeq k[X_0^{q_0},\dots,X_r^{q_r}] \hookrightarrow k[X_0,\dots,X_r]$$

The corresponding morphism of the quasicones $A^{r+1} \longrightarrow A^{r+1}$ is flat, actually is free with base given by the monomials $\left\{\prod x_i^{\alpha_i} \;/\; 0 \leqslant \alpha_i \leqslant q_i\right\}$.

Therefore $p_* \mathcal{O}_{\mathbb{P}^r} = \oplus\, \mathcal{O}_{\mathbb{P}(Q)}(-\sum_0^r i\alpha_i)$. However p need not be flat, indeed p is flat iff $\mathbb{P}(Q)$ is regular and we shall see (Theorem 1.13 g)) that $\mathbb{P}(Q)$ is regular iff $\mathbb{P}(Q)$ is $\mathbb{P}^r$.

<u>Reduction of Q</u>

Let $Q = (q_0,\dots q_r)$, $Q' = (aq_0,\dots,aq_r)$. Then $\mathbb{P}(Q') = \mathrm{Proj}(S(Q')) \simeq \mathrm{Proj}(S(Q')^{(a)})$.

But it is easy to see that $S(Q')^{(a)} \simeq S(Q)$, hence $\mathbb{P}(Q) \simeq \mathbb{P}(Q')$.

So we may divide the weights by their G.C.D. without changing $\mathbb{P}(Q)$ (up to isomorphisms). If $\mathrm{G.C.D}(q_0,\dots q_r) = 1$, Q is said to be <u>reduced.</u>

<u>Normalization of Q</u>

Put $d_i = \mathrm{G.C.D}(q_0,\dots,\hat{q}_i,\dots,q_r)$ $\quad a_i = \mathrm{l.c.m}(d_0,\dots,\hat{d}_i,\dots d_r)$

$a = \mathrm{l.c.m}(d_0,\dots d_r)$

We say that Q is <u>normalized</u> if $d_i = 1$ $i=0,\dots r$.

In general it turns out that $a_i \mid q_i$ and if we put $\overline{Q} = (\frac{q_0}{a_0},\dots,\frac{q_r}{a_r})$, then $\overline{Q}$ is normalized.

<u>Proposition</u> 1.11. a) $a_i d_i = a$

b) $S(Q)^{(a)} = k[T_0^{d_0},\dots,T_r^{d_r}] \simeq k[T'_0,\dots,T'_r]$ <u>where</u> $\deg(T'_i) = \frac{q_i}{a_i}$

c) $\mathbb{P}(Q) \simeq \mathbb{P}(\overline{Q})$

Pf. a) Easy exercise b) The first equality is an easy computation and the equivariant isomorphism comes from a) since $\frac{d_i q_i}{a} = \frac{q_i}{a_i}$.

c) Clear from b).

Proposition 1.12. Assume Q to be reduced; then the isomorphism $\mathbb{P}(Q) \simeq \mathbb{P}(\bar{Q})$ induces an isomorphism of sheaves

$$\mathcal{O}_{\mathbb{P}(Q)}(n) \simeq \mathcal{O}_{\mathbb{P}(\bar{Q})}((n - \sum b_i(n) q_i)/a)$$

where the b_i's are uniquely determined by the relations $n = b_i(n) q_i + c_i(n) d_i$ $0 \leq b_i(n) < d_i$ and $(n - \sum b_i(n) q_i)/a$ is an integer.

Pf. See [Del] p. 205.

Theorem 1.13. Let $m = \mathrm{l.c.m}(q_0, \dots, q_r)$ then

a) $\mathcal{O}_{\mathbb{P}(Q)}(m)$ is invertible and ample.

b) $\mathcal{O}(rm) \otimes \mathcal{O}(s) \simeq \mathcal{O}(rm+s)$ for every r, s.

c) $\mathcal{O}(s) \simeq \underline{\mathrm{Hom}}(\mathcal{O}(rm), \mathcal{O}(rm+s))$

d) There exists an integer $G(Q)$ such that for every $n > G(Q)$ $\mathcal{O}_{\mathbb{P}(Q)}(n)$ is generated by global sections.

e) For every integer $n > \frac{G(Q)}{m}$ $\mathcal{O}_{\mathbb{P}(Q)}(nm)$ is very ample.

f) $\mathbb{P}(Q)$ is projective.

g) If Q is normalized, then $\mathbb{P}(Q)$ is regular iff $Q = (1,1,\dots,1)$.

Pf. See [Del].

Remark. If $Q = (1,1,2)$ it is easy to see that $\mathcal{O}(1)$ is not invertible and $\mathcal{O}(1) \otimes \mathcal{O}(1) \neq \mathcal{O}(2)$. If $Q = (1,6,10,15)$ it is easy to see that $Q = \bar{Q}$ and $m = 30$ but $\mathcal{O}(30)$ is not very ample. For further remarks on pathologies of the sheaves $\mathcal{O}_{\mathbb{P}(Q)}(n)$ see [Do] and [B-R].

Let now recall some results from [Mo 1].

As usual let $Q = (q_0, \dots q_r)$, $d = \mathrm{G.C.D}(q_0, \dots, q_r)$, $m = \mathrm{l.c.m}(q_0, \dots, q_r)$ and for every prime number p put $\nu(p) = \#\{i \mid p \nmid q_i\}$ and $\nu(Q) = \min \nu(P)$.

Then it is easy to see that $\nu(Q) > 0$ iff Q is reduced, $\nu(Q) > 1$ iff Q is normalized, $\nu(Q) > r-1$ iff $(q_i, q_j) = 1$ for every i,j $i \neq j$, $\nu(Q) = r+1$ iff $Q = (1,\dots,1)$.

<u>Definition</u>. If k is a natural number, we define S_k to be the closed subscheme of $\mathbb{P}(Q)$ defined by the ideal generated by those T_i such that $k \nmid q_i$.

<u>Lemma</u> 1.14. a) $\bigcup_k S_k = \bigcup_{\substack{p \text{ prime} \\ p \mid m}} S_p$

b) $\operatorname{codim}_{\mathbb{P}(Q)} (\bigcup_k S_k) = \nu(Q)$

<u>Definition</u>. $\mathbb{P}^{\circ}(Q) = \mathbb{P}(Q) - \bigcup_k S_k$

<u>Proposition</u> 1.15. a) $\nu(Q) = 0$ <u>iff</u> $\mathbb{P}^{\circ}(Q) = \emptyset$

b) $\nu(Q) = 1$ <u>iff</u> $\mathbb{P}^{\circ}(Q)$ <u>is quasi affine, not empty</u>

c) <u>If</u> $\nu(Q) \geqslant 1$ <u>then</u> $\mathbb{P}^{\circ}(Q)$ <u>contains a complete subscheme of dimension</u> $\nu(Q) - 1$ <u>and does not contain any complete subscheme of dimension greater than or equal to</u> $\nu(Q)$.

<u>Proposition</u> 1.16. a) $\mathcal{O}_{\mathbb{P}(Q)}(1) \neq 0$ <u>iff</u> $d = 1$ <u>iff</u> $\nu(Q) > 0$ <u>iff</u> $\mathbb{P}(Q) \neq \emptyset$

b) $\mathcal{O}_{\mathbb{P}^{\circ}(Q)}(n)$ <u>is invertible for every</u> n

c) $\mathcal{O}_{\mathbb{P}(Q)}(n)$ <u>is reflexive</u>, C-M, <u>of rank</u> 1 (if $\neq 0$)

d) $\mathcal{O}_{\mathbb{P}^{\circ}(Q)}(1)^{\otimes n} \simeq \mathcal{O}_{\mathbb{P}^{\circ}(Q)}(n)$ <u>for every</u> n.

e) $\mathbb{P}^{\circ}(Q)$ <u>is the largest open set such that</u> b) <u>and</u> d) <u>hold and if</u> $\nu(Q) > 1$, i.e. <u>if</u> Q <u>is normalized</u>, $\mathbb{P}^{\circ}(Q)$ <u>is the largest open set such that</u> b) <u>holds</u>.

<u>Proposition</u> 1.17. <u>If we denote by</u> $U^{\circ} = p^{-1}(\mathbb{P}^{\circ}(Q))$, <u>where</u> $p: U \longrightarrow \mathbb{P}(Q)$, <u>we get</u>

a) $U^{\circ} \xrightarrow{p} \mathbb{P}^{\circ}(Q)$ <u>is a</u> $\mathbb{G}_m$ - <u>bundle</u> i.e; <u>it is locally the product of the base and</u> $\mathbb{G}_m$.

b) $\mathcal{O}_{\mathbb{P}^{\circ}(Q)}(1)$ <u>generates</u> $\operatorname{Pic}(\mathbb{P}^{\circ}(Q))$, <u>and if</u> Q <u>is normalized</u> $\operatorname{Pic}(\mathbb{P}^{\circ}(Q)) = \mathbb{Z}$.

§2 Rings associated to divisors on weighted projective spaces

Let A be a finitely generated graded (over $\mathbb{N}$) k-algebra and let $\{t_0,\dots t_r\}$ be a minimal set of homogeneous elements generating A as a k-algebra and assume $r>0$. Then of course $A \simeq k[T_0,\dots,T_r]/J$ where $\deg(T_i) = q_i$.
Therefore we get a canonical embedding of $X = \mathrm{Proj}(A)$ into $\mathbb{P}(Q)$ as a closed subscheme. If we recall that $U = \mathrm{Spec}(S(Q))-\{\mathfrak{m}\}$, $\mathfrak{m}= (T_0,\dots,T_r)$, we get the following commutative diagram

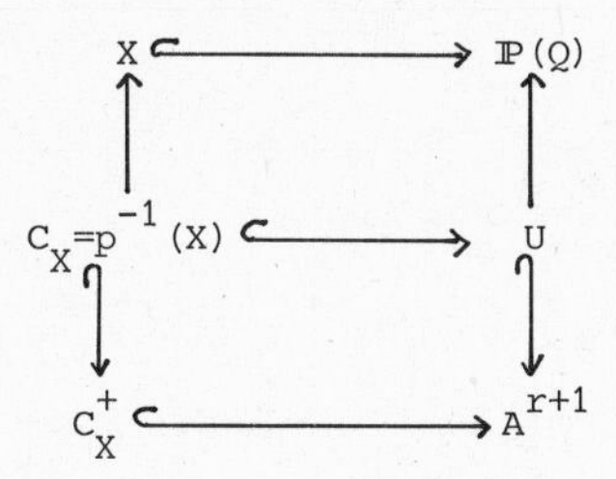

where C_X^+ is the schematic closure of $p^{-1}(X)$ in $\mathbb{A}^{r+1}$ i.e. $i_* \mathfrak{J}_{p^{-1}(X)} = \mathfrak{J}_{C_X^+}$
It is easy to see that $U = \underset{\sim}{\mathrm{Spec}}\ (\bigoplus_{n\in\mathbb{Z}} \mathcal{O}_{\mathbb{P}(Q)}(n))$ hence $p_*: \mathcal{O}_U \longrightarrow \bigoplus_{n\in\mathbb{Z}} \mathcal{O}_{\mathbb{P}(Q)}(n)$
$p^*: \bigoplus_{n\in\mathbb{Z}} \mathcal{O}_{\mathbb{P}(Q)}(n) \longrightarrow \mathcal{O}_U$ are isomorphisms inverse to each other.
Now $\Gamma(U,\mathcal{O}_U) = S(Q)$ since $\mathbb{A}^{r+1}$ is S_2 $(r>0)$, whence we get
$\Gamma(\mathbb{P}(Q),\mathcal{O}_{\mathbb{P}(Q)}(n)) = S(Q)_n$
Let us now consider the exact sequence

$$0 \longrightarrow \bigoplus_{n\in\mathbb{Z}} \mathfrak{J}(n) \longrightarrow \bigoplus_{n\in\mathbb{Z}} \mathcal{O}_{\mathbb{P}(Q)}(n) \longrightarrow \bigoplus_{n\in\mathbb{Z}} \mathcal{O}_X(n) \longrightarrow 0$$

It defines the embedding of $C_X = p^{-1}(X)$ in U, hence $i_*(\bigoplus_{n\in\mathbb{Z}} \mathfrak{J}(n)) = \tilde{I}$
where I is an ideal of $S(Q)$. We say that I is <u>the</u> ideal of the embedding of C_X^+ in A^{r+1}.

<u>Proposition</u> 2.1. a) <u>There is a canonical embedding</u> $\lambda: S(Q)/I \longrightarrow \bigoplus_{n\geqslant 0} H^0(\mathcal{O}_X(n))$

b) <u>If</u> depth $(S(Q)/I) = t>1$, <u>then</u> λ <u>is an isomorphism and</u> $H^i(X,\mathcal{O}_X(n)) = 0$ <u>for every</u> $i \leqslant t-2$

c) <u>If</u> $X = \mathbb{P}(Q)$, $H^i(\mathcal{O}_{\mathbb{P}(Q)}(n)) = 0$ <u>for every</u> n <u>and every</u> $i \leqslant r-1$.

d) $H^r(\mathcal{O}_{\mathbb{P}(Q)}(n)) \simeq S_{-r-|Q|}$

<u>Pf</u>. a), b), c) Apply the standard exact sequence of local cohomology to the triple $C_X^+, C_X, \{0\}$ $\quad 0 \longrightarrow H^0_{\mathfrak{m}}(S(Q)/I) \longrightarrow S(Q)/I \longrightarrow H^0(C_X,\mathcal{O}_{C_X}) \longrightarrow H^1_{\mathfrak{m}}(S(Q)/I) \longrightarrow \dots$

d) See [Do] p. 39-40

<u>Corollary</u> 2.2. <u>If</u> A <u>is normal, then</u> J = I <u>and</u> $A \simeq \bigoplus_{n\geq 0} H^0(X, \mathcal{O}_X(n))$

This means that from A we may construct X and every $\mathcal{O}_X(n)$, and from X and every $\mathcal{O}_X(n)$ we may reconstruct A, if it is normal. Moreover, if A is normal and generated in degree 1, then we may choose an element $x_0 \in A_1$ and consider the divisor $H = \mathrm{div}(x_0)$. Then $A \simeq \bigoplus_{n\geq 0} H^0(X, \mathcal{O}_X(nH))x_0^n$, where $\mathcal{O}_X(nH)(U) = \{f \in K(X) \,/\, (\mathrm{div}(f) + D)|_U \geq 0\}$ Therefore in this case we see that we actually reconstruct A from the pair (X,H). However, in the general case some difficulties arise; namely we saw that $\mathcal{O}(n)$ need not be isomorphic to $\mathcal{O}(1)^{\otimes n}$ and $\mathcal{O}(1)$ itself need not be invertible (see Remark after Theorem 1.13).

So we need something new, and this is provided by the beautiful theory of Demazure (see [Dem]). So let me recall the main points of this theory.

Let $(X, \mathcal{O}_X)$ be a normal noetherian scheme defined over k; denote by

W-div(X) = set of Weil divisors on X

C-div(X) = set of Cartier divisors on X

W-div(X,$\mathbb{Q}$) = W-div(X) $\otimes_{\mathbb{Z}} \mathbb{Q}$ i.e. the set of "rational coefficient Weil divisors".

If $D \in$ W-div(X,$\mathbb{Q}$), $\lfloor D \rfloor$ denotes the integral part of D i.e. $\sup\{\Delta \in \text{W-div}(X) \,/\, \Delta \leq D\}$.

If $\mathcal{O}_X(D)$ is the sheaf defined by $\mathcal{O}_X(D)(U) = \{f \in K(X) \,/\, (\mathrm{div}(f) + D)|_U \geq 0\}$, then $\mathcal{O}_X(D) = \mathcal{O}_X(\lfloor D \rfloor)$. I want to recall now that if $\mathcal{L}$ is a subsheaf of K(X), the constant sheaf of rational functions, then $\mathcal{L}$ is invertible iff $\mathcal{L} = \mathcal{O}(D)$ for some $D \in$ C-div(X); moreover, if U = Reg(X) and $j\colon U \longrightarrow X$ is the canonical embedding then the following conditions are equivalent

a) $\mathcal{L}$ is of finite type, $\mathrm{rk}\,\mathcal{L} = 1$ and $\mathcal{L}$ is reflexive

b) $\mathcal{L} = \mathcal{O}_X(D)$ for some $D \in$ W-div(X)

c) $j^*\mathcal{L}$ is invertible and $\mathcal{L} = j_*j^*\mathcal{L}$.

Let now $D = \sum p_V/q_V\, V \in$ W-div(X,$\mathbb{Q}$) with $(p_V, q_V) = 1$, $q_V > 0$ and V prime divisors. Denote by $\mathfrak{A}(X,D) = \bigoplus_{n\in\mathbb{Z}} \mathcal{O}_X(nD)\, T^n \subset K(X)(T)$ and by $C(X,D) = \mathrm{Spec}\,\mathfrak{A}(X,D)$ and assume that $ND \in$ C-div(X) for some positive N. Then C(X,D) turns out to be normal, there is a canonical projection $p\colon C(X,D) \longrightarrow X$ and $\mathrm{div}(T) = p^*(D) = \sum p_V F_V$ where $F_V = p^{-1}(V)_{red}$

Assume now that X <u>is projective and</u> ND <u>is an ample Cartier divisor for some</u> $N > 0$.

Put $A = A(X,D) = \bigoplus_{n \geq 0} H^0(X, \mathcal{O}_X(nD))T^n$, $A^{\natural} = \bigoplus_{n \geq 0} A_{\geq n} T^n$ where $A_{\geq n} = \bigoplus_{k \geq n} A_k$, $C^+(X,D) = \underset{\sim}{\mathrm{Spec}}(\bigoplus_{n \geq 0} \mathcal{O}_X(nD)T^n)$ and keep these assumptions and notations in the following theorems.

<u>Theorem</u> 2.3. a) A <u>is a finitely generated normal</u> k-<u>algebra</u>

b) <u>There is a canonical projective equivariant morphism</u> $\mathrm{Proj}(A^{\natural}) \simeq C^+(X,D) \longrightarrow \mathrm{Spec}(A)$

c) <u>This morphism induces an equivariant isomorphism</u> $C(X,D) \simeq \mathrm{Spec}(A) - \{\mathfrak{m}\}$ <u>hence an isomorphism on the geometric quotients</u> $X \xrightarrow{j} \mathrm{Proj}(A)$ <u>such that</u>

$j_* \mathcal{O}_X(nD) = \mathcal{O}_{\mathrm{Proj}(A)}(n)$ $\quad j^* \mathcal{O}_{\mathrm{Proj}(A)}(n) = \mathcal{O}_X(nD)$

d) <u>The quotient field of</u> A <u>is</u> K(X)(T).

<u>Pf</u>. See [Dem] p. 14, 15, 16.

<u>Theorem</u> 2.4. <u>Let</u> A <u>be a graded normal</u> k-<u>algebra of finite type over</u> k, $X = \mathrm{Proj}(A)$, T <u>a homogeneous element of degree</u> 1 <u>in</u> K(A). <u>Then there exists a unique</u> $D \in W\text{-}\mathrm{div}(X,\mathbb{Q})$ <u>such that</u> $A_n = H^0(X, \mathcal{O}_X(nD))T^n$, $\quad \mathcal{O}_X(n) = \mathcal{O}_X(nD)T^n$

<u>Pf</u>. See [Dem] p. 17, 18.

Therefore from A we may construct X and D and from the pair (X,D) we may reconstruct A. From the proof of Theorem 2.4 one can also get a procedure to construct D in the following way. Consider $\mathrm{Spec}(A) - \{\mathfrak{m}\}$ and assume T to be a fractional monomial on a minimal set $\{t_0, \ldots, t_r\}$ of homogeneous generators of A as a k-algebra, such that $T = \prod t_i^{s_i}$ $\deg(T) = \sum s_i q_i = 1$ (this is possible since we may assume that Q is reduced (see Reduction of Q after Theorem 1.10)).

Then on $\mathrm{Spec}(A) - \{\mathfrak{m}\}$ we have $\mathrm{div}(T) = \sum s_i \, \mathrm{div}(t_i)$ and $\mathrm{div}(t_i) = \sum r_{ij} F_{ij}$ where F_{ij} are prime divisors, which are stable under the action of $\mathbb{G}_m$ since t_i is homogeneous. Then we put $D_{ij} = p(F_{ij})$ where $p: \mathrm{Spec}(A) - \{\mathfrak{m}\} \longrightarrow X$ is, as usual, the canonical projection, we put as usual $d_i = \mathrm{G.C.D}(q_0, \ldots, \hat{q_i}, \ldots, q_r)$.

Then $D = \sum_i s_i (\sum_j s_{ij}/d_i \, D_{ij})$.

Now we come to the following application, which is the most important for our purposes.

<u>Theorem</u> 2.5. <u>Let</u> $D = \Sigma p_V/q_V \; V \in W\text{-div}(X,Q)$ <u>where</u> V <u>are prime</u>, $(p_V,q_V)=1$, $q_V > 0$ <u>and</u> ND <u>is an ample Cartier divisor for some</u> N <u>positive</u>. <u>Let</u> $A = A(X,D)$ <u>and</u> $L_D = \text{l.c.m}(q_V)$ <u>Put</u> $\alpha : \mathbb{Z} \longrightarrow \oplus \mathbb{Z}/q_V\mathbb{Z}$

$$1 \rightsquigarrow (p_V \bmod q_V)_V$$

<u>Then</u> $X = \text{Proj}(A)$ <u>and there exists an exact sequence</u>

$$0 \longrightarrow \mathbb{Z} \longrightarrow Cl(X) \longrightarrow Cl(A) \longrightarrow \text{Coker}\,\alpha \longrightarrow 0$$

$$1 \rightsquigarrow [L_D \cdot D]$$

<u>Pf</u>. See [Wa].

Now we recall some well-known definitions

<u>Definitions</u>. A normal ring A is said to be <u>factorial</u> or UFD if $Cl(A) = 0$.

A normal ring A is said to be <u>almost factorial</u> or AFD if Cl(A) is torsion.

For details on these notions see for instance [St] and [Fo].

<u>Corollary</u> 2.6. <u>Let</u> A <u>be a graded normal k-algebra of finite type over</u> k, $X = \text{Proj}(A)$, D <u>a rational coefficient Weil divisor such that</u> $A \simeq A(X,D)$. <u>Then</u>

A <u>is</u> UFD <u>iff</u> $Cl(X) = \mathbb{Z}$ <u>generated by</u> $[L_D \cdot D]$ <u>and the</u> q_V<u>'s are pairwise coprime</u>.

A <u>is</u> AFD <u>iff</u> $rk(Cl(X)) = 1$

Now we can compute the class group and the Picard group of every w.p.s.

<u>Theorem</u> 2.7. <u>Assume</u> Q <u>to be reduced; then</u>

a) $Cl(\mathbb{P}(Q)) = \mathbb{Z}$ <u>generated by</u> $\mathcal{O}(a)$ <u>where</u> $a = \text{l.c.m}(d_0,\dots,d_r)$

b) <u>If</u> $Q = \overline{Q}$, <u>then</u> $Cl(\mathbb{P}(Q)) = \mathbb{Z}$ <u>generated by</u> $\mathcal{O}(1)$

c) $\text{Pic}(\mathbb{P}(Q)) = \mathbb{Z}$ <u>generated by</u> $\mathcal{O}(m)$ <u>where</u> $m = \text{l.c.m}(q_0,\dots,q_r)$

d) $\mathbb{P}(Q)$ <u>is locally</u> AFD <u>and it is locally</u> UFD <u>iff</u> $\overline{Q} = (1,\dots,1)$.

<u>Pf</u>. Since $\mathbb{P}(Q) = \text{Proj}(S(Q))$, we can use Theorem 2.5 if we know the divisor D such that $S(Q) = A(\mathbb{P}(Q), D)$; the computation of D can be achieved by using the remarks following Theorem 2.4. Since Q is reduced, there exist integers r_i such that $\Sigma r_i q_i = 1$, hence there exists a homogeneous element $T = \prod T_i^{r_i} \in K(S(Q))$ such that $\deg(T) = 1$. So, if we denote by $H_i = \text{div}(T_i)$ on $\mathbb{P}(Q)$, we get $D = \Sigma\, r_i/d_i \; H_i$. By Theorem 2.5 we get the exact sequence

$$0 \longrightarrow \mathbb{Z} \longrightarrow Cl(\mathbb{P}(Q)) \longrightarrow 0$$

$$1 \rightsquigarrow [L_D \cdot D]$$

and by Theorem 2.4, $\mathcal{O}_{\mathbb{P}(Q)}(n) = \mathcal{O}_{\mathbb{P}(Q)}(nD)T^n$. Therefore if $Q = \bar{Q}$, $d_i = 1$ for every i, $L_D = 1$ and b) is proved. Suppose now that Q is not normalized; if $r_i = 0$ it means that $\sum r_i q_i = 1$ can be obtained with the q_j's $j \neq i$; hence all the d_i's which are different from 1 appear as denominators in D. Therefore L = a and a) is proved.

c) Clearly $\mathrm{Pic}(\mathbb{P}(Q))$ is $\mathbb{Z}$, being a subgroup of $Cl(\mathbb{P}(Q))$. Now, using the results of Delorme, we know that m is the smallest positive integer such that $\mathcal{O}(m)|_{D_+(T_i)}$ is free of rk 1 for every i. On the other hand $D_+(T_i) = \mathrm{Spec}(k[T_0,\ldots,\hat{T}_i,\ldots,T_r]^{\mu_{q_i}})$ by Theorem 1.7 b) and $k[T_0,\ldots\hat{T}_i,\ldots,T_r]^{\mu_{q_i}} = \bigoplus_n k[T_0,\ldots,\hat{T}_i,\ldots,T_r]_{nq_i}$. Therefore $D_+(T_i)$ is the affine scheme associated to a ring which can be naturally graded over $\mathbb{N}$ and whose part of degree 0 is k. By [Fo] 10.4 p. 43 its Pic is trivial, hence we can say that m is the smallest positive integer such that $\mathcal{O}(m)|_{D_+(T_i)}$ is invertible for every i. Now if Q is not normalized, $\mathcal{O}_{\mathbb{P}(Q)}(m) \simeq$ $\simeq \mathcal{O}_{\mathbb{P}(\bar{Q})}(m/a)$ by Proposition 1.12, and $m/a = \mathrm{l.c.m}(q_0/a_0,\ldots,q_r/a_r)$, therefore we may assume $Q = \bar{Q}$. With this assumption, if $\mathcal{O}_{\mathbb{P}(Q)}(r) \simeq \mathcal{O}_{\mathbb{P}(Q)}(s)$, then $\mathcal{O}_{\mathbb{P}^\circ(Q)}(r) \simeq \mathcal{O}_{\mathbb{P}^\circ(Q)}(s)$ hence r = s by Proposition 1.17 b). Now if $[\mathcal{O}(D)]$ generates $\mathrm{Pic}(\mathbb{P}(Q))$, it follows that $\mathcal{O}(D) \simeq \mathcal{O}(r)$ for a suitable positive integer r since $Cl(\mathbb{P}(Q))$ is generated by $\mathcal{O}(1)$. But m is the smallest positive integer such that $\mathcal{O}(m)$ is invertible; then $r \geq m$ and $\mathcal{O}(m) \simeq \mathcal{O}(sr)$, whence m = sr; in conclusion s = 1 and we are done.

d) By [B-O] Proposition 2.1 and by a), b), c) we get that $\mathbb{P}(Q)$ is locally AFD. Clearly it is locally UFD iff a = m. But $\mathbb{P}(Q) \simeq \mathbb{P}(\bar{Q})$ and $a(\bar{Q}) = 1$, so $\mathbb{P}(Q)$ is locally UFD iff $m(\bar{Q}) = 1$ iff $\bar{Q} = (1,\ldots,1)$.

§ 3 Applications to factorial and almost factorial rings

In the following, X is a normal projective scheme and we denote by $\mathcal{D}_X$ or simply by $\mathcal{D}$ the set of the rational coefficient Weil divisors D such that there exists a positive integer N with ND Cartier and ample.

Proposition 3.1. The following conditions are equivalent

a) There exists $D \in \mathcal{D}$ such that $A(X,D)$ is AFD

b) For every $D \in \mathcal{D}$, $A(X,D)$ is AFD

c) $rk(Cl(X) = 1$

Pf. For every $D \in \mathcal{D}$, $Proj(A(X,D) \simeq X$ by Theorem 2.3. The conclusion is now an easy consequence of the exact sequence of the Theorem 2.5.

Definition. If $D \in \mathcal{D}$ we say that X is normally embedded by D or that the embedding given by D is normalized if $A(X,D)$ can be minimally generated by homogeneous elements $t_0, \ldots t_r$ of degrees $q_0, \ldots, q_r$ such that, if $Q = \{q_0, \ldots, q_r\}$, then $Q = \overline{Q}$. The embedding is said to be UFD (resp. AFD,...) if $A(X,D)$ is UFD (resp. AFD,...)

Definition. $D \in \mathcal{D}$ is said to be a pairwise coprime divisor (p.c. div.) if $D = \sum p_V/q_V\, V$, where V are prime divisors, $(p_V, q_V) = 1$, $q_V > 0$ for every V and the q_V's are pairwise coprime. In particular if $D \in (W\text{-div}(X)) \cap \mathcal{D}$ then it is clear that it is a p.c. div. We denote by $L_D = \text{l.c.m}(q_V)$.

Proposition 3.2. Let $D, D' \in \mathcal{D}$, then T.F.A.E.

a) There exists an equivariant isomorphism between $A(X,D)$ and $A(X,D')$

b) There exists an automorphism λ of X such that $\lambda(D)$ is linearly equivalent to D'.

Pf. See [Dem] 4.2 p. 20-21.

Proposition 3.3. Let $D \in \mathcal{D}$. Then T.F.A.E.

a) X is normally embedded by D

b) $L_D = 1$ i.e. $D \in W\text{-div}(X)$.

Pf. Let $D \in \mathcal{D}$ and write $A(X,D) \simeq k[T_0, \ldots, T_r]/I$ where, if we denote by t_i the residue class of T_i mod I, $\{t_0, \ldots, t_r\}$ corresponds to a minimal set of homogeneous generators of $A(X,D)$ as a k-algebra. By Theorem 2.3 a), d), $A(X,D)$ is normal and $K(A) = K(X)(T)$ where $\deg(T) = 1$. Then if $Q = \{q_0,\dots q_r\}$ is the set of weights, Q is forced to be reduced. Therefore there exist integers r_i such that $T = \prod t_i^{r_i}$ is a homogeneous element of $K(A)$ of degree 1. On $C(X,D) \simeq \operatorname{Spec}(A) - \{\mathfrak{m}\}$ we have $\operatorname{div}(T) = \sum r_i \operatorname{div}(t_i)$ and then there exists a unique $\Delta \in W\text{-div}(X,Q)$ such that $\operatorname{div}(T) = p^*(\Delta)$ where $p : \operatorname{Spec}(A) - \{\mathfrak{m}\} \longrightarrow X$ is the canonical projection (See Theorem 2.4 and the following discussion). If we put $H_i = \operatorname{div}(t_i)$ on X, we know that $\Delta = \sum r_i/d_i \, H_i$. Therefore we get a canonical equivariant isomorphism $A(X,D) \simeq A(X,\Delta)$ hence, by Proposition 3.2, an automorphism $\lambda \in \operatorname{Aut}_k(X)$ such that $\lambda(D) \sim \Delta$. But now we have clearly the following equivalences: X is normally embedded by D iff X is normally embedded by Δ iff $d_i = 1$ for every i iff $L_\Delta = 1$ iff $\Delta \in W\text{-div}(X)$ iff $D \in W\text{-div}(X)$ iff $L_D = 1$.

Theorem 3.4. *Let X be a normal projective scheme; then T.F.A.E.*

a) $Cl(X) = \mathbb{Z}$

b) *There exists* $D \in \mathfrak{D}$ *such that* $A(X,D)$ *is* UFD.

c) *There exists a unique* *(up to linear equivalence)* $D \in \mathfrak{D} \cap W\text{-div}(X)$ *such that* $A(X,D)$ *is* UFD.

d) *There exists a unique normalized* UFD *embedding of* X *in* $\mathbb{P}(Q)$ *(up to automorphisms of* $\mathbb{P}(Q)$*)*.

Pf. a) $\Rightarrow$ c) Let $D \in W\text{-div}(X)$ be such that $[D]$ generates $Cl(X)$. First we observe that either D or $-D$ is in $\mathfrak{D}$; for, let $X \longrightarrow \mathbb{P}^N$ be any embedding and H an hyperplane section of X. Then $[nD] = [H]$ and we are done. Then denote by D a divisor such that $D \in \mathfrak{D} \cap W\text{-div}(X)$ and $[D]$ generates $Cl(X)$ and consider $A(X,D)$. It is normal by Theorem 2.3 a) and there is an exact sequence

$$0 \longrightarrow \mathbb{Z} \longrightarrow Cl(X) \longrightarrow Cl(A(X,D)) \longrightarrow 0$$
$$1 \rightsquigarrow [D]$$

by Theorem 2.5. Therefore $Cl(A(X,D)) = 0$. Suppose now that there is another $D' \in W\text{-div}(X) \cap \mathfrak{D}$ such that $A(X,D')$ is UFD. Then $[D] = [D']$ whence D is linearly equivalent to D'.

c) $\Rightarrow$ d) If we replace D by $D' = D + \operatorname{div}(f)$ we get $A(X,D') = \bigoplus_{n \geq 0} H^0(X, \mathcal{O}_X(nD))(fT)^n$ hence we may only change the embedding by choosing different bases of the

$H^0(X, \mathcal{O}_X(nD))$'s. But this means exactly that we change the embedding of X by an automorphism of $\mathbb{P}(Q)$. The fact that the embedding is normalized follows from Proposition 3.3.

d) $\Rightarrow$ b) Obvious.

b) $\Rightarrow$ a) It is a consequence of Theorem 2.5 since $\mathrm{Proj}(A(X,D)) \simeq X$ by Theorem 2.3 c).

If X is as usual a normal projective scheme and if we assume that $\mathrm{Cl}(X) = \mathbb{Z}$ generated by $[D]$, where D is a prime divisor in $W\text{-div}(X) \cap \mathcal{D}$ then to every $\Delta \in W\text{-div}(X)$ we associate an integer $d(\Delta)$ which is defined by the relation $\Delta \sim d(\Delta)D$.

Theorem 3.5. Let X be a normal projective scheme such that $\mathrm{Cl}(X) = \mathbb{Z}$ generated by $[D]$ where D is a prime divisor in $W\text{-div}(X) \cap \mathcal{D}$ and let $\Delta \in \mathcal{D}$. Then T.F.A.E.

1) $A(X, \Delta)$ is UFD

2) $\Delta = \sum_1^s{}_i (p_i/q_i) V_i$ where

a) V_i are distinct prime divisors

b) $(p_i, q_i) = 1$ for every i

c) $q_1, \ldots, q_s$ are pairwise coprime positive integers

d) $\sum (p_i \cdot d(V_i))/q_i = 1/\prod q_i$.

Pf. 1) $\Rightarrow$ 2) Let $\Delta = \sum (p_i/q_i) V_i$ where V_i are distinct prime divisors such that $(p_i, q_i) = 1$ and $q_i > 0$ for every i. Then c) follows from Corollary 2.6 and $L_\Delta = \prod q_i$. Again by Corollary 2.6 we get $L_\Delta \cdot \Delta = \sum_i (p_i \cdot \prod_{i \neq j} q_j) V_i \sim D$.

But $V_i \sim d(V_i)D$ whence $\sum (p_i \cdot d(V_i) \cdot \prod_{i \neq j} q_j) = 1$ and also d) is checked.

2) $\Rightarrow$ 1) a) and c) say that Δ is a p.c. div. By a), b), c) we get that $L_\Delta = \prod_i q_i$, hence $L_\Delta \cdot \Delta = \sum_i (p_i \cdot \prod_{i \neq j} q_j) V_i$ by d). But this divisor is equivalent to $(\sum (p_i \cdot d(V_i) \cdot \prod_{i \neq j} q_j))D = D$ by d) and we conclude by Corollary 2.6.

§4 Further remarks and examples

After the discussion of the previous sections it is clear that, if X is a normal projective variety, then either $rk(Cl(X)) > 1$ and then from X we cannot construct AFD rings or $rk(Cl(X)) = 1$ and for every $D \in \mathcal{D}$, $A(X,D)$ is AFD (Proposition 3.1). Moreover if $Cl(X) = \mathbb{Z}$ then there exists a unique (up to linear equivalence) $D \in W\text{-div}(X) \cap \mathcal{D}$ with $A(X,D)$ UFD (Theorem 3.4) and other UFD rings can be obtained by using Theorem 3.5. We want to show now some explicit examples.

Example 1. Let $A = k[X,Y,Z]/(X^2+Y^3+Z^5)$ Put $Q = (15, 10, 6)$ where $15 = \deg X$, $10 = \deg Y$, $6 = \deg Z$. Then A is a normal graded k-algebra, where, if we put $T = yz/x$, T is a homogeneous element of degree 1. Now we normalize Q (See "normalization of Q" after Theorem 1.10)

$d_1 = 2,\ d_2 = 3,\ d_3 = 5 \qquad a_1 = 15,\ a_2 = 10,\ a_3 = 6$

Then $\overline{Q} = (1,1,1)$ and $A^{(30)} = k[X^2, Y^3, Z^5]/(X^2+Y^3+Z^5)$, which is equivariantly isomorphic to $k[x,y,z]/(x+y+z)$ where $\deg x = \deg y = \deg z = 1$

Then $\mathrm{Proj}(A) \simeq \mathbb{P}^1$ whence A is AFD by Proposition 3.1.

Since $p^*(D) = \mathrm{div}(T) = \mathrm{div}(y) + \mathrm{div}(z) - \mathrm{div}(x)$ on $\mathrm{Spec}(A) - \{m\}$ and div(y) is the pull-back of $P = (1,0)$, div(z) is the pull-back of $R = (1,1)$, div(x) is the pull-back of $Q = (0,1)$, we get $D = 1/3\ P + 1/5\ R - 1/2\ Q$.

Since $1/3 + 1/5 - 1/2 = 1/30$ we get that A is UFD by Theorem 3.5.

Remark. This is a particular example of the family of the normal graded UFD k-algebras $(k = \overline{k})$ of dimension 2. If A is such an algebra, then Proj(A) is a normal (hence smooth) curve whose divisor class group is $\mathbb{Z}$; therefore $\mathrm{Proj}(A) \simeq \mathbb{P}^1$ and so Proposition 3.2 and Theorem 3.5 allow us to give an exact description of the family. This was done by Mori in [Mo 2] and by Watanabe in [Wa]. For instance all the members of the family, which can be generated as k-algebras by 3 homogeneous elements are equivariantly isomorphic to one of the following rings: $k[X,Y,Z]/(X^p+Y^q+Z^r)$ where p, q, r are pairwise coprime positive integers.

Example 2. (See [Sa] and [Ro])

Let $A = k[X,Y,Z,U]/(F)$ where $F = X^2 + Y^3 + Z^6U$ Put $Q = (12, 8, 3, 6)$ where $12 = \deg X$, $8 = \deg Y$, $3 = \deg Z$, $6 = \deg U$. Then A is a normal graded k-algebra, where, if we put $T = x/yz$, T is a homogeneous element of degree 1; now we normalize Q: $d_1 = 1$, $d_2 = 3$, $d_3 = 2$, $d_4 = 1$ $a_1 = 6$, $a_2 = 2$, $a_3 = 3$, $a_4 = 6$

Then $\bar{Q} = (2,4,1,1)$ and $A^{(6)} \simeq k[X, Y^3, Z^2, U]/(F)$ which is equivariantly isomorphic to $k[x,y,z,u]/(x^2 + y + z^3u) \simeq k[x, z, u]$ where $\deg x = 2$, $\deg y = 4$, $\deg z = 1$, $\deg u = 1$. Then $\mathrm{Proj}(A) \simeq \mathbb{P}(2,1,1)$.

Now $p^*(D) = \mathrm{div}(T) = \mathrm{div}(x) - \mathrm{div}(y) - \mathrm{div}(z)$. Denote by D_x, D_y, D_z, D_u the divisors on $\mathbb{P}(2,1,1)$ corresponding to $\mathrm{div}(x)$, $\mathrm{div}(y)$, $\mathrm{div}(z)$, $\mathrm{div}(u)$.

Since $\mathrm{Cl}(\mathbb{P}(2,1,1)) = \mathbb{Z}$ generated by $[\mathcal{O}(1)]$ (See Theorem 2.7), we get that D_x, D_z, D_u are prime divisors and $d(D_x) = 2$, $d(D_z) = 1$, $d(D_U) = 1$, while D_y is defined by $x^2 + z^3u = 0$ hence it is also a prime divisor and $d(D_y) = 4$.

Therefore $D = D_x - 1/3\, D_y - 1/2\, D_z$ and $2 - 4/3 - 1/2 = 1/6$, whence A is UFD by Theorem 3.5.

Example 3. Let $A = k[X,Y,Z,U]/(F)$ where $F = X^2 + Y^3 + Z^6U^2$ Put $Q = (12, 8, 3, 3)$

Let $T = x/yz$ and normalize Q; $\bar{Q}$ turns out to be $(4, 8, 1,1)$

$A^{(3)} \simeq k[X, Y^3, Z, U]/(F) \simeq k[x, y, z, u]/(x^2 + y + z^6u^2)$. Then $\mathrm{Proj}(A) \simeq \mathbb{P}(4,1,1)$ and A is AFD by Proposition 3.1, and Theorem 2.7.

Let D_x, D_y, D_z, D_u denote the divisors of x, y, z, u on $\mathbb{P}(4,1,1)$. Then D_x, D_z, D_u are prime divisors of degree 4, 1, 1 respectively, while D_y is defined by $x^2 + z^6u^2 = 0$. Then $D_y = D'_y + D''_y$ where D'_y is defined by $x - z^3u = 0$, and D''_y is defined by $x + z^3u = 0$. Then D'_y and D''_y are prime divisors such that $d(D'_y) = 4$, $d(D''_y) = 4$ and $D = D_x - 1/3\, D'_y - 1/3\, D''_y - D_z$. Then $L_D = 3$ and $[L_D \cdot D]$ generates $\mathrm{Cl}(\mathbb{P}(4,1,1))$. To compute $\mathrm{Cl}(A)$ we need computing α of Theorem 2.5; but $\alpha : \mathbb{Z} \longrightarrow \mathbb{Z}_3 \oplus \mathbb{Z}_3$, $1 \rightsquigarrow (-1, -1)$ Therefore $\mathrm{Cl}(A) = \mathbb{Z}_3$.

Let us now recall the following remarkable

Theorem 4.1. Let X be a projective variety of dimension ≥ 3, which is a complete intersection in a weighted projective space $\mathbb{P}(Q)$ and such that $X \subset \mathbb{P}^{\circ}(Q)$. Then $\mathrm{Pic}(X) = \mathbb{Z}$ generated by $\mathcal{O}_X(1)$.

Pf. See [Mo 1] .

It is the "weighted" version of the classical Lefschetz Theorem on complete intersections and the proof is based on the classical version, by using the canonical projection $p: \mathbb{P}^r \longrightarrow \mathbb{P}(Q)$.

Corollary 4.2. Let X be a locally UFD projective variety, such that $\dim(X) \geq 3$, which is a complete intersection in $\mathbb{P}(Q)$ with $Q = \overline{Q}$ and such that $X \subset \mathbb{P}^{\circ}(Q)$. Then its projective coordinate ring $A = S(Q)/I$ is UFD.

Pf. X is locally UFD hence $\mathrm{Pic}(X) = \mathrm{Cl}(X) = \mathbb{Z}$ generated by $\mathcal{O}_X(1)$. Let T be a homogeneous element of degree 1 in K(A) and put $H = \mathrm{div}(T)$ on X; Then $H \in W\text{-div}(X)$ since $Q = \overline{Q}$ (See Proposition 3.3), and $A \simeq A(X,H)$; moreover $\mathcal{O}(H) \simeq \mathcal{O}(1)$ on $\mathbb{P}^{\circ}(Q)$ hence $\mathcal{O}_X(H) \simeq \mathcal{O}_X(1)$ and the conclusion follows from Corollary 2.6.

Example 4. (this example was shown to me by L. Badescu)

Let $B = k[T_0, T_1, T_2, T_3, T_4]$ with $Q = (1,1,1,1,n)$ $n > 1$ and $F = T_0^{2n} + T_1^{2n} + T_2^{2n} + T_3^{2n} + T_4^{2}$, $A = B/(F)$. Then $X = \mathrm{Proj}(A)$ has a natural embedding in $\mathbb{P}(1,1,1,1,n)$ and it is easy to see that $X \subset \mathbb{P}^{\circ}(1,1,1,1,n) =$ $= \mathbb{P}(1,1,1,1,n) - \{(0,0,0,0,1)\}$. Then Corollary 4.2 tells us that A is UFD. Moreover $Q = \overline{Q}$, therefore, by Theorem 3.4 d) there is no embedding of X into any $\mathbb{P}^N$ such that the corresponding projective coordinate ring is UFD.

Example 5. Let $X = \mathbb{P}^1$ $P = (0,1)$ $Q = (1,0)$, u, v coordinates on $\mathbb{P}^1$, $D = 1/2\, P + 1/3\, Q$. Then

$\lfloor D \rfloor = 0$ hence $H^0(\mathbb{P}^1, \mathcal{O}(D))$ is generated by 1

$\lfloor 2D \rfloor = P$ " $H^0(\mathbb{P}^1, \mathcal{O}(2D))$ is " by 1, v/u

$\lfloor 3D \rfloor = P + Q$ " $H^0(\mathbb{P}^1, \mathcal{O}(3D))$ is " by 1, v/u, u/v

and it is easy to see that these sections are enough to generate $A(\mathbb{P}^1, D)$.
Therefore $A(\mathbb{P}^1, D) = k[T, v/u\, T^2, u/v\, T^3] \simeq k[X,Y,Z]/(X^5 - YZ)$ where $1 = \deg X$, $2 = \deg Y$, $3 = \deg Z$. Since $L_D = 6$ and $6D = 3P + 2Q$, from the exact sequence of Theorem 2.5 we get $Cl(A(\mathbb{P}^1, D)) = \mathbb{Z}_5$.

Let me now recall the "weighted" version of the adjunction formula

<u>Theorem</u> 4.3. <u>Let</u> X <u>be a complete intersection in</u> $\mathbb{P}(Q)$ <u>of multidegree</u> $(d_1, \ldots d_s)$, <u>such that its projecting punctured quasicone</u> C_X <u>(see the beginning of section</u> 2) <u>is smooth</u>. <u>Then</u> X <u>has a dualizing sheaf</u> ω_X <u>and</u> $\omega_X = \mathcal{O}_X(\Sigma d_i - |Q|)$

<u>Pf</u>. See [Do] 3.3.4.

<u>Example</u> 6. (<u>Description of the affine surfaces in</u> A^3_k <u>of type</u> $x^\alpha + y^\beta + z^\gamma = 0$ <u>which are</u> AFD).

Let $A = k[X, Y, Z]/(F)$ where $F = X^\alpha + Y^\beta + Z^\gamma$.

Put $d = (\alpha,\beta,\gamma)$ $\alpha' = \alpha/d$ $\beta' = \beta/d$ $\gamma' = \gamma/d$ and $r = (\alpha',\beta')$, $s = (\alpha',\gamma')$ $t = (\beta',\gamma')$. Then it is clear that $(r,s) = (r,t) = (s,t) = 1$ and we may write $\alpha = drsa$, $\beta = drtb$, $\gamma = dstc$ where $(a,b,c) = 1$

Put $N = drstabc$ and $Q = (tbc, sac, rab)$ where $tbc = \deg X$, $sac = \deg Y$, $rab = \deg Z$. Then F becomes homogeneous of degree N.

Let us normalize Q: $d_1 = a$, $d_2 = b$, $d_3 = c$ $a_1 = bc$, $a_2 = ac$, $a_3 = ab$ and $A^{(abc)} = k[X^a, Y^b, Z^c]/(F) \simeq k[x, y, z]/(G)$ where $G = x^{drs} + y^{drt} + z^{dst}$, and $t = \deg x$, $s = \deg y$, $r = \deg z$, hence $drst = \deg G$.

Now $X = \mathrm{Proj}(A) \simeq \mathrm{Proj}(k[x,y,z]/(G)$ and X is a smooth curve by Corollary 1.9.

By Theorem 4.3 $\omega_X = \mathcal{O}_X(n_0)$ where $n_0 = drst - r - s - t$.

<u>Claim</u>: If $n_0 \geq 0$ then $(k[x,y,z]/(G))_{n_0} \neq 0$

For, it is sufficient to show that if $n_0 \geq 0$ then $n_0 \in r\mathbb{N} + s\mathbb{N} + t\mathbb{N}$.

Assume $r \leq s \leq t$; if $r = 1$ it is clear; if $r = s = t = 2$ it is also clear; in the remaining cases it is sufficient to observe that

$n_0 = [(d-1)st + (s-1)(t-1) - 2]r + (r-1)s + r - 1)t$

and $(d-1)st$, $(s-1)(t-1) - 2$, $r - 1$ are non-negative.

Therefore, by Proposition 3.1 and the Claim, A is AFD iff the genus of X is 0 iff $n_0 < 0$.

Now it is easy to check that $n_0 < 0$ iff one of the following two conditions holds

1) $d = 2 \quad r = s = t = 1$

2) $d = 1 \quad r = s = 1$ any t.

For instance the following surfaces are AFD: $F = X^4 + Y^6 + Z^{10}$, $F = X^2 + Y^{15} + Z^{21}$

while the following surfaces are not AFD: $F = X^6 + Y^9 + Z^{15}$, $F = X^4 + Y^6 + Z^9$.

In conclusion, A _is_ AFD _if and only if either_ $d = 2$, $r = s = t = 1$, _or_ $d = 1$ _and two_ 1's _are in the triple_ (r, s, t).

References

[B-O] Beltrametti, M. - Odetti, F. On the projectively almost factorial varieties. Ann. Math. Pura Appl. Vol CXIII (1967)

[B-R] Beltrametti, M. - Robbiano, L. Introduction to the theory of weighted projective spaces (in preparation)

[Del] Delorme, C. Espaces projectifs anisotropes. Bull. Soc. Math. France 103 (1975)

[Dem] Demazure, M. Anneaux gradues normaux. Seminaire Demazure, Giraud, Teissier Singularites des surfaces Ecole Polytechnique (1979)

[Do] Dolgachev, I. Weighted projective varieties Proceedings, Vancouver 1981 Lecture Notes in Mathematics 956

[Ke] Kempf, G. Some quotient varieties have rational singularities Michigan Math. J. 24 (1977)

[Fl] Flenner, H. Rationale quasihomogene Singularitäten Archiv der Math. 36 (1981)

[Fo] Fossum, R. The Divisor Class Group of a Krull Domain Ergebnisse der Mathematik 74 Springer (1973)

[Mo 1] Mori, S. On a generalization of complete intersections J. Math. Kyoto Univ. 15 (1975)

[Mo 2] Graded factorial domains Japan J. Math. 3 (1977)

[Mu 1] Mumford, D. Abelian Varieties
Oxford Univ. Press (1970)

[Mu 2] Mumford, D. Geometric Invariant Theory
Springer Verlag Berlin (1965)

[Ro] Robbiano, L. Some properties of complete intersections in "good" projective varieties. Nagoya Math. J. 61 (1976)

[Sa] Salmon, P. Su un problema posto da Samuel
Rend Accad. Naz. Lincei Sez VIII Vol XL (1966)

[Se] Serre, J. P. Groupes algébriques et corps de classes
Hermann 1959

[St] Storch, U. Fastfaktorielle Ringe
Schriftenreihe des Math. Inst. der Univ. Münster 36 (1967)

[Wa] Watanabe, Keiichi Some remarks concerning Demazure's construction of normal graded rings. Nagoya Math. J. 83 (1981).

Lorenzo ROBBIANO
Istituto di Matematica dell'Università di Genova
Via L. B. Alberti 4
16132 GENOVA
ITALIA

ON SET-THEORETIC COMPLETE INTERSECTIONS

Giuseppe Valla
Istituto di Matematica
Università di Genova
Genova, Italia

These are the notes for some lectures given by the author at the C.I.M.E. Session on "Complete intersections". They aim at providing a framework, certainly defective, for studying the problem of algebraic varieties which are set-theoretic complete intersections. Focusing on the special case of affine algebraic curves, the first section of this note containes a detailed study of the work of Serre, Ferrand, Suslin Kumar and Boratynski which leads to the proof that every smooth affine curve is set-theoretic complete intersection. Unfortunately very little is known if we admit the curve to have singularities or to lie in the projective space. But if we restrict ourselves to locally Cohen-Macaulay subschemes of $\mathbb{P}^n$ or A^n of codimension two, then the particular determinantal structure of their defining ideals, given by the Hilbert Burch theorem, enables us to discuss interesting examples of algebraic varieties which are set-theoretic complete intersections. This is the argument of the second part of this note. A final section of my lectures has been devoted to make some hints at the case of projective varieties which are set-theoretic complete intersections, by means of the theory of "Grobner bases" of ideals in polynomial rings. Meanwhile this material has evolved out and now is the subject of a joint paper with L.Robbiano. (see [19]).

This paper was supported by C.N.R. (Consiglio Nazionale delle Ricerche).

1. SMOOTH AFFINE CURVES

Assume throughout that R is a Noetherian ring. For a finitely generated R-module M we shall write $v(M)$ for the minimal number of generators of M; further for any ideal I of R, $h(I)$ will denote the height of I and $rad(I)$ the radical of I. If we put $a(I) = \min\{v(J) \mid rad(I)=rad(J)\}$, then the Krull Hauptidealsatz says that: $h(I) \leq a(I) \leq v(I)$.

<u>Definition</u> 1.1. <u>The ideal I is a complete intersection</u> (c.i.) <u>if $h(I)=v(I)$</u>.

It is well known that if I is generated by a regular sequence then I is a c.i.; on the other hand, if $I=rad(I)$ or if R is Cohen-Macaulay, the converse holds.

This notion has clearly a geometric content: if $V \subset \mathbb{A}^n_k$ (or $\mathbb{P}^n_k$) is an algebraic variety over a field k, we say that V is a complete intersection, if $I(V)$ is a c.i. ideal in $k[X_1,\ldots,X_n]$ (or $k[X_0,\ldots,X_n]$). For example any linear variety in $\mathbb{P}^n$ is a c.i.. But if we consider the affine space curve with parametric equations $X=t^3$, $Y=t^4$, $Z=t^5$, then $I=I(C)$ is the ideal generated by the maximal minors of the matrix

$$\begin{bmatrix} X & Y & Z \\ Y & Z & X^2 \end{bmatrix}$$

; thus if $\underline{m}=(X,Y,Z)$ then $I \subset \underline{m}^2$, hence $I/\underline{m}^3$ is a k-vector space of dimension three, which proves that $v(I)=3$ and the curve is not a complete intersection. On the other hand it is clear that $C=H_1 \cap H_2$ where H_1 and H_2 are the surfaces: $XZ=Y^2$ $X^5+Z^3=2X^2YZ$ respectively. In this case we have $2=h(I)=a(I)< v(I)=3$.

This leads to a new and cruder definition.

<u>Definition</u> 1.2. <u>The ideal I is set-theoretic complete intersection</u> (s.t.c.i.) <u>if</u> $h(I)=a(I)$.

If $V \subset \mathbb{A}^n_k$ (or $\mathbb{P}^n_k$) is an algebraic variety, we say that V is a set-theoretic complete intersection if $I(V)$ is a s.t.c.i. ideal in $k[X_1,\ldots,X_n]$ (or $k[X_0,\ldots,X_n]$).

The problem of identify a s.t.c.i. is always a difficult matter. Even in the case of affine or projective curves there is a number of big problems still unsolved. For example it is a classical open problem the following (see[13]).

<u>Is every curve in $\mathbb{A}^3$</u> <u>(respectively connected curve in $\mathbb{P}^3$)</u> a s.t.c.i.?

Here the main difficulty is now concentrate in the case of a field of characteristic zero, because Cowsik and Nori (see[5]) have proved that every curve in A_k^n is s.t.c.i. if char(k)=p> 0.

The starting point for the recent developping of the research in this area is certainly the paper by Serre (see[21]). There, for the first time, the "Complete intersection Problem" has been linked to the famous "Serre Conjecture" on the freeness of finitely generated projective modules over the polynomial ring $k[X_1,...,X_n]$ (see [17] and [22]). The main result is the following theorem, where $h.d._R(M)$ denotes the projective dimension of the R-module M.

Theorem 1.3. *Let M be a finitely generated R-module with* $h.d._R(M) \leq 1$. *Suppose that* $Ext^1_R(M,R)$ *is a cyclic R-module. Then there is an exact sequence*

$$0 \to R \to P \to M \to 0$$

with P a projective module.

Proof. Let η be a generator of $Ext^1_R(M,R)$ and let $0 \to R \to X \to M \to 0$ be the corresponding extension. Then we have an exact sequence

$$Hom_R(R,R) \overset{\partial}{\to} Ext^1_R(M,R) \to Ext^1_R(X,R) \to 0$$

where the connecting homomorphism ∂ sends 1_R to η . Thus ∂ is surjective, hence $Ext^1_R(X,R)=0$. We have $h.d._R(X) \leq \max(h.d._R(M),h.d._R(R))$ hence $h.d._R(X) \leq 1$. Let $0 \to P_1 \to P_0 \to X \to 0$ be a projective resolution of X; since $P_1 \oplus Q = R^t$ for some t and some R-module Q, we get $0= \oplus Ext^1_R(X,R)=Ext^1_R(X,R^t)=Ext^1_R(X,P_1) \oplus Ext^1_R(X,Q)$, hence $Ext^1_R(X,P_1)=0$. This implies that the sequence $0 \to P_0 \to P_1 \to X \to 0$ splits, hence X is a projective module.

In order to exhibit an important class of ideals to which the above result applies we need the local version of the definition of complete intersection.

Definition 1.4. *The ideal I is locally a complete intersection (l.c.i.) of height r if* $IR_{\underline{m}}$ *is generated by a regular sequence of length r, for all maximal ideals* $\underline{m} \supset I$.

For eaxmple if V is a smooth algebraic variety of codimension r then I(V) is l.c.i. of height r.

If I is l.c.i. of height r we get, by using the Koszul complex, that $\text{h.d.}_{R_{\underline{m}}}(IR_{\underline{m}})=r-1$ for every maximal ideal $\underline{m} \supseteq I$, hence $\text{h.d.}_R(I)=r-1$. Further I/I^2 is a projective R/I-module of rank r-1, since, if J is an ideal generated by a regular sequence $x_1,\ldots,x_n$ then $\bar{x}_1,\ldots,\bar{x}_n$ is a base of the module J/J^2. Less trivial properties of l.c.i. ideals are the following.

Theorem 1.5. a) *If I is a l.c.i. ideal of height r then there is a canonical isomorphism* $\text{Ext}_R^r(R/I,R)=\text{Hom}_{R/I}(\Lambda^r(I/I^2),R/I)$

b) *If V is a smooth algebraic variety of codimension r in* A_k^n *then* $\text{Ext}_R^r(R/I,R)=$ $=\Lambda^{n-r}\Omega_k(R/I)$ *where* $\Omega_k(R/I)$ *denotes the module of k-differentials of R/I.*

We omit the proof of this result since is not relevant to our pourposes (for a proof see [1]or [15]).

Remark 1.6. Recall that for any ideal I of height r in the polynomial ring R, we have $\text{Ext}_R^i(R/I,R)=0$ for $i< r$ and $\text{Ext}_R^r(R/I,R) \neq 0$: in fact this is the homological characterisation of the grade of an ideal. The module $\text{Ext}_R^r(R/I,R)$ is called the canonical module for R/I and is denoted by $K_{R/I}$ (see[9]).

The main application of the above results is the following theorem.

Theorem 1.7. *Let C be a smooth irreducible curve of genus 0 or 1 in* A_k^3, *where k is algebraically closed. Then C is a complete intersection.*

Proof. Let I be the defining prime ideal of C in $R=k[X,Y,Z]$. Then $\text{Ext}_R^1(I,R)=$ $=\text{Ext}_R^2(R/I,R)=\Omega_k(R/I)$; by a classical result, every curve of genus 0 or 1 has cyclic module of differentials. Thus we can apply theorem 1.3 and we get an exact sequence $0 \to R \to P \to I \to 0$ with P projective. Since R is a domain we can define the rank of any R-module M as $\text{rk}(M)=\dim_{K(R)}[M\otimes K(R)]$, where K(R) is the field of fractions of R. Now it is clear that $\text{rk}(I)=1$, hence P is a projective module of rank 2. The relevance of Serre Conjecture now comes into the picture: the result of Quillen and Suslin says that P is free of rank 2 and the conclusion follows.

Remark 1.8. In Theorem 1.7. the assumption that C is a smooth curve cannot be deleted. In 1916 Macaulay has proved that for every n there exists a rational irreducible curve C in A^3 such that $v(I(C)) \geq n$.

More recently Ferrand has found a brilliant argument which yields the possibility to use Serre Theorem 1.3 for the problems of curves which are set-theoretic complete intersections (see [8]). Again the starting point of Ferrand construction is a theorem of Serre. Here projective modules are always finitely generated.

Theorem 1.9. Let R be a noetherian ring and P a projective R-module. Then $P=F \oplus Q$ where Q is projective of rank $\leq$ dim R and F is free. (If $rk(P) \leq$ dim R, F=0).

For the proof of this result see [20].

Corollary 1.10. Let R be a noetherian ring and let dim R=1. If P is a projective module of rank $r \geq 1$, then $P=F \oplus \Lambda^r P$ for some free module F of rank r-1.

Proof. Using Theorem 1.9. we get $P=F \oplus Q$ where Q is projective of rank 1 and F is free of rank r-1. Hence we get $\Lambda^r P=\Lambda^r(F \oplus Q)=\Lambda^{r-1}F \otimes_R \Lambda^1 Q=R \otimes_R Q=Q$.

We recall that if P is a projective module of rank 1 over the ring R and if we put $P^{\wedge}=Hom_R(P,R)$, then $P \otimes_R P^{\wedge}=R$. In the following, if n is any integer, $P^{\otimes(n)}$ means $P \otimes_R P \otimes_R \ldots \otimes_R P$ n times if $n>0$, $P^{\wedge} \otimes_R P^{\wedge} \otimes_R \ldots \otimes_R P^{\wedge}$ (-n) times if $n<0$ and R if n=0. Thus $P^{\wedge}=P^{\otimes(-1)}$ and $P^{\otimes(n)} \otimes_R P^{\otimes(m)}=P^{\otimes(n+m)}$ for every integers n and m.

Corollary 1.11. Let R be a noetherian ring and let dim R=1. If P is a projective module of rank $r>1$, then $(\Lambda^r P)^{\wedge}$ is a direct summand of P.

Proof. Let $K=(\Lambda^r P)^{\wedge}$; then $K^{\otimes(n)}$ is projective of rank 1 for every n. Let $N=R^{r-2} \oplus K \oplus K^{\otimes(-2)}$; then N is projective of rank r and we have $\Lambda^r N= R \otimes_R K \otimes_R K^{\otimes(-2)}= =K^{\wedge}$. But since P is projective, $\Lambda^r P$ is reflexive, hence $\Lambda^r P=K^{\wedge}$. Thus P is isomorphic to N by Corollary 1.10 and the conclusion follows.

Theorem 1.12. Let I be an ideal of the ring $R=k[X_1,\ldots,X_n]$ $(n \geq 3)$. If I is l.c.i. of height n-1, then there exists an ideal $J \subseteq I$ such that $I^2 \subseteq J$, J is l.c.i. and J/J^2 is a free module of rank n-1 over R/J.

Proof. Let $P=I/I^2$; since I is l.c.i. of height n-1, P is a projective module of rank $n-1>1$ over the one-dimensional ring R/I=A. Let $K=(\Lambda^{n-1}P)^{\wedge}$; then by Corollary 1.11, we have an exact sequence $0 \to K^{\otimes(-2)} \oplus A^{n-3}=J/I^2 \to I/I^2 \to K \to 0$ where J is an ideal of R such that $I^2 \subseteq J \subseteq I$. Clearly K=I/J and J/I^2 are projective A-module of rank 1 and n-2 respectively. Thus if $\underline{m} \supseteq J$ is a maximal ideal in R, there exist elements $x \in I$, $x_1, \dots, x_{n-2} \in J$ such that $IR_{\underline{m}} \subseteq (x+J)R_{\underline{m}}$ and $JR_{\underline{m}} \subseteq (x_1, \dots, x_{n-2}, I^2)R_{\underline{m}}$. It follows that $JR_{\underline{m}} \subseteq (x_1, \dots, x_{n-2}, x^2, \underline{m}J)R_{\underline{m}}$ and, by Nakayama Lemma, $JR_{\underline{m}} = (x^2, x_1, \dots, x_{n-2})R_{\underline{m}}$. This proves that J is l.c.i. of height n-1 so that J/J^2 is a projective module of rank n-2 over R/J. We have an exact sequence of projective A-modules

$$0 \to I^2/IJ \to J/IJ \to J/I^2 \to 0$$

Since $J/IJ = (J/J^2) \otimes_{R/J} A$, we have $rk(J/IJ) = n-1$, $rk(J/I^2) = n-2$, so that $rk(I^2/IJ) = 1$. But we have a canonical epimorphism $K^{\otimes(2)} = (I/J) \otimes_A (I/J) \to I^2/IJ$, hence we get $K^{\otimes(2)} = I^2/IJ$. It follows that $\Lambda^{n-1}(J/IJ) = \Lambda^{n-1}((J/I^2) \oplus (I^2/IJ)) = \Lambda^{n-2}(K^{\otimes(-2)} \oplus A^{n-3}) \otimes K^{\otimes(2)} = K^{\otimes(-2)} \otimes K^{\otimes(2)} = A$. Since J/IJ is a projective A-module of rank n-1, this proves by Corollary 1.10, that J/IJ is free of rank n-1 over A. As $(I/J)^2 = 0$, we get $v(J/J^2) = n-1$; but J/J^2 is R/J-projective of rank n-1, so it is free.

As a first corollary of this result one can prove that every smooth curve in A^3 is s.t.c.i.. In fact, more generally, we have

Theorem 1.13. Let $C \subseteq A^3$ be a curve which is l.c.i.. Then C is s.t.c.i..

Proof. Let I be the defining ideal of C in R=k[X,Y,Z] . By Ferrand construction there exists an ideal $J \subseteq I$ such $I^2 \subseteq J$, J/J^2 is a free module of rank 2 over R/J and $h.d._R(J) = 1$. Thus we get, by Theorem 1.5 a), $Ext^1_R(J,R) = Ext^2_R(R/J,R) = Hom_{R/J}(\Lambda^2(J/J^2), R/J) = R/J$. The conclusion follows again by Theorem 1.3 and the fact that every projective module over R is free.

When we want to tackle the same question for curves in A^n with n>3, the problem is to get a substitute for Theorem 1.3. Boratynski and Kumar have found the following method to get a projective module to map onto an ideal (see [3]and[11]).

For a commutative ring R, $Gl_n(R)$ will denote the group of invertible n×n matrices with elements in R and $E_n(R)$ the subgroup generated by $E_{ij}(a) = I + ae_{ij}$ where $e_{ij} = (\delta_{ij})$ and δ_{ij} is the usual Kroneker symbol. If $E \in E_n(R)$ we say that E is an elementary matrix. A vector $u = [a_1, \dots, a_n] \in R^n$ is called "unimodular" if there is a vector $v \in R^n$ such that $u^t v = 1$ or, which is the same, if $a_1, \dots, a_n$ generate the ideal (1).

It is clear that we have an action of $Gl_n(R)$ on the set of unimodular vectors; the orbits, under this action, correspond to the isomorphism classes of projective modules P such that $P \oplus R = R^n$. In fact if $u=[a_1,\dots,a_n]$ is an unimodular row we get a surjective map which we call again $u:R^n \to R$; the kernel $P(u)$ of this map is a projective module such that $P(u) \oplus R = R^n$. Now if $v=[b_1,\dots,b_n]$ is another unimodular vector and $P(u)$ is isomorphic to $P(v)$ then we get a diagram

$$\begin{array}{ccccccc} 0 \to & P(u) & \overset{h}{\leftrightarrows} & R^n & \overset{u}{\to} & R & \to 0 \\ & \psi \downarrow & & \vdots & & \| & \\ 0 \to & P(v) & \underset{i}{\to} & R^n & \overset{v}{\underset{t}{\rightleftarrows}} & R & \to 0 \end{array}$$

The map $\phi = i\psi h + tu : R^n \to R^n$ makes the diagram commute, hence it is an isomorphism. Thus $u = v\phi$ and u,v are in the same orbit. Conversely if $u=v\phi$ with $\phi \in Gl_n(R)$ then clearly $P(u)$ is isomorphic to $P(v)$.

Definition 1.14. A unimodular vector u is called "completable" if u is in the same orbit as $v=[1,0,\dots,0]$.

Since $P(v)=R^{n-1}$, a unimodular vector $u=[a_1,\dots,a_n]$ is completable if and only if $P(u)=R^{n-1}$, or equivalently if $[a_1,\dots,a_n]$ is the top row of an invertible matrix.

It is clear that if $n=2$ every unimodular vector is completable; but if $R=Z[X_1,X_2,X_3]/(X_1^2+X_2^2+X_3^2-1)=k[x_1,x_2,x_3]$ then $[x_1,x_2,x_3]$ is a unimodular vector which is not completable (see for example [14]).

A fundamental result of Suslin (see [23]) says that if $u=[a_1,\dots,a_n]$ is a unimodular vector, then $[a_1,a_2,a_3^2,\dots,a_n^{n-1}]$ is completable.

In the following if M and N are R-modules and if $f \in Hom_R(M,P)$, $g \in Hom_R(N,P)$ we denote by (f,g) the map in $Hom_R(M \oplus N,P)$ defined by $(f,g)(x,y)=f(x)+g(y)$, while if $h \in Hom_R(P,M)$, $k \in Hom_R(P,N)$ we write $\binom{h}{k}$ for the map in $Hom_R(P,M \oplus N)$ defined by $\binom{h}{k}(x)=(h(x),k(x))$. Hence we get the formula $(f,g)\binom{h}{k}=fh+gk$. Further if $f \in Hom_R(M,P)$, $h \in Hom_R(M,Q)$, $g \in Hom_R(N,P)$, $k \in Hom_R(N,Q)$ we write $\begin{pmatrix} f & g \\ h & k \end{pmatrix}$ for the map in $Hom_R(M \oplus N, P \oplus Q)$ defined by $\begin{pmatrix} f & g \\ h & k \end{pmatrix}(x,y)=(f(x)+g(y),h(x)+k(y))$. We have the usual rule for matrix multiplication $\binom{h}{k}(f,g)=\begin{pmatrix} hf & hg \\ kf & kg \end{pmatrix}$ and $\begin{pmatrix} f & g \\ h & k \end{pmatrix}\begin{pmatrix} l & m \\ p & q \end{pmatrix}=\begin{pmatrix} fl+gp & fm+gq \\ hl+kp & hm+kq \end{pmatrix}$.

Lemma 1.15. Let P be a projective module and let $a \in R$ such that P/aP is a free module over R/aR of rank m with base $\overline{p}_1,\dots,\overline{p}_m$. Let $p:R^m \to P$ be the map defined by $p(a_1,\dots,a_m)=\Sigma a_ip_i$ and $\psi=\begin{pmatrix} a1_m \\ p \end{pmatrix}\in \mathrm{Hom}_R(R^m,R^m\oplus P)$. Then ψ is injective and $\mathrm{Coker}\psi\simeq P$.

Proof. We have a commutative diagram

$$\begin{array}{ccccccc} P & \overset{q}{\dashrightarrow} & R^m & \overset{p}{\to} & P & \to & 0 \\ \pi\downarrow & & \tau\downarrow & & \pi\downarrow & & \\ P/aP & \underset{k}{\to} & (R/aR)^m & \underset{h}{\to} & P/aP & & \end{array}$$

where π and τ are the canonical projections, q exists since P is projective, h is the isomorphism induced by p and k its inverse. We get $\pi(1_P-pq)=\pi-\pi pq=\pi-h\tau q=$ $=\pi-hk\pi=0$ and $\tau(1_m-qp)=\tau-\tau qp=\tau-k\pi p=\tau-kh\tau=0$ hence $\mathrm{Im}(1_P-pq)\subseteq aP$ and $\mathrm{Im}(1_m-qp)\subseteq aR^m$. It follows that there exist maps $r:P\to P$, $s:R^m\to R^m$ such that $1_P-pq=ar$ and $1_m-qp=as$. Let $u=\begin{bmatrix} a1_m & -q \\ p & r \end{bmatrix}$ $v=\begin{pmatrix} s & q \\ -p & a1_P \end{pmatrix}\in \mathrm{Hom}_R(R^m\oplus P,R^m\oplus P)$; then $uv=\begin{bmatrix} 1_m & 0 \\ ps-rp & 1_P \end{bmatrix}$ $vu=\begin{bmatrix} 1_m & qr-sq \\ 0 & 1_P \end{bmatrix}$. This proves that u is an isomorphism and the conclusion follows by the commutative of the diagram

$$\begin{array}{ccccccccc} 0 & \to & R^m & \xrightarrow{\begin{bmatrix} 1_m \\ 0 \end{bmatrix}} & R^m\oplus P & \to & P & \to & 0 \\ & & \downarrow & & u\downarrow\wr & & \wr\downarrow & & \\ 0 & \to & R^m & \underset{\psi}{\to} & R^m\oplus P & \to & \mathrm{Coker}\psi & \to & 0 \end{array}$$

Lemma 1.16. Let P be a projective module and let $a\in R$ such that P/aP is a free R/aR-module of rank m. If $p_m\in P$ is such that $\bar{p}_m\in P/aP$ can be extended to a base of P/aP and if $\phi=\begin{pmatrix} a \\ p_m \end{pmatrix}\in \mathrm{Hom}_R(R,R\oplus P)$, then ϕ is injective and $\mathrm{Coker}\phi\simeq P$.

Proof. Let $u:R\to R^m$ be the map defined by $u(1)=(0,\dots,1)$, $h:R^m\to R^m$ the map associated to the matrix $\begin{bmatrix} I_{m-1} & 0 \\ 0 & a_m \end{bmatrix}$ and $k:R^m\to P$ the map defined by $k(x_1,\dots,x_m)=x_mp_m$. Let $\psi=\begin{pmatrix} h \\ k \end{pmatrix}\in \mathrm{Hom}_R(R^m,R^m\oplus P)$; then we have a commutative diagram

$$\begin{array}{ccccccc} 0 & \dashrightarrow & R & \overset{\phi}{\to} & R\oplus P & \longrightarrow & \mathrm{Coker}\,\phi \to 0 \\ & & u\downarrow & & \downarrow v & & \vdots\downarrow \\ 0 & \longrightarrow & R^m & \underset{\psi}{\to} & R^m\oplus P & \longrightarrow & \mathrm{Coker}\,\psi \to 0 \end{array}$$

where $v=\begin{pmatrix} u & 0 \\ 0 & 1_P \end{pmatrix}$. If we prove that ψ is injective and $\mathrm{Coker}\psi\simeq P$, then ϕ is injective and since $\bar{\psi}:\mathrm{Coker}\ u\simeq R^{m-1}\to \mathrm{Coker}\ v\simeq R^{m-1}$ is an isomorphism, we get by the Snake Lemma, that $\mathrm{Coker}\ \phi\simeq P$. Let $p_1,\dots,p_{m-1}\in P$ be such that $\bar{p}_1,\dots,\bar{p}_m$ form a base of P/aP. Then $\Lambda^m(P/aP)=\Lambda^mP/a\Lambda^mP$ is a free R/aR-module with base $\overline{p_1\wedge\dots\wedge p_m}$. Since Λ^mP is a projective module, we get a map $\lambda:\Lambda^mP\to R$ such that the following diagram is commutative

$$\begin{array}{ccccccc} & & \Lambda^m P & & & & \\ & \swarrow^{\lambda} & & \searrow & & & \\ R & \to & R/aR & \to & \Lambda^m P/a\Lambda^m P & \to & 0 \end{array}$$

Let $b=\lambda(p_1 \ldots p_m)$ and for every $i=1,\ldots,m$ let $\psi_i:P \to R$ be the map defined by $\psi_i(x)=\lambda(p_1\wedge\ldots\wedge x\wedge\ldots\wedge p_m)$ (x in place of p_i). Then $\psi_i(p_j)=b\delta_{ij}$. Further it is clear that $\bar{a}$ is a unit in the ring R/bR. But if x is a unit in a ring A, by Whitehead's Lemma (see [2]Cor.1.8, pg.227), the matrix $\begin{bmatrix} xI_{m-1} & 0 \\ 0 & x^{1-m} \end{bmatrix}$ is in $E_m(A)$ and, of course, $\begin{bmatrix} xI_{m-1} & 0 \\ 0 & x^{1-m} \end{bmatrix}\begin{bmatrix} I_{m-1} & 0 \\ 0 & x^{m} \end{bmatrix}=xI_m$. In our situation, there exists $\alpha\in E_m(R)$ $\beta \in M_m(R)$ such that $aI_m-b\beta = \alpha\begin{bmatrix} I_{m-1} & 0 \\ 0 & a^{m} \end{bmatrix}$, hence $\alpha h+b\beta=a1_m$.

Let $p:R^m \to P$ be the map defined by $p(x_1,\ldots,x_m)=\Sigma\, x_i p_i$, let $f=p-k$ and $g=\beta\begin{bmatrix} \psi_1 \\ \vdots \\ \psi_m \end{bmatrix}$

Then it is clear that $gp=p\beta$ anf $fh=f$. It follows that if we define

$\sigma = \begin{bmatrix} 1_m & 0 \\ f & 1_P \end{bmatrix}$, $\tau = \begin{bmatrix} \alpha & g \\ 0 & 1_P \end{bmatrix} \in \mathrm{Hom}_R(R^m\oplus P,R^m\oplus P)$ then σ and τ are isomorphisms

and $$\tau\sigma\psi = \begin{bmatrix} \alpha & g \\ 0 & 1_P \end{bmatrix}\begin{bmatrix} 1_m & 0 \\ f & 1_P \end{bmatrix}\begin{bmatrix} h \\ k \end{bmatrix} = \begin{bmatrix} \alpha & g \\ 0 & 1_P \end{bmatrix}\begin{bmatrix} h \\ p \end{bmatrix} = \begin{bmatrix} \alpha h+gp \\ p \end{bmatrix} = \begin{bmatrix} a1_m \\ p \end{bmatrix}$$

We can apply Lemma 1.15 to the map $\eta=\tau\sigma\psi$ and get a commutative diagram

$$\begin{array}{ccccccc} 0 \to & R^m & \xrightarrow{\psi} & R^m\oplus P & \to & \mathrm{Coker}\,\psi & \to 0 \\ & \| & & \downarrow\tau\sigma & & \downarrow\wr & \\ 0 \to & R^m & \xrightarrow{\eta} & R^m\oplus P & \to & \mathrm{Coker}\,\eta \simeq P & \to 0 \end{array}$$

from which the conclusion follows.

<u>Theorem</u> 1.17. <u>Let</u> $u=[a_1,\ldots,a_n]$ <u>be an unimodular vector of</u> R^n. <u>Then the unimodular vector</u> $[a_1,a_2,a_3^2,\ldots,a_n^{n-1}]$ <u>is completable</u>.

Proof. Let $P=R^{n-1}$, $p_{n-1}=(a_1,a_2,a_3^2,\ldots,a_{n-1}^{n-2})$ and $a=a_n$. By induction $\overline{p}_{n-1}$ can be extended to a base of the free R/a_nR-module R^{n-1}/a_nR^{n-1}. Hence, by Lemma 1.16, $\psi =$

$$= \begin{bmatrix} a_n^{n-1} \\ p_{n-1} \end{bmatrix} \in \operatorname{Hom}_R(R, R\oplus R^{n-1})$$

is injective and Koker $\psi \approx R^{n-1}$. This proves that the vector $[a_1,a_2,a_3^2,\ldots,a_n^{n-1}]$ is completable.

Theorem 1.18. *Let J be an ideal in R and let $a_1,\ldots,a_n, s\in J$, $t\in R$ such that $tJ \subseteq (a_1,\ldots,a_n)$ and $(s,t)=R$ so that the vector $[a_1,\ldots,a_n]$ is unimodular in $(R_{st})^n$. If $[a_1,\ldots,a_n]$ is completable over $(R_{st})^n$, there exists a projective R-module P of rank n mapping onto J.*

Proof. We have $JR_s=R_s$ since $s\in J$; hence the map $f:(R_s)^n \xrightarrow{[1,0,\ldots,0]} JR_s=R_s$ is surjective. On the other hand $JR_t=(a_1,\ldots,a_n)R_t$ hence the map $g:(R_t)^n \to JR_t$ defined by the vector $[a_1,\ldots,a_n]$ is surjective. Since $[a_1,\ldots,a_n]$ is completable over R_{st} we have a commutative diagram

$$\begin{array}{ccccc} [(R_t)^n]_s & \xrightarrow{g_s} & (JR_t)_s=R_{ts} & \longrightarrow & 0 \\ \psi\,\vdots & & \| & & \\ [(R_s)^n]_t & \xrightarrow{f_t} & (JR_s)_t=R_{ts} & \longrightarrow & 0 \end{array}$$

where ψ is an isomorphism of R_{st}-modules. Since $(s,t)=R$ we know that J is the pullback of the canonical diagram

$$\begin{array}{ccccc} & \overset{\alpha}{\nearrow} & JR_s & \overset{i_t}{\searrow} & \\ J & & & & (JR_s)_t=JR_{st}=R_{st} \\ & \underset{\beta}{\searrow} & JR_t & \underset{i_s}{\nearrow} & \end{array}$$

(see [12]). Let P be the pullback of the diagram

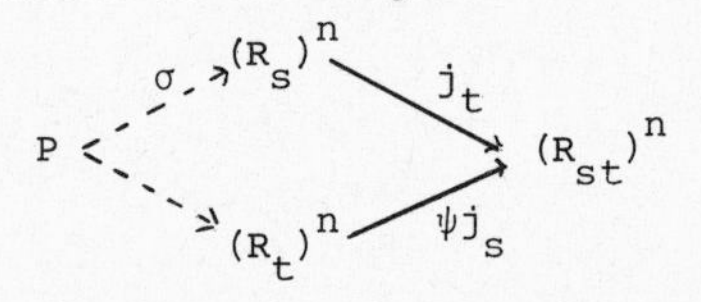

then we know that $\sigma_s:P_s \to (R_s)^n$ and $\tau_t:P_t \to (R_t)^n$ are isomorphisms. Further it is clear that we have commutative diagrams

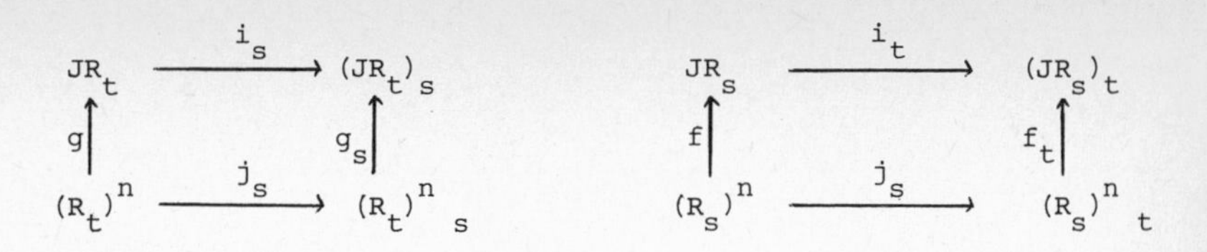

It follows that in the diagram

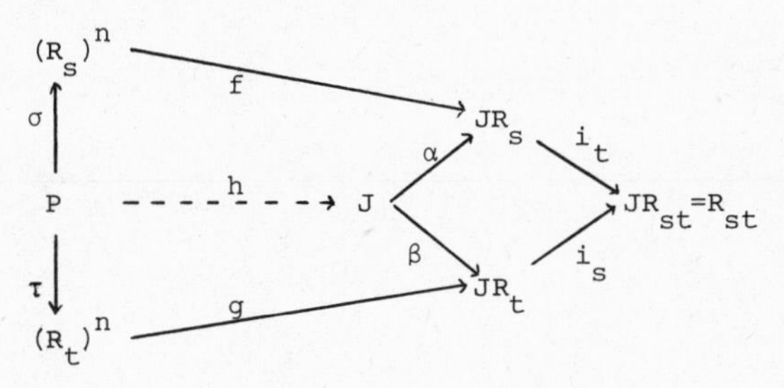

we have $i_t f\sigma = f_t j_t \sigma = f_t \psi j_s \tau = g_s j_s \tau = i_s g\tau$. Thus, by the functorial properties of the pullback we get a map $h: P \to J$ such that the above diagram commutes. Since σ_s and τ_t are isomorphisms, we get $h_s = f$ and $h_t = g$, hence h is surjective. Finally $P_s \approx R_s^n$ and $P_t \approx R_t^n$ so that P is a projective module of rank n (see [4]).

In order to apply the above result we need the following easy Lemma.

Lemma 1.19. *Let I be an ideal of R such that $v(I/I^2) = n$. Then there exist elements $a_1, \dots, a_n, s \in I$, $t \in R$ such that, if $J = (a_1, \dots, a_n)$,*

1) $tJ \subseteq J$

2) $(s,t) = R$.

In particular $I = (a_1, \dots, a_n, s)$ and $v(I/I^2) \leq v(I) \leq v(I/I^2)+1$.

Proof. Let $\bar{a}_1, \dots, \bar{a}_n$ be a system of generators of I/I^2. Then $I = I^2 + J$, hence $I(I/J) = I/J$. It follows by Nakayama Lemma (non local version) that there exists $t \in R$ such that $1-t \in I$ and $t(I/J) = 0$. The conclusion follows with $s = 1-t$.

Now we can prove Boratynski result.

Theorem 1.20. *If I is an ideal in $R = k[X_1, \dots, X_n]$ such that I/I^2 is free over R/I, then I is s.t.c.i.*

Proof. Let $v(I/I^2) = p$ and, accordingly to Lemma 1.19, $a_1, \dots, a_p, s \in I$, $t \in R$ such that $tI \subseteq (a_1, \dots, a_p)$ and $(s,t) = R$. Let $J = (a_1, a_2, a_3^2, \dots, a_p^{p-1}) + I^{p(p-1)}$; then $JR_t = (a_1, a_2, a_3^2, \dots, a_p^{p-1})R_t + (IR_t)^{p(p-1)} = (a_1, a_2, a_3^2, \dots, a_p^{p-1})R_t$, hence for some positive integer q, $t^q J \subseteq (a_1, a_2, a_3^2, \ldots, a_p^{p-1})$. Further rad(I)=rad(J), $s^{p(p-1)} \in J$ and $(t^q, s^{p(p-1)})=R$. By Theorems 1.17 and 1.18 we get a surjective map $P \to J \to 0$ where P is a projective module of rank p; thus P is free, hence v(J)=p. But, since I/I^2 is a free module of rank p and $R_{\underline{m}}$ is a regular local ring for every maximal ideal $\underline{m}$ ofR, we get $gr(JR_{\underline{m}})=p$, hence gr(J)=p. Thus v(J)=gr(J) and this implies that J can be generated by a regular sequence (see [10], Th.125).

Corollary 1.21. Let C be an affine curve in A^n which is l.c.i.. Then C is s.t.c.i.

Proof. Apply Theorem 1.12 and Theorem 1.20.

Remark 1.22. In [13] Murthy asked if every ideal I of $R=k[X_1, \ldots, X_n]$ with I/I^2 free over R/I is a complete intersection. Thus Boratynski result gives a partial answer to this problem. Using similar arguments as in the proof of Theorem 1.18, M.Kumar (see [11]) settles the question in several important cases. In fact he proved that if $v(I/I^2) \geq \dim(R/I)+2$ then $v(I)=v(I/I^2)$.

We notice that one can deduce Corollary 1.21 also by Kumar result.

2. CODIMENSION 2 CASE.

In this section I is perfect ideal of the noetherian ring R with gr(I)=2. This means that $h.d._R(R/I)=2$ so that, by the Hilbert-Burch Theorem, I can be generated by the maximal minors of an $(n+1)\times n$ matrix A. Now there is an exact sequence

$$0 \to R^n \xrightarrow{d_2} R^{n+1} \xrightarrow{d_1} R \xrightarrow{\varepsilon} R/I \longrightarrow 0$$

where A is the matrix of d_2 relative to the canonical base of R^n and R^{n+1} respectively. This means that if $A=(a_{ij})$ then $d_2(e_i)=\Sigma\, a_{ji}e_j$. Further by the well known result of Northcott and Eagon we have h(I)=2 (see [6]). Now it is clear that $\mathrm{Ext}^i_R(R/I,R)=0$ for i=0,1 so that, taking Hom, we get a free resolution for the canonical module $K_{R/I}=\mathrm{Ext}^2_R(R/I,R)$ of R/I

$$0 \to R^{\wedge} \xrightarrow{d_1^{\wedge}} (R^{n+1})^{\wedge} \xrightarrow{d_2^{\wedge}} (R^n)^{\wedge} \xrightarrow{\pi} K_{R/I} \longrightarrow 0$$

where as usual $M^{\wedge}=\mathrm{Hom}_R(M,R)$.

Finally following [16], we say that I is "self-linked" if we can find an ideal J generated by two elements of I, such that J:I=I. This implies $I^2 \subseteq J$, so that I is s.t.c.i.. The following theorem is the main result in [24]. The proof given here is partly due to J.Herzog (see also [18]).

Theorem 2.1. Let $A=\binom{M}{m}$ where M is a square matrix such that $M={}^tM$; if $f=\det(M)$ $g=\det \begin{bmatrix} M & {}^tm \\ m & 0 \end{bmatrix}$ and $J=(f,g)$ then $J:I=I$ and I is self-linked.

Proof. It is clear that we have a commutative **diagram**

$$\begin{array}{ccccccc} 0 \longrightarrow R^n & \xrightarrow{d_2=\binom{M}{m}} & R^{n+1} & \xrightarrow{d_1} & I \longrightarrow 0 \\ \Big\downarrow u_2=\binom{I}{0} & & \Big\downarrow u_1=(I,0) & & \Big\downarrow h \\ (R^{n+1})\hat{} & \xrightarrow[d_2\hat{}=({}^tM,{}^tm)]{} & (R^n)\hat{} & \xrightarrow{\pi} & K_{R/I} \longrightarrow 0 \end{array}$$

where in fact $(I,0)\binom{M}{m}=M=({}^tM,{}^tm)\binom{I}{0}$. Thus we get a map $h:I \to K_{R/I}$ such that the above diagram commutes. Since $\pi u_1=hd_1$ is surjective, h is surjective; further, since I annihilates R/I, we have $h(I^2) \subseteq I\,\mathrm{Ext}^2_R(R/I,R)=0$. Let $J=\ker(h)$; then $I^2 \subseteq J \subseteq I$. We can apply the snake Lemma to the morphism of short exact sequences

$$\begin{array}{ccccccc} 0 \longrightarrow & R^n & \xrightarrow{d_2} & R^{n+1} & \xrightarrow{d_1} & I \longrightarrow 0 \\ & \Big\downarrow pu_2 & & \Big\downarrow u_1 & & \Big\downarrow h \\ 0 \longrightarrow & (R^{n+1})\hat{}\,/\ker(d_2\hat{}) & \xrightarrow{\bar{d}_1\hat{}} & (R^n)\hat{} & \xrightarrow{\pi} & K_{R/I} \longrightarrow 0 \end{array}$$

where $p:(R^{n+1})\hat{} \to (R^{n+1})\hat{}/\ker(d_2\hat{})$ is the canonical projection. We infer that the sequence $\ker(u_1) \xrightarrow{d_1} J=\ker(h) \xrightarrow{\partial} \mathrm{Coker}(pu_2) \longrightarrow \mathrm{Coker}(u_1)$ is exact. But u_1 is surjective, hence $\mathrm{Coker}(u_1)=0$; further $\ker(u_1)$ is a cyclic module with base e_{n+1} while $\mathrm{Coker}(pu_2)=(R^{n+1})\hat{}/\ker(d_2\hat{})+\mathrm{Im}(u_2)$ is a cyclic module generated by $\overline{e\hat{}_{n+1}}$. Moreover $d_1(e_{n+1})=\det(M)=f$ and $g=d_1(\Sigma\, m_i e_i)$ if $m=(m_1,\dots,m_n)$; but $u_1(\Sigma\, m_i e_i)=d_2\hat{}(e\hat{}_{n+1})$ so that $g \in J$ and $\partial g=\overline{e\hat{}_{n+1}}$. It follows that if $a \in J$, then $\partial a=x\overline{e\hat{}_{n+1}}$ for some $x \in R$, hence $a-xg=d_1(ye_{n+1})=yf$ for some $y \in R$; thus J is generated by f and g. Since $gr(J)=2$, by the depth sensitivity of the Koszul complex, we have an exact sequence

$$0 \longrightarrow R \xrightarrow{\binom{-f}{g}} R^2 \xrightarrow{(g,f)} R \longrightarrow R/J \longrightarrow 0$$

Let $f_1,\dots,f_{n+1}=f$ be the generators of I; then it is very easy to check that we have a map of free complexes

$$\begin{array}{lccccccc}
\mathbb{L}: & 0 \longrightarrow R^n & \xrightarrow{\binom{M}{m}} & R^{n+1} & \xrightarrow{(f_1,\dots,f_{n+1})} & R \longrightarrow 0 \\
& \Big\uparrow \begin{bmatrix} f_1 \\ \vdots \\ f_n \end{bmatrix} & & \Big\uparrow \begin{bmatrix} t_m & 0 \\ 0 & 1 \end{bmatrix} & & \Big\| \\
\mathbb{K}: & 0 \longrightarrow R & \xrightarrow{\binom{-f}{g}} & R^2 & \xrightarrow{(g,f)} & R \longrightarrow 0
\end{array}$$

If $\mathbb{E}$ denotes the mapping cone of the map $\mathbb{L}\hat{}\to\mathbb{K}\hat{}$ we have an exact sequence

$$0 \to H_1(\mathbb{E}) \to H_0(\mathbb{L}\hat{}) \to H_0(\mathbb{K}\hat{}) \to H_0(\mathbb{E}) \to 0$$

But $H_0(\mathbb{L}\hat{})=\mathrm{Ext}^2_R(R/I,R)$; $H_0(\mathbb{K}\hat{})=\mathrm{Ext}^2_R(R/J,R)$; now it is well known (see[10]) that there is a natural equivalence of functors $\mathrm{Ext}^2_R(R/I,R)=\mathrm{Hom}_{R/J}(R/I,R/J)$ and $\mathrm{Ext}^2_R(R/J,R)=\mathrm{Hom}_R(R/J,R/J)$ so that the map $H_0(\mathbb{L}\hat{}) \to H_0(\mathbb{K}\hat{})$ is the natural inclusion $(J:I)/J \to R/J$. This implies that $H_0(\mathbb{E})=R/(J:I)$ and $H_i(\mathbb{E})=0$ for $i \geq 1$, so that $\mathbb{E}$ is a resolution of $R/(J:I)$; by the definition of the boundary map in the mapping cone we get $J:I=(f_1,\dots,f_n,f,g)=I$ and the conclusion follows.

We want to prove now that the above theorem has a partial converse.

<u>Theorem</u> 2.4. <u>Let I be a perfect ideal of codimension two of the local ring R. If I is self-linked, then I can be generated by the maximal minors of an $(n+1)\times n$ matrix $A=\binom{M}{m}$ where $M={}^tMU$ for some invertible matrix U.</u>

<u>Proof</u>. Let $0 \to R^n \xrightarrow{d_2} R^{n+1} \xrightarrow{d_1} I \to 0$ be a minimal projective resolution of I and let $J \subseteq I$ be an ideal generated by a regular sequence of two elements such that $I=J:I$. We have $K_{R/I}=\mathrm{Ext}^2_R(R/I,R)=\mathrm{Hom}_R(R/I,R/J)=(J:I)/J=I/J$, hence we get a surjective map $I \xrightarrow{u} K_{R/I}$. This induces a chain map

$$\begin{array}{ccccccccc}
& 0 \longrightarrow R^n & \xrightarrow{d_2} & R^{n+1} & \xrightarrow{d_1} & I \longrightarrow 0 \\
& \Big\downarrow u_2 & & \Big\downarrow u_1 & & \Big\downarrow \\
0 \longrightarrow & R\hat{} \xrightarrow{\hat{d_1}} (R^{n+1})\hat{} & \xrightarrow{\hat{d_2}} & (R^n)\hat{} & \xrightarrow{\pi} & K_{R/I} \longrightarrow 0
\end{array}$$

Now it is clear that the bottom row is a minimal resolution of $K_{R/I}$ (in fact $\hat{d_1}$ and $\hat{d_2}$ are minimal since such are d_1 and d_2, while $\ker(\pi)=\mathrm{Im}(\hat{d_2}) \subseteq \underline{m}(R^n)\hat{}$ since a matrix of d_2 has all its entries in $\underline{m}$). This implies that $\pi\otimes 1_k$ and $d_1\otimes 1_k$ are isomorphisms so that, since u is surjective, $u_1 \otimes 1_k$ is surjective and, by Nakayama, u_1 is surjective. It follows that u_1 is split, hence u_1 can be represented by the matrix $(I,0)$. Let $\binom{M}{m}$ and $\binom{S}{s}$ represent d_2 and u_2 respectively. Then we have $(I,0)\binom{M}{m}=$ $=({}^tM, {}^tm)\binom{S}{s}$ from which we infert $M={}^tMS+{}^tms$. We obtain

$\det(M)=\det(M)\det(S)+\Sigma m_i s_j$ [(ij)thcofactor of ${}^tMS$]$=\det(M)\det(S)+\Sigma\, m_i s_j(\,\Sigma M_{li} S_{lj})$

(see[14]pg.6). Now $\Sigma m_i M_{li}$ give the minimal generators of I toghether with det(M); hence det(S) is a unit in R and S is invertible. Now we have

$\begin{bmatrix} {}^tM \\ m \end{bmatrix} = \begin{bmatrix} {}^tSM+{}^tsm \\ m \end{bmatrix}$ and it is clear that, by elementary operations, we obtain the matrix $\begin{bmatrix} {}^tSM \\ m \end{bmatrix}$. Thus $\begin{bmatrix} {}^tM \\ m \end{bmatrix}$ is equivalent to $\begin{bmatrix} {}^tSM \\ m \end{bmatrix}$ (see [14], pg.8); further we have $\begin{bmatrix} {}^tS & 0 \\ 0 & 1 \end{bmatrix}\begin{bmatrix} M \\ m \end{bmatrix} = \begin{bmatrix} {}^tSM \\ m \end{bmatrix}$ so that $\binom{M}{m}$ is equivalent to $\begin{bmatrix} {}^tM \\ m \end{bmatrix}$ and they have the same determinantal ideal. Hence we get an isomorphism of short exact sequences

$$\begin{array}{ccccccc} 0 \longrightarrow & R^n & \xrightarrow{\begin{bmatrix} {}^tM \\ m \end{bmatrix}} & R^{n+1} & \longrightarrow & J & \longrightarrow 0 \\ & \Big\downarrow w & & \Big\downarrow v & & \Big\| & \\ 0 \longrightarrow & R^n & \xrightarrow{\binom{M}{m}} & R^{n+1} & \longrightarrow & J & \longrightarrow 0 \end{array}$$

where the matrix of v is of the form $\begin{bmatrix} V & 0 \\ v & 1 \end{bmatrix}$. Let W be the matrix of w, then

$\binom{M}{m}W=\begin{bmatrix} V & 0 \\ v & 1 \end{bmatrix}\begin{bmatrix} {}^tM \\ m \end{bmatrix}$ hence $MW=V{}^tM$ and $V^{-1}M={}^t(V^{-1}M)\,{}^tVW^{-1}$. Let $U={}^tVW^{-1}$; then U is invertible and since $\begin{bmatrix} V^{-1} & 0 \\ 0 & 1 \end{bmatrix}\begin{bmatrix} M \\ m \end{bmatrix} = \begin{bmatrix} V^{-1}M \\ m \end{bmatrix}$, the conclusion follows.

Remark. One can ask if every matrix M such that $M={}^tMU$ for some invertible matrix U, is equivalent to a symmetric matrix. This would imply that every self-linked ideal can be generated by the maximal minors of a matrix $\binom{M}{m}$ with $M={}^tM$. There is an unpublished manuscript of Ferrand (see [7]) in which a proof of this result is given. But the arguments are quite difficult to follow, so it could be of some interest the following easy proof in the case $v(I)=3$.

Proposition 2.5. If the perfect codimension two ideal I of the local ring R is self linked and v(I)=3, then I can be generated by the 2×2 minors of a matrix $\binom{M}{m}$, where $M={}^{t}M$.

Proof. By the preceding remark we need only to prove that if M is a 2×2 matrix such that $M={}^{t}MU$ with U invertible, then M is equivalent to a symmetrix matrix. Let

$M=\begin{bmatrix} a & b \\ c & d \end{bmatrix}$, $U=\begin{bmatrix} x & y \\ z & t \end{bmatrix}$; then we get a=ax+cz, b=ay+ct, c=bx+dz and d=by+dt. If x is a unit, then we have $\begin{bmatrix} a & b \\ bx+dz & d \end{bmatrix}\begin{bmatrix} 1 & 0 \\ -z & x \end{bmatrix}=\begin{bmatrix} . & bx \\ bx & . \end{bmatrix}$; otherwise yz is a unit and

we have $\begin{bmatrix} a & b \\ c & d \end{bmatrix}\begin{bmatrix} 1-xt & xy \\ -zt & yz \end{bmatrix}=\begin{bmatrix} . & axy+byz \\ c(1-xt)-dzt & . \end{bmatrix}$. But yz(1-xt)+ytxz=yz

and axy+byz+dzt+cxt-c=bx+dz-c=0 so the conclusion follows.

For much more informations about the properties of self linked curves, one can see the paper of A.P.Rao (see [18]).

REFERENCES

[1] Altman A. and Kleiman S."Introduction to Grothendieck duality theory" Lecture Notes in Math., 146, Springer, New York,1970.

[2] Bass H.,"Algebraic K-theory" Benjamin, new York, 1968.

[3] Boratynski M., A note on set-theoretic complete intersection ideals, J. of Al. 54 (1978), 1-5.

[4] Bourbaki N.,"Algebre Commutative", Ch.1, 2, Hermann, Paris, 1961.

[5] Cowsik C.P. and Nori, Affine Curves in characteristic p are set-theoretic complete intersections, Inv. Math. 45 (1978), 111-114.

[6] Eagon J.A. and Northcott D.G., Ideals defined by matrices and a certain complex associated with them,Proc. Roy. Soc. A 269 (1962), 188-204.

[7] Ferrand D., Un critere explicite pour les intersections completes ensemblistes d'ordre de nilpotence deux, (unpublished).

[8] Ferrand D., Courbes gauches et fibres de rang deux, C.R. Acad. Sci. Paris, 281 (1975), Ser.A, 345-347.

[9] Herzog J. und Kunz E.,"Der kanonische Modul eines Cohen-Macaulay Rings", Lecture Notes in Math. 238, Springer, New York, 1971.

[10] Kaplansky I;, "Commutative Rings", Allyn and Bacon, Boston, 1972.

[11] Kumar M., On two conjectures about polynomial rings, Inv.Math. 46 (1978), 225-236.

[12] Kunz E.,"Einfuhrung in die Kommutative Algebra und Algebraische Geometrie" F. Vieweg und Sohn, Braunschweig/Wiesbaden, 1980.

[13] Murthy M.P., Complete intersections,Conf. in Comm.alg., 1974, Queen's Papers in Pure and applied Math., 42.

[14] Northcott D.G., "Finite free resolutions", Cambr.Tracts in Math. 71, Cambr. Univ.Press, London, 1976.

[15] Ohm J., "Space curves as ideal theoretic complete intersections" (preprint).

[16] Peskine C. et Szpiro L., Liaison des varietes algebriques, Inv.Math.26 (1974), 271-302.

[17] Quillen D., Projective modules over polynomial rings, Inv.Math. 36 (1976),166-172.

[18] Rao A.P., On self-linked curves, Duke Math.Jou. 49,n.2 (1982), 251-273.

[19] Robbiano L. and Valla G., On set theoretic complete intersections in the projective space, (to appear).

[20] Serre J.P., Modules Projectifs et espaces fibres a fibre vectorielle, Sem. Dubreil Pisot, Algebre et Theorie des nombres, t.11, 1957/58,n.23.

[21] Serre J.P., Sur les modules projectifs, Sem. Dubreil-Pisot, 14 (1960-61).

[22] Suslin A.A., Projective modules over a polynomial ring are free,Sov.Math.Dokl. 17 (1976), 1160-1164.

[23] Suslin A.A.,On stably free modules, Math.U.S.S.R. Sbornik 102 (1977),537-550.

[24] Valla G., On determinantal ideas which are set-theoretic complete intersections, Comp.Math. 42 (1981), 3-11.

THE CLASSIFICATION OF QUOTIENT SINGULARITIES WHICH ARE COMPLETE INTERSECTIONS

by

Haruhisa NAKAJIMA and Kei-ichi WATANABE

Department of Mathematics

Tokyo Metropolitan University

Tokyo, 158, JAPAN

and

Department of Mathematics

Nagoya Institute of Technology

Nagoya, 466, JAPAN(*)

This note is an expansion of the talk which the second-named author gave at the C.I.M.E. Conference on Complete Intersections at Acireale (13-21, June 1983). The first part of this note is aimed to introduce the leaders to some basic facts about the quotient singularities which are complete intersections and meant to be rather expository.

At the occasion of his talk at Acireale, the second-named author talked about some results of the first-named author and said, "The classification of quotient singularities which are complete intersections will be achieved in near future", and so it was achived after only few months after the conference. The second half of this note includes the classification accomplished by the first-named author.

(*) This author attended the Acireale Conference while he was staying at Politecnico di Torino. He is grateful for the hospitality of Politecnico di Torino and for the finantial support from C.N.R. of Italy.

1. Introduction and preliminaries.

Let V be a vector space of dimension n over the complex number field $\mathbb{C}$ and G be a finite subgroup of GL(V). The action of G on V induces the action of G on S := S(V), the symmetric algebra of V over $\mathbb{C}$. In this note we will investigate the following

Problem. *When is the invariant subring S^G a complete intersection ?* and we will give the complete answer to this problem by classifying all the groups G whose invariant subring is a complete intersection (C.I., for short).

To begin with, let us notice the following hierarchy of conditions for commutative rings;

regular $\Longrightarrow$ hypersurface $\Longrightarrow$ C.I. $\Longrightarrow$ Gorenstein $\Longrightarrow$ Cohen-Macaulay and concerning these properties we have the following theorems.

Theorem A. ([S-T], [Ch]) S^G is a polynomial ring if and only if G is generated by its pseudo-reflections.

We say that $\sigma \in G$ is a pseudo-reflection if rank$(\sigma - I) = 1$, where I is the identity matrix of GL(V). We say that G is a reflection group if G is generated by its pseudo-reflections. Finite reflection groups are completely classified by Shephard and Todd (cf. [S-T] or [Co]).

Theorem B. ([H-E]) If R is a Cohen-Macaulay ring and if G is a finite group acting on R whose order is a unit in R, then the invariant subring R^G is again a Cohen-Macaulay ring.

This shows that in our case, S^G is always a Cohen-Macaulay ring.

Theorem C. ($[W_1]$) If G does not contain pseudo-reflections, S^G is a Gorenstein ring if and only if $G \subset SL(V)$.

The condition "G does not contain pseudo-reflections" is not so serious. If G does contain psudo-reflections, let H be the subgroup of G generated by all the psudo-reflections of G. Then S^H is a polynomial ring by Theorem A and the action of G/H on S^H is linear and does not contain pseodo-reflections. So we can apply Thorem C to this action.

As a C.I. is a Gorenstein ring, hereafter *we will always suppose that* G *is a finite subgroup of* $SL(V)$.

If $n = \dim V = 2$, the finite subgroups of $SL(V)$ are classified into five types (cf. [Sp_2]) and their invariant subrings are hypersurfaces which are the celebrated "rational double points" of type (A_m), (D_m), (E_6), (E_7), (E_8).

But for $n \geqq 3$, there are rather few subgroups of $SL(V)$ whose invariant subrings are C.I. and our problem becomes complicated. To show the difficulty of giving the answer to our problem by some group-theoretic conditions, let us note the following examples.

Example 1. Let e_m denote a primitive m-th root of unity and let G be the finite subgroup of $SL(3,\mathbb{C})$ of order 3m generated by

$$\begin{pmatrix} e_m & 0 & 0 \\ 0 & e_m^a & 0 \\ 0 & 0 & e_m^{a^2} \end{pmatrix} \quad \text{and} \quad \begin{pmatrix} 0 & 1 & 0 \\ 0 & 0 & 1 \\ 1 & 0 & 0 \end{pmatrix},$$

where m and a satisfy the condition $a + a^2 + 1 \equiv 0 \pmod{m}$. If $m = 7$ and $a = 2$, then (putting $S = \mathbb{C}[X,Y,Z]$), $S^G = \mathbb{C}[X^7+Y^7+Z^7,\ XY^4+YZ^4+ZX^4,\ X^3Y^2+Y^3Z^2+Z^3X^2,\ X^5Y+Y^5Z+Z^5X,\ XYZ]$, which is a C.I., but if $m = 13$ and $a = 3$, it will be shown that S^G is not C.I. (cf. § 2).

In our classification, the reflection groups are very important and if $G = H \cap SL(V)$ for some finite reflection group H, the condition for S^G to be C.I. was given by Stanley [St_1]. He also conjectured in [St_3] that if S^G is C.I., then there is a reflection group H such that $H \supset G \supset H'$ (where H' is the commutator subgroup of H). Although his conjecture turned out to be false (for example, the group G in Example 1 for $m = 7$ is a counterexample and also D. Rotillon noticed that there is a group G in $SL(3,\mathbb{C})$, whose invariant subring is a hypersurface and there is no finite reflection group containing G), study of the groups satisfying the condition of his conjecture is very important (cf. § 4).

2. Fundamental theorems.

In this section, we represent two theorems which are essential in our classification.

Theorem 1. ([K-W]) Let k be any field and G be a finite subgroup of $GL(n,k)$ acting on $S = k[X_1,\ldots,X_n]$. If S^G is a C.I., then G is generated by $\{\sigma\varepsilon G \mid \text{rank}(\sigma-I) \leqq 2\}$.

To prove this theorem, the concept of "purity" is essential.

Definition. ([SGA 2, exp. 10]) Let $(A,\underline{m})$ be a Noetherian local ring and put $X = \text{Spec}(A)$ and $X' = X - \{\underline{m}\}$. A is a pure local ring if the functor $\underline{\text{Et}}(X) \longrightarrow \underline{\text{Et}}(X')$ induced by inclusion $X' \longrightarrow X$ is a category equivalence (where, in general, $\underline{\text{Et}}(Y)$ is the category of étale coverings of a scheme Y).

This definition means that if there is an étale covering over X' and if A is a pure local ring, then we can extend this étale covering to that over X in a unique manner. In particular, we have the following

Corollary. Let A be a normal pure local ring and let B be a normal finite A-algebra. Assume that $\text{Spec}(B) \longrightarrow X = \text{Spec}(A)$ is étale over X'. Then B is an étale A-algebra.

Theorem of purity ([SGA 2, exp. 10]). (i) If A is a regular local ring of $\dim(A) \geqq 2$, then A is pure.

(ii) If A is a C.I. and if $\dim(A) \geqq 3$, then A is pure.

Now we can state the proof of Theorem 1.

(Proof of Theorem 1) Assume that S^G is a C.I. Then by the purity theorem, $(S^G)_{\underline{p}}$ is a pure local ring for every prime ideal of height $\geqq 3$ of S^G. Let H be the subgroup of G generated by $\{\sigma\varepsilon G \mid \text{rank}(\sigma-I) \leqq 2\}$. It is clear that H is a normal subgroup of G. We put $Z = \text{Spec}(S)$, $Y = \text{Spec}(S^H) = Z/H$, $X = \text{Spec}(S^G) = X/G = Y/(G/H)$ and put L to be the union of linear subspaces of codimension $\geqq 3$ of S which is fixed pointwise by some element $\sigma \neq I$ of G. Also we put $M = L/H$ (resp. $N = L/G$) to be the image of L in Y (resp. in Z). Then by our assumption, the canonical covering $Y - M \longrightarrow X - N = (Y - M)/(G/H)$ is étale. Then by the purity theorem and by the assumption that S^G is a C.I., $Y \longrightarrow X$ should be étale. But if $H \neq G$, the covering $Y \longrightarrow X$ is necessarily ramified at the point of X which is the image of the otigin of S. Thus we have $G = H$.

Now, we will state the second fundamental theorem. We denote by $\mathrm{emb}(S^G)$ the minimal number of generators of S^G over k (= embedding dimension of S^G).

Theorem 2 ([G-W]). Let k be a field and G be a finite subgroup of $GL(n,k)$ whose order is $\neq 0$ in k. If S^G is a C.I., then $\mathrm{emb}(S^G) \leq 2n-1$.

This theorem is a direct corollary of the following theorem if we note that every regular local ring is pseudo-rational by [L-T].

Theorem 2'. ([G-W]) Let B be a pseudo-rational local ring and A be a subring of B such that B is integral and pure over A. We assume that $\dim A = n$ and the residue field of A is infinite. If A is a C.I., then $\mathrm{emb}(A) \leq 2n-1$.

We will omit the definition of pseudo-rational local rings as it is rather complicated. But for local rings which are essentially of finite type over a field of characteristic 0 or for complex analytic local rings, "pseudo-rational" is equivalent to "rational singularity" and it is known that quotient singularities of $\mathbb{C}^n$ by a finite group or a reductive algebraic group are rational singularities by Boutot's theorem ([Bo]). The following theorem is essential in our proof of Theorem 2.

Theorem ([L-T]). Let $J \subset I$ be ideals in a pseudo-rational local ring A of $\dim A = n$ and assume that I is integral over J. Then $I^n \subset J$.

(Proof of Theorem 2) Let B (resp. A) be the localization of S (resp. S^G) at the irrelevant maximal ideal and J be a minimal reduction of the maximal ideal $\underline{m}$ of A (we may assume that k is an infinite field). Then $\underline{m}B$ is integral over JB and by the Theorem above, $(\underline{m}B)^n \subset JB$. As B is pure over A, $\underline{m}^n = (\underline{m}B)^n \cap A \subset (JB) \cap A = J$. As A/J is an Artinian ring with $\mathrm{emb}(A/J) = \mathrm{emb}(A) - n$, it suffices to prove the following

Lemma. If $(A,\underline{m})$ is an Artinian local ring which is a C.I. and if $\underline{m}^n = 0$, then $\mathrm{emb}(A) < n$.

(Proof) Let $A = B/(f_1,\dots,f_s)$, where B is a regular local ring with $\dim B = \mathrm{emb}(A) = s$ and $(f_1,\dots,f_s)$ is a B-regular sequence. Let $\underline{x} = (x_1,\dots,x_s)$ be a regular parameter system of B. Then we can write $f_j = \sum_{i=1}^{s} g_{ij}$, where g_{ij}'s are in the maximal ideal of B. By the following Sublemma (which can be proved by the mappimg $K.(\underline{f},B) \longrightarrow K.(\underline{x},B)$ between two Koszul complexes), we can prove $s < n$.

Sublemma. The socle of A is generated by the image of $\det(g_{ij})$ in A.

Remark. The conditions of Theorem 1 or Theorem 2 are not sufficient condition for S^G to be a C.I. For example, the groups in Example 1 in § 1 are generated by $\{\sigma\in G \mid \operatorname{rank}(\sigma-I) = 2\}$. But if $m \geqq 13$, then S^G is not a C.I. As for Theorem 2, for $n = 3$, we can claim;

Corollary. If G is a finite subgroup of $SL(3,\mathbb{C})$, then S^G is a C.I. if and only if $\operatorname{emb}(S^G) \leqq 5$.

(Proof) A Gorenstein ring of embedding codimension $\leqq 2$ is a C.I. ([Se]).

But if $n = 4$, S^G need not be a C.I. even if $\operatorname{emb}(S^G) = 7$.

Example 2. Let ω be a cube root of unity and G be the subgroup of $SL(4,\mathbb{C})$ of order 6 generated by $\begin{pmatrix} \omega & 0 & 0 & 0 \\ 0 & \omega^2 & 0 & 0 \\ 0 & 0 & \omega & 0 \\ 0 & 0 & 0 & \omega^2 \end{pmatrix}$ and $\begin{pmatrix} 0 & 1 & 0 & 0 \\ 1 & 0 & 0 & 0 \\ 0 & 0 & 0 & 1 \\ 0 & 0 & 1 & 0 \end{pmatrix}$. Then S^G is generated by X_1X_2, X_3X_4, $X_1X_4+X_2X_3$, $X_1^3+X_2^3$, $X_3^3+X_4^3$, $X_1^2X_3+X_2^2X_4$, $X_1X_3^2+X_2X_4^2$ and thus $\operatorname{emb}(S^G) = 7$. The simplest way to prove that S^G is not a C.I. will be to compute the Poincare series of S^G by the aid of Molien's Theorem which we will state below. In our case,

$$P(S^G,t) = \frac{1}{6}\left[\frac{1}{(1-t)^4} + \frac{2}{(1+t+t^2)^2} + \frac{3}{(1-t^2)^2}\right] = \frac{1+t^2+2t^3+t^4+t^6}{(1-t^2)^2(1-t^3)^2} \quad (*).$$

But if $S^G \cong \mathbb{C}[Y_1,\dots Y_{n+r}]/(F_1,\dots,F_r)$ is a C.I., then putting $\deg(Y_i) = d_i$ and $\deg(F_j) = e_j$,

$$P(S^G, t) = \frac{(1 - t^{e_1})\dots\dots(1 - t^{e_r})}{(1 - t^{d_1})\dots(1 - t^{d_{n+r}})}.$$

S^G is not a C.I. because the polynomial in the numerator of (*) is not a product of cyclotomic polynomials.

Molien's Theorem ([Mo], [St_3]). If G is a finite subgroup of order g of $GL(n,\mathbb{C})$, then,

$$P(S^G, t) = \frac{1}{g} \sum_{\sigma \in G} \frac{1}{\det(I - \sigma t)} .$$

3. The case of Abelian groups.

In this section, we assume that G is a finite Abelian subgroup of $GL(n,\mathbb{C})$. As G is diagonalizable, we use the following notation;

$(\alpha; i)$ (resp. $(\alpha,\beta; i,j)$) is the diagonal matrix whose (i,i)-entry is α (resp. whose (i,i)-entry is α and (j,j)-entry is β) and the other diagonal entries are 1.

The finite Abelian subgroups of $GL(n,\mathbb{C})$ whose invariant subring is a C.I. can be classified by the aid of the following "data".

Definition. (I) A special datum is a couple $\mathbb{D} = (D,w)$, where D is a set of subsets of $[n] := \{1,\dots,n\}$ and w is a mapping of D into the set of positive integers satisfying the following conditions;

(1) For every $i \in [n]$, $\{i\} \in D$, (2) If $J, J' \in D$, then either $J \subset J'$, $J' \subset J$ or $J \cap J' = \phi$, (3) If J is a maximal element of D with respect to the implication, then $w(J) = 1$, (4) If $J \subsetneq J'$, then $w(J')$ divides $w(J)$ and $w(J') < w(J)$, (5) If $J_1, J_2 < J$, then $w(J_1)=w(J_2)$, where $J' < J$ if $J' \subsetneq J$ and there is no element $J'' \in D$ such that $J' \subsetneq J'' \subsetneq J$.

(II) A datum is a pair $(\mathbb{D} ; a_1,\dots,a_n)$, where $\mathbb{D}$ is a special datum and $a_1,\dots,a_n$ are positive integers. A special datum $\mathbb{D}$ can be identified with $(\mathbb{D} ; 1,\dots,1)$.

(III) If $\mathbb{D} = (D, w, a_1,\dots,a_n)$ is a datum, we define the subgroup $G_{\mathbb{D}}$ of $GL(n,\mathbb{C})$ and the subring $R_{\mathbb{D}}$ of S as follows;

$G_{\mathbb{D}}$ is the group generated by $\{(e_{a_j} ; j) \mid j \in [n]\}$ and $\{(e_{wa_i}, e_{wa_j}^{-1} ; i,j) \mid i \in J_1,\ j \in J_2,\ J_1, J_2, J \in D,\ J_1, J_2 < J,\ w=w(J_1)=w(J_2)\}$ and

$$R_{\mathbb{D}} = \mathbb{C}[\, X_J \mid J \in D \,], \text{ where } X_J = (\prod_{i \in J} X_i^{a_i})^{w(J)} .$$

It is easy to check that $R_{\mathbb{D}}$ is the invariant subring of $G_{\mathbb{D}}$. $G_{\mathbb{D}} \subset SL(n,\mathbb{C})$ if and only if $\mathbb{D}$ is a special datum.

Theorem 3. ([W_2]) If G is a finite Abelian subgroup of GL(n,$\mathbb{C}$) (resp. SL(n,$\mathbb{C}$)) and if S^G is a C.I., then G is conjugate to $G_{\mathbb{D}}$ for some datum (resp. special datum) $\mathbb{D}$.

Remark. Sometimes, it is convenient to express a special datum by its "graph". For example, if $R_{\mathbb{D}} = \mathbb{C}[X_1^4, X_2^4, (X_1X_2)^2, X_3^6, X_4^6, (X_3X_4)^2, X_1X_2X_3X_4]$ (resp. $R_{\mathbb{D}} = \mathbb{C}[X_1^m, X_2^m, \ldots, X_n^m, X_1X_2\ldots X_n]$), the graph of $\mathbb{D}$ is given by

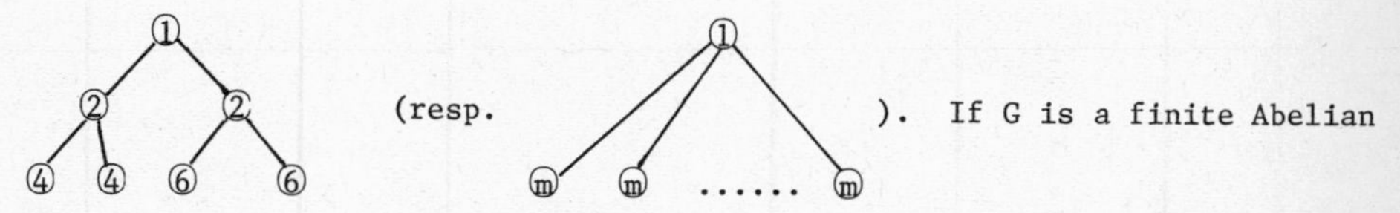

subgroup of SL(n,$\mathbb{C}$), which does not have invariant linear forms and if S^G is a hypersurface, then $S^G \cong \mathbb{C}[X_1^m, \ldots, X_n^m, X_1X_2\ldots X_n]$ for some m. (The definition of the graph of a special datum can be found in [W_2].)

4. Relative invariants of finite reflection groups.

For a finite group H, a subgroup N of H and a representation $\rho: H \to GL(V)$ of H, we adopt the following notation and terminology: H is said to be reducible (irreducible, imprimitive, prmitive, monomial) in GL(V), if ρ is so. Let V_N be the $\mathbb{C}$N-submodule of V generated by $(\sigma-1)V$ $(\sigma \in N)$ and R(V;N) the largest reflection subgroup of $\rho(N)$. A subspace U of codimension one in V is said to be a reflecting hyperplane relative to N if $\mathrm{Ker}(\sigma-1) = U$ for some σ in N. Denote by P(V,N) the set consisting of all reflecting hyperplanes relative to N and denote by $I_U(N)$ the subgroup $\{\tau \in \rho(N) : \mathrm{Ker}(\tau-1) \supset U\}$ for $U \in P(V,N)$. The cardinalities of $I_U(N)$ $(U \in P(V,N))$ are called orders of pseudoreflections in N. For each $U \in P(V,N)$, let $L_U(V,N)$ be a fixed nonzero element in $V_{I_U(N)}$. For a linear character χ of H with $\mathrm{Ker}\chi \supset \mathrm{Ker}\rho$, let $s_U(V,N,\chi)$ be the minimal natural number s satisfying $\chi(\sigma) = (\det(\sigma))^s$ for all $\sigma \in I_U(N)$ and $f_\chi(V,N)$ the product of $L_U(V,N)^{s_U(V,N,\chi)}$ $(U \in P(V,N))$. Further S_χ^N denotes the set $\{ f \in S : \sigma(f) = \chi(\sigma)f \text{ for } \sigma \in N \}$, whose elements are known as relative invariants.

Theorem ([St_1]). S_χ^N is a graded free S^N-module if and only if $f_\chi(V,N)$ belongs to S_χ^N.

$P(V,N)/N$ stands for a complete system of representatives of $P(V,N)$ modulo N under the natural action of N. The linear characters

$$\det:\langle I_{U'}(N) : N U \ni U' \rangle \to (\mathbb{C}^*)_U$$

induce the natural homomorphism

$$\Phi_{N,V}: R(V;N) \to \oplus_{U \in P(V,N)/N} (\mathbb{C}^*)_U \to GL_{|P(V,N)/N|}(\mathbb{C})$$

where $(\mathbb{C}^*)_U = \mathbb{C}^*$, $\oplus_{U\in P(V,N)/N}(\mathbb{C}^*)_U$ is diagonally embedded in $GL_{|P(V,N)/N|}(\mathbb{C})$.

For a representation $\delta: K \to GL(V)$ of a finite group K, $(R(V;N),K,V)$ is defined to be a CI-triplet, if $R(V;N) \supset \delta(K) \supset [R(V;N),R(V;N)]$ (the commutator of $R(V;N)$) and $\Phi_{N,V}(\delta(K))$ is conjugate to $G_{\mathbb{D}}$ in $GL_{|P(V,N)/N|}(\mathbb{C})$ for some datum $\mathbb{D}$. Moreover K is said to be extended to a CI-triplet in $GL(V)$, if (L,K,V) is a CI-triplet for a finite reflection group L in $GL(V)$.

Theorem 4. ([N_1, N_2]) Let K be a subgroup of a finite reflection group W in $GL(V)$ and suppose that K contains $[W,W]$. Then

(1) S^K is a complete intersection if and only if (W,K,V) is a CI-triplet.

(2) If S^K is a complete intersection, there is a CI-triplet (K^*,K,V) such that a regular system of homogeneous parameters of S^{K^*} can be extended to a minimal system of homogeneous generators of S^K.

When N is normal in H and $\rho(N)$ is a reflection group, we denote by $V(H\#N)$ a $\mathbb{C}H/N$-submodule of S^N which satisfies $S(V(H\#N)) = S^N$ and is generated by some graded elements.

5. Classification.

The slice method is well known and useful also in studying our subject.

Example 3 (Slice Method). For any v in V, $S^{(G_v)}$ is etale over S^G at the maximal ideals induced from v, and hence if S^G is a C.I., $S^{(G_v)}$ is also a C.I., where G_v is the stabilizer of v under the action of G. (Clearly this can be extended to the case where G is a reductive algebraic group, with the additional assumption on "closedness of Gv".)

To simplify matrix notation, we denote by $g((n))$ the permutation matrix of degree n associated with any element g in S_n and by $[a_1,a_2,\ldots,a_n]$ the diagonal matrix $[a_{ij}]$ with $a_{ij} = a_i$.

Example 4. Suppose that n = 4 and G is monomial in GL(V) such that the permutation group on the complete system of imprimitivities induced from G is isomorphic to <(12)(34), (13)(24)>. Then S^G is a C.I. if and only if G is conjugate to one of the following groups ; $\langle g_1,g_2,h_1,h_2\rangle$ $(4|c)$, $\langle g_3,g_4,g_7,h_1,h_2\rangle$ $(a < c/2,\ a|c/2,\ 2|c)$, $\langle g_3^2,g_5^2,g_7,h_1,h_2\rangle$ $(4|c)$, $\langle g_3^2,g_5^2,g_7,h_1',h_2\rangle$ $(a = c/4)$, $\langle g_2,g_5^2,g_6,h_1,h_2\rangle$ $(4a|c,\ b-a = c/2,\ a < c/4,\ b/a \equiv 3\ (4))$, $\langle g_2,g_5^2,g_6,h_1',h_2\rangle$ $(4a|c,\ b-a = c/2,\ a < c/4,\ b/a \equiv 3\ (4))$, $\langle g_2,g_5,g_6,h_1,h_2\rangle$ $(4a|c,\ b-a = c/2,\ a < c/4,\ b/a \equiv 3\ (4))$, $\langle g_2,g_5,g_6,h_1',h_2\rangle$ $(4a|c,\ b-a = c/2,\ a < c/4,\ b/a \equiv 3\ (4))$. Here $g_1 = [e_c,1,1,e_c^{-1}]$, $g_2 = [1,1,e_a,e_a^{-1}]$, $g_3 = [e_{c/2},e_{c/2}^{-1},1,1]$, $g_4 = [1,e_a,e_a^{-1},1]$, $g_5 = [1,e_{c/2},e_{c/2}^{-1},1]$, $g_6 = [e_c^{-b/a},e_c^{-1},e_c^{b/a},e_c]$, $g_7 = [e_c,e_c^{-1},e_c^{-1},e_c]$, $h_1 = (12)(34)((4))$, $h_2 = (13)(24)((4))$, $h_1' = [1,1,e_{2a},e_{2a}^{-1}](12)(34)((4))$ and a, b, c are natural numbers.

In the case where G is not explicitely defined (cf. Example 4), we can not compute the Poincare series of S^G and so use Theorem 2 essentially.

Example 5. Let G be a 6-dimensional representation of a nonsplitting central extension of Z/3Z by A_6 of order 1080 such that G is generated by matrices with two eigenvalues -1 and the rest 1. The Taylor series expansion of the Poincare series of S^G is

$$1+2t^3+7t^6+16t^9+38t^{12}+\ldots.$$

([H-S]). From this we can easily see that $\mathrm{emb}(S^G) > 11$, and S^G is not a C.I..

Hereafter let V_i $(1 \leq i \leq m)$ be irreducible $\mathbb{C}G$-submodules of V with $\dim V_i = n_i$ which satisfy $V = \oplus_{i=1}^m V_i$ and $p_i\colon G \longrightarrow GL(V_i)$ the representation of G afforded by the $\mathbb{C}G$-module V_i. Let G^* be the direct product of $\{g \in GL(V) : g(V_i) = V_i\ (1 \leq i \leq m),\ g = 1 \text{ on } V_i\ (j \neq i),\ g \in p_j(G) \text{ on } V_i\}$ $(1 \leq j \leq m)$, and put $R = R(V:G^*)$, $G^i = \mathrm{Ker}(\oplus_{j\neq i} p_i)$, $G_i = p_i(G)$, $(G^i)_i = p_i(G^i)$ and $R_i = p_i(R)$, respectively. Let E_m be $\langle e_m\rangle$, D_m the binary dihedral group of order 4m, T the binary tetrahedral group of order 24, O the binary octahedral group of order 48 and I the binary icosahedral group of order 120. The notations $(E_p|E_q;H|K)$, $G(p,q,n)$ and $W(F)$ for a root graph F are defined in [Co].

Example 6. Let us assume that $m > 1$, $\dim V^G = 0$, $n_i > 8$ and G is primitive in $GL(V_i)$. Moreover suppose that S^G is a hypersurface. Then $S(V_i)^G$ is a polynomial ring and S^G is a simple extension of S^R. On the other hand, by Theorem 1, we have $G_i = R_i(G^i)_i = R_i$. Since R_i can be identified with $W(A_{n_i})$, $(G^i)_i$ is equal to $R_i \cap SL(V_i)$ and we must have $G = R \cap SL(V)$, if they are nontrivial. The nontriviality of $(G^i)_i$ follows from Examples 2 and 3.

Example 7. Suppose that $G_1 = G_2 = E_{60}.I$ and $(G^1)_1 = (G^2)_2 = 1$. Let U_i ($i = 1, 2$) be two dimensional $\mathbb{C}G$-modules such that $U_1 = V_1$ and $U_2 = 1 \oplus \det^{-1}$ as $\mathbb{C}G$-modules and let A be the graded subalgebra of $S(U_1) \otimes S(U_2)$ generated by all graded elements of degree (1,1). Then G acts on A and $\mathrm{emb}(A^G) > 7$. From this we see that S^G is not a C.I., because there is a $\mathbb{C}$-epimorphism from S^G to A^G.

Example 8. Suppose that $m = 2$, $n_1 = n_2 = 2$, each G_i agrees with one of the groups $E_{2a}.T$, $E_{2a}.I$, $(E_{6a}|E_{2a};T|D_2)$, $(E_{2a}|Ea;O|T)$ in $GL_2(\mathbb{C})$ and G is generated by elements g in G satisfying $\mathrm{rank}(g-1) = 2$ or $p_1(g) = p_2(g)^{-1}$ in $GL_2(\mathbb{C})$ (we may identify $GL(V_1)$ with $GL(V_2)$). Then : (1) If $G^1 = 1$, $G^2 = 1$, G_1 is conjugate to G_2 in $GL_2(\mathbb{C})$ and the natural representation of G in GL(V) is equivalent to the representation of G associated with $(\det_{V_2}^{-1} \oplus \det_{V_2}) \otimes p$ on $Z(G) \times L$ where $p = p_1 \oplus p_2$ and $p_1^{-1}(p_1(G) \cap SL(V_1))$. Moreover if $p_1(G) \neq E_{2a}.I$, p_1 is equivalent to p_2 on L (there are exactly two inequivalent irreducible representations of SL(2,5) of degree 2). (2) If $(G^1)_1 = \langle -1 \rangle$, $(G^2)_2 = \langle -1 \rangle$, G_1 is conjugate to G_2 in $GL_2(\mathbb{C})$ and there is a normal subgroup N of G such that $N \cap G^1 = N \cap G^2 = 1$ and N satisfies the condition in (1) for $G = N$. (3) If $(G^1)_1 = D_2$ and $G_1 \neq (E_{6a}|E_{2a};T|D_2)$, $(G^2)_2 = D_2$, G_1 is conjugate to G_2 in $GL_2(\mathbb{C})$ and agrees with one of $E_{2a}.T$, $E_{2a}.O$, $(E_{4a}|E_{2a};O|T)$.

Theorem 5. S^G is a C.I. if and only if the following conditions are satisfied :

(1) G is generated by $\{g \in G : \mathrm{rank}(g-1) = 2\}$.

(2) $(R,R\cap G,V)$ is a CI-triplet.

(3) $f_\chi(V_i,G_i)$ $(1 \leqq i \leqq m)$ are invariants of G^i for all $\chi\in\mathrm{Hom}(G^*,\mathbb{C}^*)$ such that $\chi(G) = 1$.

(4) All $S(V_i)^G$ are C.I..

This was essentially proved in $[N_4]$.

Theorem 6. S^G is a C.I. if and only if the following conditions are satisfied :

(1) G is generated by $\{g\in G : \mathrm{rank}(g-1) = 2\}$.

(2) $(R,R\cap G,V)$ is a CI-triplet.

(3) For each $1 \leqq i \leqq m$:

Case A "R is irreducible in $GL(V_i)$". If $R_i \neq G_i$, up to conjugacy, the groups G_i, R_i, $(G^i)_i$ are respectively listed in one of the following triplets ; 1) $G_i = \langle[e_{2^b}, e_{2^b}^{-1}], R_i\rangle$, $R_i = G(u,u,2)$, $(G^i)_i = G_i\cap SL(V_i)$ $(b > 1,\ 2^{b-1}||u)$; 2) $E_4.T$, $E_4.D_2$, $G_i\cap SL(V_i)$; 3) $E_6.O$, $E_6.T$, $G_i\cap SL(V_i)$; 4) $\langle -1,R_i\rangle$, $G(u,u,3)$, $G_i\cap SL(V_i)$ (u odd) ; 5) $\langle[e_{3u}^{-2},e_{3u},e_{3u}],R_i\rangle$, $G(u,v,3)$, $\langle[e_{3u}^{-2},e_{3u},e_{3u}],G(u,u,3)\cap SL(V_i)\rangle$ $(u > 1)$; 6) $\langle W(L_3)\cap SL(V_i),R_i\rangle$, $G(3,3,3)$, $G_i\cap SL(V_i)$; 7) $E_9.R_i$, $W(L_3)$, $G_i\cap SL(V_i)$; 8) $E_9.R_i$, $W(M_3)$, $G_i\cap SL(V_i)$; 9) $E_{18}.R_i$, $W(M_3)$, $G_i\cap SL(V_i)$; 10) $E_3.R_i$, $W(J_3(4))$, $G_i\cap SL(V_i)$; 11) $\langle[e_{2^b},e_{2^b},e_{2^b}^{-1},e_{2^b}^{-1}],R_i\rangle$, $G(u,v,4)$, $\langle[e_{2^b},e_{2^b},e_{2^b}^{-1},e_{2^b}^{-1}],G(u,u,4)\cap SL(V_i)\rangle$ $(2^{b-1}||u)$; 12) $E_4.R_i$, $W(F_4)$, $G_i\cap SL(V_i)$; 13) $E_2.R_i$, $W(A_4)$, $G_i\cap SL(V_i)$; 14) $E_{12}.R_i$, $W(L_4)$, $G_i\cap SL(V_i)$; 15) $E_8.R_i$, $EW(N_4)$, $G_i\cap SL(V_i)$; 16) $E_2.R_i$, $W(A_5)$, $G_i\cap SL(V_i)$; 17) $E_6.R_i$, $W(K_5)$, $G_i\cap SL(V_i)$; 18) $E_2.R_i$, $W(E_6)$, $G_i\cap SL(V_i)$.

Case B "R_i is reducible in $GL(V_i)$ and not abelian". (i) $n_i = 4$. (ii) G_i/R_i is conjugate in $GL(V_i(G^*\#R))$ to one of the groups in Example 4 or can be extended to a CI-triplet in $GL(V_i(G^*\#R))$. (iii) For any element x in V_i with $\dim\,(V_i)_K = 3$, in $GL((V_i)_K)$, K is extended to a CI-triplet or conjugate to one of the groups listed in [W-R, Sect. 3], where K denotes the stabilizer of x under the action of G^i. (iv) If, for an irreducible $\mathbb{C}R$-submodule U of V_i, $(G^i)_U$ is not contained in R, up to conjugacy, the groups $(G_i)_U$, $(R_i)_U$ and $((G^i)_i)_U$ agree, in $GL((V_i)_{R_U})$, respectively with G_i, R_i, and $(G^i)_i$ of $n_i = 2$ listed in Case A.

<u>Case</u> C "R_i is reducible in $GL(V_i)$ and nontrivial abelian". For each g in $(G^i)_i$, the product of nonzero entries of the matrix $[g_{ij}]$ of g is equal to one, where $[g_{ij}]$ of g is afforded by a $\mathbb{C}$-basis on which R_i is represented as a diagonal group, and G^i is conjugate in $GL(V_i)$ to one of $G(u,u,n_i)\cap SL(V_i)$ ($(u,u,n_i) \neq (2,2,2)$), $\langle G(u,u,4)\cap SL(V_i),[e_{2^b},e_{2^b},e_{2^b}^{-1},e_{2^b}^{-1}]\rangle$ ($2^{b-1}||u$), the groups in Example 4, $\langle G(u,u,3)\cap SL(V_i),[e_{3u}^{-2},e_{3u},e_{3u}]\rangle$ ($u > 1$), $\langle G(u,u,3)\cap SL(V_i),[e_{7u},e_{7u}^{2},e_{7u}^{-3}]\rangle$.

<u>Case</u> D "R_i is trivial". $G_i = (G^i)_i$, and G_i can be extended to a CI-triplet in $GL(V_i)$ or conjugate to one of the following groups ; 1) $(G^i)_i$ in Case A ; 2) the groups in Example 4 ; 3) $n_i = 4$ and there is a system of imprimitivities W_j ($j = 1,2$) of G_i in V_i with $\dim W_j = 2$. Let H be the subgroup of G_i consisting of all elements g satisfying $g(W_j) = W_j$ ($j = 1, 2$), $q_j: H \longrightarrow GL(W_j)$ the natural representations of H, L the subgroup generated by Ker q_j ($j = 1, 2$) and elements g in G_i such that both $q_j(g)$ are pseudo-reflections and put $L^* = q_1(L)\times q_2(L)$ ($q_j(L)$ is naturally regarded as a subgroup of $GL(V_i)$). L and W_j satisfy the conditions in Case A of (3) for $G = L$, $m = 2$ and $n_1 = n_2 = 2$ (i.e., $S(V_i)^L$ is a C.I.). Moreover $\bigcap_{\chi\in X}\mathrm{Ker}\,\chi\cap L^* = L$ for a subset X in $\mathrm{Hom}(L^*G_i,\mathbb{C}^*)$, and, in $GL(V_i(L^*G_i\#R(V_i:L^*G_i)))$, L^*G_i is extended to a CI-triplet or conjugate to one of the groups in Example 4 ; 4) $\langle A(u,u,4),[e_{2^b},e_{2^b},e_{2^b}^{-1},e_{2^b}^{-1}],(123)((4)),(234)((4))\rangle$ ($2^{b-1}||u$) ; 5) $E_4.I^{\otimes 2}$, $E_4.O^{\otimes 2}$, $(E_4|E_2;O|T)^{\otimes 2}$, $I^{\otimes 2}$, $O^{\otimes 2}$, $T^{\otimes 2}$; 6) the groups classified in [W-R].

Remark. The conditions in Case B and of 3) in Case D can be replaced by concrete classifications of some subgroups in $GL(V_i)$. However it is rather complicated.

<u>Theorem</u> 7. ([N_2, N_3]) S^G is a hypersurface if and only if $G = H\cap SL(V)$ for a finite reflection group H in GL(V) in which all orders of pseudo-reflections are equal to the index $[H : G]$ or G is conjugate to one of the groups classified in Case D of Theorem 6 whose algebras of invariants are hypersurfaces (delete the trivial subrepresentations of G).

To show Case D of Theorem 6, we use the classification of some finite linear groups obtained by Blichfeld, Huffman and Wales [B, H, Wal].

Our result is useful in classifying the representations of simple algebraic groups whose algebras of invariants are C.I.($[N_5]$).

6. Proof of Theorems 5 and 6.

In this section we will make a sketch of the proof of our classification and use the notation in Sect. 5.

Lemma 1. Suppose that S^G is a C.I.. Then G contains the commutator $[G^*, G^*]$ of G^*.

(Proof) Thanks to Theorem 1, we may assume $m = 2$, $n_1 = n_2 = 2$ and $G_i = R_i$ ($i = 1, 2$). Let G be a counter-example for this lemma.

Case 1 "both p_i ($i = 1, 2$) are primitive". Suppose $G_i = E_{2u}.I$, $G^1 = G^2 = 1$ and V_1 is not $\mathbb{C}L$-isomorphic to V_2, where $L = p_1^{-1}(I)$. Then, by the representation theory of SL(2,5), u is not divisible by 5. So u = 2, 6 or 3. We can compute the Taylor expansion of the Poincare series of S^G and see $\mathrm{emb}(S^G) > 7$, a contradiction. In the case where $G_i = E_{2u}.I$ and $(G^1)_1 = (G^2)_2 = \langle -1 \rangle$, using Example 8, we similarly get $\mathrm{emb}(S^G) > 7$. Moreover, if $G_i \neq E_{2u}.I$, by the method of Examples 7 and 8, one sees that both G^i ($i = 1, 2$) are not isomorphic to 1 and Z/2Z ($I/\langle -1 \rangle$ is simple). So G^i are isomorphic to D_2. Suppose there is an element g in G such that $p_1(g)$ is a pseudo-reflection of order 4. Let H be the stabilizer of elements in $\mathrm{Ker}(p_1(g)-1)$ under the action of G. Then, since S^H is a C.I. and $p_1(H)$ is a cyclic group of order 4, $p_2(H)$ contains a pseudo-reflection of order 4, which implies $p_2(H)$ is identified with $(E_{6u}|E_{2u};T|D_2)$ or $E_{2u}.D_2$. This conflicts with the classification in [Co]. Thus G_i is one of $(E_4|E_2;O|T)$, $E_4.O$, $(E_{12}|E_6;O|T)$, $E_{12}.O$. We easily see that S^G is not a C.I..

Case 2 "both p_i ($i = 1, 2$) are imprimitive". Let $\{X_1, X_2\}$ (resp. $\{X_3, X_4\}$) be a basis of V_1 (resp. V_2) on which G is represented as $G(u_1,v_1,2)$ (resp. $G(u_2,v_2,2)$), let A be $\{g \in G : p_i(g)$ are pseudo-reflections$\}$ and put $B = \{g \in A : p_i(g)$ are diagonal$\}$, $N = \langle G^1, G^2, B \rangle$ and $C = \{g \in G :$ only one of $p_i(g)$ is diagonal$\}$. Clearly $G = \langle A \rangle$ and G^i are diagonal. Suppose C is nonempty. Exchanging the indices of V_i, we can choose $g = [a,a^{-1},-1,1](12)((4))$, $h = [b,b^{-1},1,-1](12)((4))$ from A. For some natural numbers e, w, t, $S^{<B>} = \mathbb{C}[X_1^e,X_2^e,X_3^e,X_4^e,X_1X_2X_3X_4]$ and $S^N = \mathbb{C}[X_1^{ew},X_2^{ew},X_3^{et},X_4^{et},(X_1X_2)^e,(X_3X_4)^e,X_1X_2X_3X_4]$. The natural representation $\underline{p}_1$: $G/N \longrightarrow GL(\mathbb{C}X_1^{ew}\oplus\mathbb{C}X_2^{ew})$ is faithful and the image of this map can be identified $G(\underline{u}_1,\underline{v}_1,2)$ on this basis. Since $\underline{p}_1$ has two distinct systems of imprimitivities, by [Co], $(\underline{u}_1,\underline{v}_1) = (2,1)$, $(4,4)$ or $(4,2)$. Consequently the action of G/N on S^N may be given by one of the following rules ; 1) $G/N = \langle gN,hN,kN\rangle$, $\underline{p}_1(G) = G(4,4,2)$, $\underline{p}(gN) = [1,1,-1,1](12)((4))$, $\underline{p}(hN) = [-1,-1,1,-1](12)((4))$, $\underline{p}(kN) = [e_4^{-1},e_4,1,1](12)(34)((4))$, $g((X_1X_2)^e) = h((X_1X_2)^e) = k((X_1X_2)^e) = (X_1X_2)^e$, $g((X_3X_4)^e) = h((X_3X_4)^e) = -(X_3X_4)^e$, $k((X_3X_4)^e) = (X_3X_4)^e$, $g(X_1X_2X_3X_4) = h(X_1X_2X_3X_4) = -X_1X_2X_3X_4$, $k(X_1X_2X_3X_4) = X_1X_2X_3X_4$; 2) $G/N = \langle gN,hN,kN,fN\rangle$, $\underline{p}_1(G) = G(4,2,2)$, the action of g, h, k is the same one as in Case 1, $\underline{p}(f) = [-1,1,e_4,e_4^{-1}](34)((4))$, $f((X_1X_2)^e) = -(x_1X_2)^e$, $f((X_3X_4)^e) = (X_3X_4)^e$, $f(X_1X_2X_3X_4) = -X_1X_2X_3X_4$; 3) $G/N = \langle gN,hN,k'N\rangle$, $\underline{p}_1(G) = G(2,1,2)$, the action of g, h is the same one as in Case 1, $\underline{p}(k') = [-1,1,1,1](34)((4))$, $k'((X_1X_2)^e) = (X_1X_2)^e$, $k'((X_3X_4)^e) = (X_3X_4)^e$, $k'(X_1X_2X_3X_4) = X_1X_2X_3X_4$; where $\underline{p}$ is the natural representation of G/N in $GL(\mathbb{C}X_1^{ew}\oplus\mathbb{C}X_2^{ew}\oplus\mathbb{C}X_3^{et}\oplus\mathbb{C}X_4^{et})$. We can compute the invariants of G and see $\mathrm{emb}(S^G) > 7$, which is a contradiction. Thus C is empty. Moreover B is nonempty and $X_1X_2X_3X_4$ is not contained in a minimal system of monomial generators of S^D, where D is the subgroup consisting of all diagonal matrices in G. So $G^1 = G^2 = 1$ and by Example 2, $e > 1$. In this circumstance, we have $\mathrm{emb}(S^G) > 7$.

Case 3 "p_1 is primitive and p_2 is imprimtive". By Clifford's theorem, $(G^1)_1$ can be identified with one of D_2, $\langle -1\rangle$ and 1. Then , using the classification of 2-dimensional irreducible reflection groups ([Co]), we easily get an elementary group theoretic contradiction. Thus the proof of Lemma 1 is completed.

Lemma 2. Suppose that $f_\chi(V,G^*)$ is a χ-invariant of G^* for every linear character χ of G^* satisfying $\chi(G) = 1$. Then S^G is a C.I. if and only if $(R,R\cap G,V)$ is a CI-triplet, G is generated by elements g with $\mathrm{rank}(g-1) = 2$ and all $S(V_i)^G$ are C.I..

(Proof) By Stanley's theorem stated in Sect. 4, (in case of "if" part or in case of "only if" part of this lemma,) we always see that the closed fibre of the flat morphism $(S_{SV})^{G^*} \longrightarrow (S_{SV})^G$ is isomorphic to one of the flat morphism $(S_{SV})^R \longrightarrow (S_{SV})^{G\cap R}$ and hence the assertion follows.

(Proof of Theorem 5) It suffices to show that if S^G is a C.I. condition (3) holds. Because $f_\chi(V_i,G_i)$ may be regarded as a product of relative invariants $f_\chi(V_i,G_x)$ for some elements x in $\oplus_{j\neq i}V_j$, in order to prove (3), we may assume that $m = i = n_2 = 2$, $n_1 = 1$. But this case had already been treated in [W-R].

(Proof of Theorem 6) In the case where G is essentialy reducible in GL(V), the assertion follows from Theorem 5 and [Sp_1, B-E, St_2] (the proof of Case B is similar to the proof of 3) in Case D). So we suppose that G is irreducible.

Case 1 "G is imprimitive and non-monomial". Assume that S^G is a C.I.. By Theorem 1, there is a block diagonal subgroup H of G such that the dimension of each irreducible $\mathbb{C}H$-submodule of V is 2. Let W_i be irreducible $\mathbb{C}H$-submodules of V which satisfy $V = \oplus_{i=1}^{n/2}W_i$ and q_i be the natural representation of H in $GL(W_i)$. There are elements t_i in G such that $t_i(W_{i-1}) = W_i$, $t_i(W_i) = W_{i-1}$, $t_i(W_j) = W_j$ $(j \neq i-1,i)$, $t_i = 1$ on W_j $(j \neq i-1, i)$, $ord(t_i) = 2$ and $G = \langle H,t_2,\ldots,t_{n/2}\rangle$. Let $\{X_i,Y_i\}$ be a $\mathbb{C}$-basis of W_i. We may assume that the restriction of t_i to $W_{i-1}\oplus W_i$ is equal to (13)(24)((4)) on this basis. Then, if $n > 5$, the stabilizer of elements $X_4, \ldots, X_{n/2}$ is isomorphic to S_3, and hence S^H is not a C.I.. Hence we must have $n = 4$. As in the proof of Lemma 1, by the method of Examples 7 and 8, $emb(S^G) > 7$ if H does not contain the commutator of L^*G (we use the notation defined in 3) in Case D of Theorem 6) or if $q_1(H)$ does not contain a pseudo-reflection. Because $f_\chi(W_1,L^*)$ is a product of $f_\chi(V_1,G_x)$ for some elements x in W_2, $f_\chi(V,L^*)$ is an invariant of Ker q_1 and Ker q_2 if χ is a linear character of L^* satisfying $\chi(L) = 1$. This implies $S^G = \oplus_{\chi\varepsilon X}S^{L^*G}f_\chi(V,L^*)$. Thus the closed fibre of the flat morphism $(S_{SV})^{L^*G} \longrightarrow (S_{SV})^G$ is isomorphic to one of the flat morphism $(S_{SV})^{L^*} \longrightarrow (S_{SV})^L$, and S^L is a C.I.. Because L^*G contains a direct product of 2-dimensional irreducible reflection groups, by Example 4, we easily infer the remainder of the assertion. In order to show the converse of our theorem, we need only to check the proof of "only if" part.

Case 2 "G is monomial". By the slice method and [H], we immediately obtain our classification, except for the case where $n = 4$ and the permutation group associated with the monomial action of G on a basis is isomorphic to <(12),(34),(13)(24)> or <(12)(34),(13)(24)>. But, in the last case, the method used in the proof is one of in the proof in Case 1.

Case 3 "G is primitive". If G is one of the groups in our theorem (cf. Case D), using [Se], we can show that S^G is a C.I., and so suppose that G is not listed and S^G is a C.I.. (1) $n = 4$. Let us use the notation in [B]. If G is one of the groups (A)...(K) (some of them are listed and should be deleted), we easily compute the Taylor expansion of the Poincare series of S^G and get a contradiction. Suppose that G has a normal intransitive imprimitive subgroup. In this case G is obtained as a tensor product of 2-dimensional primitive irreducible groups, and, by the similar method as in Example 7, $\mathrm{emb}(S^G) > 7$ (it should be noted that G leaves a hypersurface in $\mathbb{C}^4$). If G is the groups $8^o...12^o$, we apply Theorem 1. Thus G is one of the groups $13^o...21^o$. Let f_i be homogeneous polynomials in S which minimally generate S^K such that $\oplus_{i=1}^{6}\mathbb{C}f_i$ is a $\mathbb{C}G$-module and let $A = \mathbb{C}[T_1,\ldots,T_6]$ be a graded polynomial algebra defined by $\deg T_i = \deg f_i$. Then Ker h is generated by two graded elements F_j as an ideal where h is the natural graded epimorphism defined by $h(T_i) = f_i$. As G-modules, we identify $\oplus\,\mathbb{C}T_i$ with $\oplus\mathbb{C}f_i$ and need only to compute the invariants of A^G (note that S_6 does not have a 2-dimensional irreducible representation). (2) $n = 5$. Brauer's classification is well known and essential. Suppose that G is isomorphic to PSL(2,11) and on a $\mathbb{C}$-basis $\{X_1,\ldots,X_5\}$, G contains $g = [e_{11}^4,e_{11}^5,e_{11}^9,e_{11}^3,e_{11}]$. $g^4 = g'gg'^{-1}$ for an element g' in G. Then S^H is not a C.I., where H is the stabilizer of $X_1+\ldots+X_5$ under the action of G. Hence G is associated with some extra 2-special groups. On $\{X_1,\ldots,X_5\}$, the 5-Sylow intersection group D is monomial and generated by an abelian group A and (12345)((5)). Let H be the stabilizer of G at $X_1+\ldots+X_5$. H_5 (5-Sylow group) = <(12345)((5))>, and S^H is not a C.I.. (3) $n > 5$. We use the classification of [Wal]. Clearly G is not the groups (A), (B), (F), (I) and

(M) (for the last two groups, the Taylor expansions of the Poincare series of S^G are known). The slice method is useful. For example : G = (C) ; Let x and y be elements respectively in Y and Y_1 such that A_y is trivial and K_y is a 2-dimensional irreducible reflection group. Then $S^{G_{x\otimes y}}$ is not a C.I.. G = (K) ; Since $SL(2,5)oSL(2,5)ol_2$ is contained in H and $V = U\oplus U$ for some $\mathbb{C}SL(2,5)oSL(2,5)$-module U, we can choose an element g from $SL(2,5)oSL(2,5)ol_2$ such that dim $(g-1)U = 2$. Then the stabilizer of every elements in $U^{\langle g \rangle}$ under the action of G does not satisfy Theorem 1.

REFERENCES

[B] H. F. Blichfeld, Finite Collineation Groups, University of Chicago Press, Chicago, 1917.

[B-E] D. Buchsbaum and D. Eisenbud, Algebra structure for finite free resolutions, and some structure theorems for ideals of codimension 3, Amer. J. Math. 99 (1977), 447-485.

[Bo] J.-F. Boutot, Singularites rationelles et quotient par les groupes reductifs, preprint.

[Ch] C. Chevalley, Invariants of finite groups generated by reflections, Amer. J. Math. 67 (1955), 778-782.

[Co] A. M. Cohen, Finite complex reflection groups, Ann. Sci. Ecole Norm. Sup. 9 (1976), 379-436.

[G-W] S. Goto and K.-i. Watanabe, The embedding dimension and multiplicity of rational singularities which are complete intersections, preprint.

[H] W. G. Huffman, Imprimitive linear groups generated by elements containing an eigenspace of codimension two, J. Algebra 63 (1980), 499-513.

[H-E] M. Hochster and J. Eagon, Cohen-Macaulay rings, invariant theory, and generic perfection of determinantal loci, Amer. J. Math. 93 (1971), 1020-1058.

[H-S] W. C. Huffman and N. J. A. Sloane, Most primitive groups have messey invariants, Advances in Math. 32 (1979), 118-127.

[K-W] V. Kac and K.i. Watanabe, Finite linear groups whose ring of invariants is a complete intersection, Bull. Amer. Math. Soc. 6 (1982), 221-223.

[L-T] J. Lipman and B. Teissier, Pseudo-rational local rings and a theorem of Briancon-Skoda about integral closures of ideals, Michigan Math. J. 28 (1981), 97-116.

[Mo] T. Molien, Uber die Invarianten der linearen Substitutionsgruppen, Sitzungsber. Konig Preuss. Akad. Wiss. (1897), 1152-1156.

[N_1] H. Nakajima, Relative invariants of finite groups, J. Algebra 79 (1982), 218-234.

[N_2] H. Nakajima, Rings of invariants of finite groups which are hypersurfaces, J. Algebra 80 (1983), 279-294.

[N_3] H. Nakajima, Rings of invariants of finite groups which are hypersurfaces, II, Advances in Math., to appear.

[N_4] H. Nakajima, Quotient complete intersections of affine spaces by finite linear groups, preprint, 1982.

[N_5] H. Nakajima, Representations of simple Lie groups whose algebras of invariants are complete intersections, preprint, 1983.

[Se] J.-P. Serre, Sur les modules projectifs, Sem. Dubreil-Pisot, 1960/1961.

[SGA 2] A. Grothendieck, Cohomologie locale des faisceaux coherents et Theoremes de Lefschetz locaux et globaux (SGA 2), North-Holland, Amsterdam, 1968.

[Sp_1] T. A. Springer, Regular elements of finite reflection groups, Invent. Math. 25 (1974), 159-198.

[Sp_2] T. A. Springer, Invariant Theory, Lect. Notes in Math. No. 585, Springer, Berlin, 1977.

[St_1] R. Stanley, Relative invariants of finite groups generated by pseudo-reflections, J. Algebra 49 (1977), 134-148.

[St_2] R. Stanley, Hilbert functions of graded algebras, Advances in Math. 28 (1978), 57-83.

[St_3] R. Stanley, Invariants of finite groups and their applications to combinatorics, Bull. Amer. Math. Soc. 1 (1979), 475-511.

[S-T] G. C. Shephard and J. A. Todd, Finite unitary reflection groups, Canad. J. Math. 6 (1954), 274-304.

[Wal] D. B. Wales, Linear groups of degree n containing an involution with two eigenvalues -1, II, J. Algebra 53 (1978), 58-67.

[W_1] K.-i. Watanabe, Certain invariant subrings are Gorenstein, II, Osaka J. Math. 11 (1974), 379-388.

[W_2] K.-i. Watanabe, Invariant subrings which are complete intersections, I (Invariant subrings of finite Abelian groups), Nagoya Math. J. 77 (1980), 89-98.

[W-R] K.-i. Watanabe and D. Rotillon, Invariant subrings of $\mathbb{C}[X,Y,Z]$ which are complete intersections, Manuscripta Math. 39 (1982), 339-357.

COMPRESSED ALGEBRAS

by

R. Fröberg and D. Laksov
University of Stockholm

0. INTRODUCTION.

The purpose of the following work is twofold. Firstly we want to give a presentation of A. Iarrobino's construction of graded compressed artinian algebras (Theorem 14 below and [Ia] Thm III), that is algebras of maximal length among those of a given socle type. Secondly we want to generalize the notion of compressed algebras, introduced by Iarrobino for artinian rings, to the class of graded Cohen-Macaulay algebras and to generalize Iarrobino's results to this class of rings. We feel that our presentation of the existence result adds substantially to the understanding of the subject. It introduces the compressed algebras in a natural way that brings out the reason why these algebras are important extremal cases among the algebras of a given socle type. Our approach is completely within the framework of algebras and avoids the duality, used by Iarrobino, between graded algebras and the corresponding algebra of differential operators. Finally our presentation points out that the crucial and most interesting part of the work lies in the construction of certain "generic" compressed algebras. Our main mathematical contribution is the method for constructing such algebras which is natural, illustrative and characteristic free. The elegance of the construction was achieved thanks to ideas of J. Boman (private communication).

For readers familiar with (and appreciating) the dual viewpoint alluded to above, we have included an appendix giving the connections between our work and that of Iarrobino. In particular we show how our approach gives results of Grace-Young, Iarrobino and Jordan about powers of linear forms.

In his article [Ia] Iarrobino also announces a result of Buchsbaum, Eisenbud and himself about the resolutions of compressed artinian algebras satisfying an additional condition (Prop. 4.1 a and b, also c.f. [Bu-Ei]). We generalize their result and give a useful extension to the Cohen-Macaulay case. We prove that the compressed algebras satifying this additional condition, we call them extremely compressed algebras, comprise extremal Cohen-Macaulay rings and extremal Gorenstein rings. In particular they contain interesting classes of rings like rings of surfaces with rational or elliptic singularities, Cohen-Macaulay and Gorenstein rings of maximal embedding dimension, some extremal Stanley-Reisner rings and certain classes of determinantal and pfaffian coordinate rings. Thus our generalization of the Buchsbaum-Eisenbud-Iarrobino result implies a series of results on resolutions by Schenzel, Sally and Wahl. We also prove that the coordinate ring of points in general position in projective space are all compressed.

We have also included a simplified version of Iarrobino's classification of compressed algebras that are complete intersections without reference to the dual viewpoint mentioned above.

1. NOTATION AND DEFINITIONS.

Let k be a field. In the following we shall study finitely generated graded k-algebras that can be generated by elements of degree one and shall tacitly assume that all algebras are of this type. Given such an algebra A and a graded A-module M we denote by M_d the vector space of elements of degree d and we denote by

$$\mathrm{Hilb}_M(z) = \sum_{c=0}^{\infty} \dim_k(M_c) z^c$$

the Hilbert series of M. For an artinian algebra A we define the *socle type* $S(A)$ to be the polynomial $\mathrm{Hilb}_{\mathrm{Soc}(A)}(z)$ where $\mathrm{Soc}(A) = \{f \in A;\ fg = 0 \text{ for every } g \text{ in } \bigoplus_{i=1}^{\infty} A_i\}$ is the socle of A. Later we shall define the *type* of a graded Cohen-Macaulay algebra, which coincides with the socle type when the algebra is artinian, and we shall be interested in constructing Cohen-Macaulay algebras of a prescribed type. In this generality the problem of construction is too difficult. Therefore we shall limit ourselves to certain "generic" cases that pose rather severe restrictions on the type. To display these limitations it is convenient to introduce the following notation;

Given a positive integer s, we let

$$N(s,c) = \binom{c+s-1}{c}$$

and denote by P_s the set of polynomials

$$\{\sum_{c=0}^{e} s_c z^c;\ e \text{ and } s_c \text{ non-negative integers and } s_c \leq N(s,c) \text{ for all } c\}.$$

For each element $p = \sum_{c=0}^{e} s_c z^c$ in P_s with $s_e \neq 0$ we define integers $r_d(c)$ for all $0 \leq c \leq d$ by

$$r_d(c) = N(s,c) - N(s,d-c)s_d - N(s,d+1-c)s_{d+1} - \ldots - N(s,e-c)s_e$$

and we let $r_d = r_d(d)$. We note that $r_d = N(s,d)$ for $d > e$ and that $r_d \leq 0$ for $d \leq e/2$.

LEMMA 1. *With the notation above the following relations hold;*

(*i*) $r_d(c+1) - r_d(c) = N(s-1,c+1) + N(s-1,d-c)s_d + N(s-1,d+1-c)s_{d+1} + \ldots + N(s-1,e-c)s_e$

(*ii*) $r_{d+1}(c) - r_d(c) = N(s,d-c)s_d$

In particular we have that $r_{d+1} > r_d$.

As a consequence of Lemma 1 we can, for each polynomial $p \in P_s$ uniquely determine an integer $b \leq e$ by the inequalities $r_b \geq 0$ and $r_{b-1} < 0$. With this notation it follows from Lemma 1 that $r_d(c) \geq 0$ for all $b \leq c \leq d$.

For convenience of notation the integers $r_d(c)$, r_d or b do not refer to the number s or the polynomial p from which they are constructed. We shall however, whenever these integers appear, state explicitely from which number s and polynomial p they are formed.

When A is an *artinian* ring the integer $r = \dim_k A_1$ is called the *codimension* of A. Clearly A is the quotient of a polynomial ring $R = k[x_1, x_2, \ldots, x_r]$ and therefore if $S(A) = \sum_{c=0}^{e} s_c z^c$ we have inequalities

$$s_c \leq \dim_k R_c = N(r,c).$$

We can therefore define integers r_d and b associated to $r = \dim_k A_1$ and $S(A)$. These integers we denote by $r_d(A)$ and $b(A)$.

We shall consider algebras A that are quotients of $R = k[x_1, x_2, \ldots, x_r]$ by a graded ideal I. We denote by $(I_d : R)_c$ the sub-vector space

$$\{f \in R_c;\ fR_{d-c} \subseteq I_d\}$$

of R_c. Then $\mathrm{Soc}(A)_d = (I_{c+1} : R)_c / I_c$, that is

$$S(A) = \sum_{c=0}^{\infty} \dim_k((I_{d+1} : R)_c / I_d) z^c.$$

In our construction of algebras we shall need the following two observations;

LEMMA 2. *Let* $J \subseteq V \subseteq R_d$ *be sub-vectorspaces of* R_d. *Then the inequality*

$$\dim_k (J:R)_c \geq \dim_k (V:R)_c - N(r,d-c) \dim_k (V/J)$$

holds for all $0 \leq c \leq d$.

PROOF. Consider the map

$$\Psi:\ (V:R)_c \longrightarrow \bigoplus_M (V/J)$$

defined by $\Psi(F) = \sum_M MF$, where both sums are over the $N(r,d-c)$ monomials in R_{d-c}. Then $(J:R)_c$ is the kernel of Ψ and consequently is of dimension $\dim_k (V:R)_c - \dim_k(\mathrm{im}\Psi) \geq \dim_k (V:R)_c - N(r,d-c)\dim_k(V/J)$.

LEMMA 3. *For* $c \leq i \leq d$ *and for each subspace* V *of* R_d *we have the relation*

$$(V:R)_c = ((V:R)_i:R)_c.$$

2. CONDITIONS SATISFIED BY ARTINIAN ALGEBRAS OF A GIVEN SOCLE TYPE.

We now describe conditions that ideals I in R must satisfy in order to give an artinian quotient algebra R/I and be of a given socle type $p = \sum_{c=0}^{e} s_c z^c$ with $s_e \neq 0$. From the integer r and the polynomial p we define the numbers $r_d(c)$ and b as in section 1. First we observe that because we require that the quotient R/I shall be artinian we must have $I_c = R_c$ for $c>e$. Then we observe that I_e must be a subspace of R_e of codimension s_e. Given such a space it follows from Lemma 2 that we have an inequality

$$\dim_k(I_e:R)_c \geq N(r,c) - N(r,e-c)s_e = r_e(c) \tag{1}$$

for all $0 \leq c \leq e$. Next we observe that, for I to be an ideal with socle dimension s_{e-1} in degree e-1, we must have that I_{e-1} is a subspace of $(I_e:R)_{e-1}$ of codimension s_{e-1}. For such a space to exist we must have an inequality

$$\dim_k(I_e:R)_{e-1} \geq s_{e-1}. \tag{2}$$

From the inequality (1) it follows that (2) always holds if $r_e(e-1) - s_{e-1} = r_{e-1} \geq 0$. In particular, if $e-1 \geq b$ we can always find such a space I_{e-1}. Moreover, given I_{e-1}, it follows from Lemma 2 that we have an inequality

$$\dim_k(I_{e-1}:R)_c \geq r_e(c) - N(r,e-1-c)s_{e-1} = r_{e-1}(c) \tag{3}$$

for all $c \geq e-1$. Next we observe that, by the same reasoning as above, we must have that I_{e-2} is a subspace of $(I_{e-1}:R)_{e-2}$ of codimension s_{e-2}. For such a space to exist we must have an inequality

$$\dim_k(I_{e-1}:R)_{e-2} \geq s_{e-2}.$$

From (3) it follows that the latter inequality always holds if $r_{e-1}(e-2) - s_{e-2} = r_{e-2} \geq 0$. Inparticular, if $e-2 \geq b$, we can always find such a space I_{e-2}. Again it follows from Lemma 2 that we have an inequality

$$\dim_k(I_{e-2}:R)_c \geq \dim_k(I_{e-1}:R)_c - N(r,e-2-c)s_{e-2}.$$

Hence by the inequality (3) together with Lemma 1(ii) we have an inequality

$$\dim_k(I_{e-2}:R)_c \geq r_{e-2}(c)$$

for all $c \leq e-2$. In this manner we can always find spaces $I_e, I_{e-1},\ldots, I_b$ such that I_d is a subspace of $(I_{d+1}:R)_d$ of codimension s_d and such that the inequalities

$$\dim_k(I_d:R)_c \geq \dim_k((I_{d+1}:R)_d:R_c)-N(r,d-c)s_d=\dim_k(I_{d+1}:R)_c-N(r,d-c)s_d$$

hold and by the same reasoning as above, we obtain by descending induction on d the inequality

(4) $$\dim_k(I_d:R)_c \geq r_d(c)$$

for all $c \leq d$.

We note that the above argument can not be continued to claim the existence of an I_{b-1} contained in $(I_b:R)_{b-1}$ and of codimension s_{b-1}. Indeed, we have that $r_{b-1}<0$ so that the particular case

$$\dim_k(I_b:R)_{b-1} \geq r_b(b-1) = r_{b-1}+s_{b-1}$$

of the inequality (4) does not guarantee that $\dim_k(I_b:R)_{b-1} \geq s_{b-1}$. All we can say is that $I = 0\oplus 0\oplus\ldots\oplus 0\oplus I_b\oplus I_{b+1}\oplus\ldots\oplus I_e\oplus R_{e+1}\oplus R_{e+2}\oplus\ldots$ is an ideal of R of socle type $\dim_k(I_b:R)_{b-1}z^{b-1}+\sum_{c=b}^{e} s_c z^c$ such that R/I is artinian.

3. CONSEQUENCES IN THE ARTINIAN CASE.

The observations of section 2 lead to the following two results about ideals of a given socle type having an artinian quotient algebra;

PROPOSITION 4. *Let* A *be an artinian algebra of codimension* r *and socle type* $S(A) = \sum_{c=0}^{e} s_c z^c$ *with* $s_e \neq 0$. *Then the following assertions hold;*

(i) $\dim_k A_d \leq \min(N(r,d)-r_d, N(r,d))$ *for all* d.

(ii) $\ell(A) \leq N(r+1,b(A)-1)-s_{b(A)}-N(r+1,1)s_{b(A)+1}-\ldots-N(r+1,e-b(A))s_e$.

(iii) Equality holds in (ii) if and only if all the inequalities of (i) are equalities.

(iv) Let $r_d(c)$ *and* b *be the integers defined by the number* r *and the polynomial* $p = \sum_{c=0}^{e} s_c z^c$ *in* P_r *with* $s_e \neq 0$. *If* $s_c=0$ *for* $c=0,1,\ldots,b-1$ *then any sequence of vector spaces* J_d *in* R_d *for* $d=0,1,\ldots$ *with* $J_d \subseteq (J_{d+1}:R)_d$

and satisfying the relations $\dim_k(J_d:R)_c = \max(r_d(c),0)$ *for all* $0 \leq c \leq d$, *give rise to an ideal*

$$J = 0 \oplus 0 \oplus \ldots \oplus 0 \oplus J_b \oplus J_{b+1} \oplus \ldots \oplus J_e \oplus R_{e+1} \oplus R_{e+2} \oplus \ldots$$

such that $S(R/J) = p$ *and such that the inequalities of (i) and (ii) are equalities for the quotient algebra* R/J. *Moreover, if* $e>1$, *then* $\mathrm{codim}(R/J)=r$.

PROOF. Since A has codimension r we have that A is a quotient of R by an ideal I. We observed in section 2 that $\dim_k(I_d:R)_c \geq N(r,c) - \sum_{i=d}^{e} N(r,i-c)s_i$ for all d such that $b(A) \leq d \leq e$. Hence, for such a d the inequality $\dim_k A_d \leq N(r,d) - r_d(A)$ holds. For $d<b(A)$ and for $d>e$ the inequalities of (i) are the obvious $\dim_k A_d \leq N(r,d)$ and $\dim_k A_d \leq 0$. Summing the left and right sides of the inequalities of part (i) we obtain an inequality $\sum_{d=0}^{\infty} \dim_k A_d = \sum_{d=0}^{b-1} N(r,d) + \sum_{d=b(A)}^{e} (N(r,d) - r_d)$. Using well-known relations for binomial coefficients, we obtain the inequality (ii). Since we obtained the inequality (ii) by "summing" the inequalities (i), part (iii) is clear. Assertion (iv) is an immediate consequence of the observations of section 2 in the case when the vector spaces $(J_d:R)_c$ all have the minimal possible dimension $\max(r_d(c),0)$. In particular we have that $\dim_k(J_b:R)_{b-1} = \max(r_b(b-1),0)$ and that $\dim_k J_c = \max(r_c,0) = 0$ for $c<b$. We saw in section 2 that under these circumstances we obtain an ideal $0 \oplus 0 \oplus \ldots \oplus 0 \oplus J_b \oplus J_{b+1} \oplus \ldots \oplus J_e \oplus R_{e+1} \oplus R_{e+2} \oplus \ldots$ of socle type

$$\dim_k(J_b:R)_{b-1} z^{b-1} + \sum_{c=b}^{e} s_c z^c.$$

However $\dim_k(J_b:R)_{b-1} = \max(r_b(b-1),0)$ and by Lemma 1 we have that $r_b(b-1) = r_{b-1} + s_{b-1} = r_{b-1} < 0$ so that we have $\dim_k(J_b:R)_{b-1} = 0$.

Moreover, if $e>1$ then we have observed that $r_c = 0$ for $c \leq e/2$. In particular we have that $r_1 = 0$, so that $\dim_k J_1 = 0$. That means $\mathrm{codim}_k A = \dim_k R_1 = r$.

Notice that Proposition 4 does <u>not</u> assert that given an element $p = \sum_{c=0}^{e} s_c z^c$ in P_r such that $s_c = 0$ for $c<b$, then there exists an algebra of socle type p, it simply states that if there exist vector spaces J_d

for $d=0,1,\dots$ satisfying the properties of part (iv) of that proposition, then there exists an algebra for the given p. One main result of this note is that such vector spaces exist for all p in P_r such that $s_c=0$ for $c<b$. We postpone the proof of this existence result to section 7.

Assume that J is a graded ideal satisfying the relations

$$\dim_k(J_d:R)_c = \max(r_d(c),0)$$

for all $0\leqq c\leqq d$ as in Proposition 4 (iv). If I_d is a subspace of R_d of codimension n, then, since the vector spaces $(I_d:R)_c$ are defined by linear equations in the coefficients of monomials in R_c, the subspaces I_d satisfying an inequality of the form $\dim_k(I_d:R)_c>m$ form a closed algebraic subset in the grassmannian $\mathrm{Grass}_n(R_d)$ of all codimension n subspaces of R_d. Hence, for each I_e in an open subset U_e, containing J_e, of $\mathrm{Grass}_{s_e}(R_e)$ the equalities $\dim_k(I_e:R)_c=\max(r_d(c),0)$ hold. Similarly, for each space I_{e-1} in an open subset U_{e-1}, containing J_{e-1}, of the relative grassmannian over U_e whose fiber over I_e is $\mathrm{Grass}_{s_{e-1}}((I_e:R)_{e-1})$, the equalities $\dim_k(I_{e-1}:R)_c=\max(r_d(c),0)$ hold. Continuing recursively, we obtain for each $d=e-1,e-2,\dots,b$ that for each space I_d in an open subset, containing J_d, of the relative grassmannian over U_{d+1} whose fiber over I_{d+1} is $\mathrm{Grass}_{s_d}((I_{d+1}:R)_d)$, the equalities $\dim_k(I_d:R)_c=\max(r_d(c),0)$ hold. The variety $\mathrm{Grass}_{s_d}((I_{d+1}:R)_d$ has dimension $s_d(\dim_k(I_{d+1}:R)_d-s_d)=s_d(r_{d+1}(d)-s_d)=s_dr_d$ for $d=b,b+1,\dots,e-1$. Hence, assuming the result of section 7 stating that for each p in P_r vector spaces $J_0,J_1,\dots$ satisfying the conditions of Proposition 4(iv) exist, we see that the following result hold;

PROPOSITION 5. *Let* $p=\sum_{c=0}^{e}s_cz^c$ *with* $s_e\neq 0$ *be an element of* P_r *and* r_d *and* b *the numbers defined by* r *and* p. *If* $s_c=0$ *for* $c<b$ *there is an open subset* U *of the* $r_es_e+r_{e-1}s_{e-1}+\dots+r_bs_b$*-dimensional affine space such that each point of* U *correspond in a natural one to one way to graded ideals* I *such that* R/I *is artinian of socle type* p. *Moreover, the space* U *is an open subset in a chain of relative grassmannians in such a way that* I_d *lies in an open subset of* $\mathrm{Grass}_{s_d}((I_{d+1}:R)_d)$ *for* $d=b,b+1,\dots,e-1$ *and* I_e *lies in an open subset of* $\mathrm{Grass}_{s_e}(R_e)$.

4. RECALL OF PROPERTIES OF GRADED ALGEBRAS.

Our next task shall be to extend Proposition 4 to the Cohen-Macaulay case. To accomplish this we shall need the following well-known definitions and properties of graded algebras.

Let A be a graded k-algebra of (Krull) dimension t. Then the Hilbert series of A is of the form

$$\mathrm{Hilb}_A(z) = p(z)/(1-z)^t,$$

where p(z) is a polynomial [At-Ma, thm 11.1]. The integer p(1) is called the *multiplicity* of A and is denoted m(A). When A is artinian, that is $t=0$, we clearly have that $m(A) = \ell(A)$.

(a) Let K be a field containing k. Then we have that

$$\mathrm{Hilb}_A(z) = \mathrm{Hilb}_{A\otimes_k K}(z).$$

Indeed $\dim_k A_d = \dim_K (A\otimes_k K)_d$ for all d.

(b) If y is a homogeneous non-zero divisor of degree d in A we have

$$\mathrm{Hilb}_{A/y}(z) = \mathrm{Hilb}_A(z)(1-z)^d.$$

This equation follows immediately from the exact sequence $0 \to A \xrightarrow{y} A \to A/y \to 0$.

(c) If k is an infinite field we can find a maximal A-regular sequence consisting of elements of A of degree one. Indeed, by the Noether normalization lemma, see e.g. [At-Ma, p. 69 exercise 16], there exist elements $y_1, y_2, \ldots, y_t$ in A_1 that are algebraically independent and such that A is integral over $k[y_1, y_2, \ldots, y_t]$. Then clearly $y_1, y_2, \ldots, y_t$ is a regular sequence of the desired form.

(d) The graded algebra A is *Cohen-Macaulay* if one of the following equivalent conditions is satisfied;

(i) There exists a regular sequence of homogeneous elements of positive degrees of length dimA in A.

(ii) The local ring obtained by localizing A at A_+ (the maximal graded ideal) is a local Cohen-Macaulay ring.

(iii) The local rings obtained by localizing A at any prime ideal are all

Cohen-Macaulay.

(iv) The local rings obtained by localizing A at any maximal ideal are all Cohen-Macaulay.

(v) The local rings obtained by localizing A at any graded prime ideal are Cohen-Macaulay.

(vi) The polynomial ring $A[y_1,y_2,\dots,y_u]$ is Cohen-Macaulay of dimension dimA+u.

Proofs of these equivalences can be found in [Na-v.O, ch. B III] and [Ho-Ra, thm 4.11].

Denote by R[d] the graded R-module defined by $R[d]_c=R_{c+d}$. Any graded quotient algebra A of R clearly has a resolution of the form

$$F_*:\ 0\to \bigoplus_{i=1}^{b_c} R[-n_{i,c}]\to \dots\to \bigoplus_{i=1}^{b_1} R[-n_{i,1}]\to R\to A\to 0$$

where all the maps are of degree zero and are given by forms of positive degrees on the components where they are non-zero. Such a resolution is called R-*minimal*. We see that the vector spaces $\mathrm{Tor}_j^R(A,k)$ are graded k-spaces isomorphic to the spaces $\bigoplus_{i=1}^{b_j} k[-n_{i,j}]$. In particular the numbers $n_{i,j}$ are the same for all minimal resolutions. We call them the R-*numerical characters* of A. We will always assume that the $n_{i,j}$'s are ordered so that $n_{1,k}\le n_{2,k}\le\dots\le n_{b_k,k}$ for $k=1,2,\dots,c$.

(e) If y is a homogeneous non-zero divisor of degree one in A, we have that the R/Y-numerical characters of A/y are equal to the R-numerical characters of A, where Y is a homogeneous pre-image of y in R. This is so, since Y non-zero divisor on A implies that $\mathrm{Tor}_i^R(A,R/Y)=0$ for i>0. Hence if $P_*\to A$ is an R-free resolution of A, then $P_*\otimes_R R/Y\to A/y$ is an R/Y-free resolution of A/y and $\mathrm{Tor}_i^R(A,k)=H_i(P_*\otimes_R(R/Y\otimes_{R/Y}k))\simeq H_i((P_*\otimes_R R/Y)\otimes_{R/Y}k)=\mathrm{Tor}_i^{R/Y}(A/y,k)$.

(f) If K is a field containing k, then the $R\otimes_k K$- numerical characters of $A\otimes_k K$ are the same as the R-numerical characters of A. Indeed, extension of the base field is a flat operation.

Since R is regular of dimension r, hence of global dimension r, we have that

$b_c = 0$ for $c>r$. If A is Cohen-Macaulay of dimension t we have that $b_c=0$ for $c>r-t$. Indeed by (f) we may assume that k is infinite and it then follows from (c) and repeated use of (e) that A is the quotient of a polynomial ring in r-t variables.

(g) If A is artinian there is a degree zero isomorphism $\mathrm{Soc}A \simeq \bigoplus_{i=1}^{b_r} k[r-n_{i,r}]$ of graded k-vector spaces. In particular, if $g_1, g_2, \ldots, g_u$ is a basis of SocA consisting of homogeneous elements and ordered such that $\deg g_i \leq \deg g_{i+1}$ for $i = 1,2,\ldots,u-1$, then $u = b_r$ and $\deg g_i = n_{i,r}-r$ for $i = 1,2,\ldots,u$. Indeed let K_* be the Koszul resolution of the R-algebra k with respect to the variables $x_1, x_2, \ldots, x_r$. Then K_* is an R-free resolution of k and hence $\mathrm{Tor}_i^R(A,k) \simeq H_i(K_* \otimes_R A)$. Moreover we have pointed out above that $\mathrm{Tor}_r^R(A,k) \simeq \bigoplus_{i=1}^{b_r} k[-n_{i,r}]$ and, with the obvious notation, we clearly have $H_r(K_* \otimes_R A)[r] = Z_r(K_* \otimes_R A)[r] = \{aT_1T_2\ldots T_r;\ d(aT_1T_2\ldots T_r) = \sum_{i=1}^{r}(-1)^{i-1} ax_iT_1T_2\ldots\hat{T}_i\ldots T_r = 0\}[r] = \mathrm{Soc}A$.

(h) For any algebra we have the inequalities $n_{1,1}<n_{1,2}<\ldots<n_{1,c}$. Indeed, these inequalities follow immediately from the minimality of F_*.

(i) If A is Cohen-Macaulay we have the inequalities $n_{b_1,1}<n_{b_2,2}<\ldots<n_{b_c,c}$. Indeed, we have that $\mathrm{Ext}_R^d(A,R) = 0$ for $d<r-\dim A = c$, see e.g. [Ma, thms 25 and 31]. Hence, applying the functor $\mathrm{Hom}_R(*,R)$ to the sequence F_* we obtain a sequence $G_* = \mathrm{Hom}_R(F_*,R)$ which is a resolution of $\mathrm{Ext}_R^{r-\dim A}(A,R)$. Moreover $\mathrm{Hom}_R(R[-n],R) = R[n]$ and G_* is easily seen to be minimal. Hence assertion (i) follows from assertion (h).

(j) For any algebra A we have that

$$\mathrm{Hilb}_A(z) = (1-\sum_{i=1}^{b_1} z^{n_{i,1}} + \sum_{i=1}^{b_2} z^{n_{i,2}} - \ldots + (-1)^c \sum_{i=1}^{b_c} z^{n_{i,c}})/(1-z)^r.$$

Indeed, for each fixed degree d we have that $(F_*)_d$ is an exact sequence of vector spaces and consequently we have that $\dim_k A_d = \dim_k R_d - \dim_k (F_1)_d + \ldots$, hence $\mathrm{Hilb}_A(z) = \sum_{d=0}^{\infty} \dim_k R_d z^d - \sum_{d=0}^{\infty} \dim_k (F_1)_d z^d + \sum_{d=0}^{\infty} \dim_k (F_2)_d z^d - \ldots =$
$(1-z)^{-r} - (1-z)^{-r} \sum_{i=1}^{b_1} z^{n_{i,1}} + \ldots$

5. COMPRESSED ALGEBRAS.

Let $A = R/I$ be a graded Cohen-Macaulay algebra of dimension t. We define the *type* of A to be the polynomial

$$S(A) = \sum_{c=0}^{\infty} \dim_k\left(\mathrm{Tor}^R_{r-t}(A,k)[r-t]\right)_c Z^c$$

As we have seen above the type is equal to $\sum_{i=1}^{b_{r-t}} z^{n_{i,r-t}-r+t}$ and from (g) above it follows that the type of an artinian ring equals the socle-type. Let K be any field containing k such that there is a maximal regular sequence $y_1, y_2, \ldots, y_t$ of $B = A \otimes_k K$ contained in B_1. ((c) above shows that any infinite field K containing k will do.) The artinian ring $B/(y_1, y_2, \ldots, y_t)$ we call an *artinian reduction* of A.

It follows from (e) that if A is a quotient of R then the R-numerical characters of A are equal to the $R \otimes_k K/(Y_1, Y_2, \ldots, Y_t) = S$-numerical characters of B, where $Y_1, Y_2, \ldots Y_t$ in $R_1 \otimes_k K$ are preimages of $y_1, y_2, \ldots, y_t$. Since the numerical characters of A and B respectively determine their types and since S is a polynomial ring in r-t variables wee see that the type of A is independent of the polynomial ring R and also that it is equal to the socle-type of any artinian reduction.

Moreover, we see that the number $\dim_k A_1 - \dim A$, called the *codimension* of A, is equal to the codimension of B. We define the numbers $r_d(A)$ and $b(A)$ to be those r_d and b determined by the number $\dim_k A_1 - \dim A$ and the polynomial $S(A)$. As we have seen we have then $r_d(A) = r_d(B)$ and $b(A) = b(B)$ for any artinian reduction B of A.

After all these preliminaries we are ready to generalize Proposition 4.

PROPOSITION 6. *Let* A *be a Cohen-Macaulay ring of dimension* t *and codimension* r *and let* $S(A) = \sum_{c=0}^{e} s_c z^c$ *with* $s_e \neq 0$ *be the type of* A. *Then the following four assertions hold;*

(i) $\mathrm{Hilb}_A(z) \leq (1-z)^{-t} \sum_{c=0}^{e} \min(N(r,c) - r_c(A), N(r,c)) z^c$

where the inequality holds coefficientwise.

(ii) $m(A) \leq N(r+1, b(A)-1) - s_{b(A)} - N(r+1,1)s_{b(A)+1} - \ldots - N(r+1, e-b(A))s_e$.

(iii) Equality holds in (i) if and only if it holds in (ii).

(iv) Keep the notation of Proposition 4(iv) and let $J \subseteq R$ *be an ideal such that the vector spaces* R_d *satisfy the conditions of that proposition. If* $S = R[y_1, y_2, \ldots, y_t]$ *is the polynomial ring in the variables* $y_1, y_2, \ldots, y_t$ *over* R, *then* S/JS *is a Cohen-Macaulay ring of dimension* t *such that* $\mathcal{S}(S/JS) = p$ *and that the inequalities (i) and (ii) are equalities. Moreover, if* $e > 1$ *then* $\operatorname{codim}(R/J) = r$.

PROOF. Let B be an artinian reduction of A. Then we have that $\mathcal{S}(A) = \mathcal{S}(B)$, $r_d(A) = r_d(B)$ and $b(A) = b(B)$. Moreover we have that $\mathrm{Hilb}_A(z) = (1-z)^{-t}\mathrm{Hilb}_B(z)$ by (b) and thus $m(A) = m(B) = \ell(B)$. The assertions (i), (ii) and (iii) thus follow from the corresponding assertions of Proposition 4.

As for (iv) we have that R/J is artinian of socle sequence p by Proposition 4(iv). We conclude that $S/JS = R/J[y_1, y_2, \ldots, y_t]$ is Cohen-Macaulay of dimension t by (d) (vi) of section 4. Clearly R/J is an artinian reduction of S/JS. Thus $\mathcal{S}(R/J) = \mathcal{S}(S/JS)$ and $\operatorname{codim} R/J = \operatorname{codim} S/JS$. Hence assertion (iv) follows from Proposition 4(iv).

Let SP_r be the subset of P_r consisting of polynomials $q = \sum_{c=0}^{e} s_c z^c$ such that $s_c = 0$ for $c < b$ where b is determined by r and q. We say that a Cohen-Macaulay algebra A of codimension r is *compressed* if $\mathcal{S}(A)$ belongs to SP_r and the inequalities (i) and (ii) of Proposition 6 are equalities. In other words, A is compressed if it has maximal multiplicity, or equivalently, has coefficientwise maximal Hilbert series among all Cohen-Macaulay algebras of the same dimension, codimension and type. We call a compressed algebra *extremely compressed* if $r_{b(A)} = 0$.

PROPOSITION 7. *Let* A *be a Cohen-Macaulay algebra.*

(i) Let $y \in A_1$ *be a non-zero divisor. Then* A *is compressed if and only if* A/yA *is compressed.*

(ii) A *is compressed if and only if* $A[y_1, y_2, \ldots, y_s]$ *is compressed.*

PROOF. Assertions (i) and (ii) follow from the observation that A, A/y and $A[y_1, y_2, \ldots, y_s]$ all have a common artinian reduction.

6. EXAMPLES OF COMPRESSED ALGEBRAS.

EXAMPLE 1. Some extremely compressed artinian algebras with socle in several degrees.

The ring $k[x_1,x_2,x_3]/(x_1^2x_2,x_1^2x_3,x_1x_2^2,x_1x_2x_3,x_1x_3^2,x_2^2x_3,x_2x_3^2,x_1^3-x_2^3,x_1^3-x_3^3)$ of type $3z^2+z^3$ is a simple example of an extremely compressed algebra with socle in several degrees. More generally, the factor ring of $k[x_1,x_2,\ldots,x_r]$ where one puts all x_i^d equal and all other monomials of degree d equal to zero is extremely compressed of type $(N(r,d-1)-r)z^{d-1}+z^d$ for $d\geq 3$.

EXAMPLE 2. Extremal Cohen-Macaulay rings.

Let $p(z)$ be a polynomial and t a natural number. Then it is easily seen by induction on t that the coefficients $a_i(t)$ of the power series $p(z)(1-z)^{-t}$ are given by the values $f_t(i)$ of a polynomial $f_t(x)$ in x for large values of i and that if $i_t(p)$ denotes the number $1+\max\{j;\ f_t(j)\neq a_j(t)\}$, then $i_t(p)=\deg p\ -t+1$.

When A is a t-dimensional k-algebra, then $\mathrm{Hilb}_A(z)=p(z)(1-z)^{-t}$ for a polynomial $p(z)$, see e.g. [At-Ma, ch. 11], and we denote $i_t(p)$ by $i(A)$.

Let $r=\dim_k A_1$ and write A as a quotient R/I of the polynomial ring R. We denote the number $\min\{d;\ I_d\neq 0\}$ by $d(A)$. It is clear that if $n_{i,j}$ denote the R-numerical characters of A defined in section 4 then $d(A)=n_{1,1}$.

If A is Cohen-Macaulay the following inequality holds, see [Sc, Kor. 1],

$$i(A)+\dim A\geq d(A).$$

We recall that A is called extremal if this inequality is an equality, see [Sc].

PROPOSITION 8. *Let* A *be a graded Cohen-Macaulay ring of dimension* t *and codimension* r *and let* $e=d(A)-1$. *Then the following five conditions are equivalent;*

(i) A *is extremely compressed of type* $N(r,e)z^e$.

(ii) A *is compressed of type* $N(r,e)z^e$.

(iii) A *is an extremal Cohen-Macaulay ring.*

(iv) $\mathrm{Hilb}_A(z)=(1-z)^{-t}(1+N(r,1)z+N(r,2)z^2+\ldots+N(r,e)z^e)$.

(v) A *has an artinian reduction of the form* $k[x_1,x_2,\ldots,x_r]/(x_1,\ldots,x_r)^{d(A)}$.

PROOF. The equivalence of (i) and (ii) is a trivial checking of the numbers $r_d(A)$ and $b(A)$.

If (ii) holds then $r_c(A) = N(r,c) - N(r,e-c)N(r,e) < 0$ holds for $c<e$. Hence the implication (ii) $\Rightarrow$ (iv) follows from the definition of a compressed algebra.

The Hilbert series of the ring in part (v) is equal to $1 + N(r,1)z + N(r,2)z^2 + \ldots + N(r,e)z^e$. Hence the equivalence (iv) $\Leftrightarrow$ (v) follows from the relation between a ring and an artinian reduction. Clearly, the ring in part (v) is compressed of type $N(r,e)z^e$. Moreover, a ring and an artinian reduction have the same type and one is compressed if and only if the other is. Hence we have an implication (v) $\Rightarrow$ (ii).

Finally we have that a ring A is an extremal Cohen-Macaulay ring if and only if an artinian reduction B is. Indeed the numbers $d(A)$ and $i(A)$ are clearly invariant under an extension of the field k. Moreover, if y is a non-zero divisor of degree 1 in A then it follows from (e) in section 4 that $d(A) = d(A/y)$ and from (b) in section 4 that $i(A) = i(A/y)-1$. Consequently $d(A) = d(B)$ and $i(A) + \dim A = i(B) + \dim B$. For the artinian ring B we have that $i(B) = \min\{j;\ B_j = 0\}$. Moreover, B is isomorphic to a quotient R/I and $d(B) = \min\{d;\ I_d \neq 0\}$. Hence B is an extremal Cohen-Macaulay ring, that is $d(B) = i(B)$, if and only if $I_d = 0$ for $d<i(B)$ and $I_{i(B)} = R_{i(B)}$. Hence (iii) is equivalent to (v).

REMARK. The implication (iii) $\Rightarrow$ (iv) of Proposition 8 is [Sc, Thm A (c)].

In the following four examples the Hilbert series are well-known or can easily be calculated from known results , see e.g. [Ea, Thm 1], [Go-Ta, second Cor., p. 53], [Sa, Thm 1 (i)], [Wa, Thm 1] respectively. They are, as is remarked in [Sc], all extremal Cohen-Macaulay rings. For these examples one can also obtain simple proofs of the fact that they are extremely compressed without knowledge of their Hilbert series by showing directly that they have an artinian reduction of the form of Proposition 8(v).

(2.1) Let (X_{ij}), $1 \leq i \leq m$, $1 \leq j \leq n$, be an mxn-matrix of indeterminates and let I be the ideal in $k[X_{ij}]$ generated by all maximal minors of (X_{ij}).

Then $k[X_{ij}]/I$ is an extremal Cohen-Macaulay ring.

(2.2) Let (X_{ij}), $1 \leq i,j \leq n$, be an nxn symmetric matrix of indeterminates and let I be the ideal in $k[X_{ij}]$ generated by all submaximal minors of (X_{ij}). Then $k[X_{ij}]/I$ is an extremal Cohen-Macaulay ring.

(2.3) Let (Q,m) be a d-dimensional local Cohen-Macaulay ring of multiplicity m. Then the embedding dimension of Q is $\leq m+d-1$. A ring for which this inequality is an equality is called a Cohen-Macaulay ring of maximal embedding dimension. If Q is a Cohen-Macaulay ring of maximal embedding dimension, then $gr_m Q$ is an extremal Cohen-Macaulay ring.

(2.4) If (Q,m) is the local ring of a rational surface singularity, then $gr_m Q$ is an extremal Cohen-Macaulay ring.

EXAMPLE 3. Extremal Gorenstein rings.

We keep the notation of the previous example. Assume that the k-algebra A is Gorenstein, that is Cohen-Macaulay of type z^e for some integer e. Then the following inequality holds, see [Sc, Kor 1];

$$i(A) + \dim A \geq 2d(A)-1.$$

We recall that A is called an extremal Gorenstein ring if this inequality is an equality, see [Sc].

Observe that an extremal Gorenstein ring is *not* an extremal Cohen-Macaulay ring.

PROPOSITION 9. *Let* A *be a Gorenstein ring of dimension* t *and codimension* r *and let* $e = d(A)-1$. *Then the following four conditions are equivalent;*

(i) A *is extremely compressed of type* z^{2e}.

(ii) A *is compressed of type* z^{2e}.

(iii) A *is an extremal Gorenstein ring.*

(iv) $\mathrm{Hilb}_A(z) = (1-z)^{-t}\left(\sum_{c=0}^{e} N(r,c)z^c + \sum_{c=e+1}^{2e} N(r,2e-c)z^c\right)$.

PROOF. The equivalence of (i) and (ii) is a trivial checking of the numbers $r_d(A)$ and $b(A)$.

If (ii) holds then $r_c(A) = N(r,c) - N(r,2e-c)$, hence the implication (ii)$\Rightarrow$(iv) follows from the definition of a compressed algebra. Since A is Gorenstein the opposite implication (iv)$\Rightarrow$(ii) is also a simple computation.

To prove the remaining part we see as in the proof of Proposition 8 that it suffices to do so for an artinian reduction B. Then $i(B) = \min\{j;\ j>0 \text{ and } B_j = 0\}$. Hence if (iv) holds we have that $i(B) = 2e+1$. Consequently $i(B) = 2d(B)-1$ and we have proved the implication (iv)$\Rightarrow$(iii). Conversely if (iii) holds we note that since B is Gorenstein we have that $\dim_k B_c = \dim_k B_{2e-c}$ (e.g. [St, thm 4.1]) for all c. Since $\dim_k B_c = N(r,c)$ for $0 \leq c \leq d(B)-1 = e$ it follows that $\mathrm{Hilb}_B(z)$ is of the form in (iv). Hence the implication (iii)$\Rightarrow$(iv) holds.

REMARK. The implication (iii)$\Rightarrow$(iv) in Proposition 9 is [Sc, Thm B (c)].

In the following five examples the Hilbert series are well known or can easily be calculated from known results, see e.g. [Bu-Ei, Thm 3.3], [Gu-Ne, Thm 2], [Sa, Thm 2(i)], [Wa, Thm 2] respectively. They are, as remarked in [Sc] all Extremal Gorenstein rings.

(3.1) Let (X_{ij}), $1 \leq i,j \leq 2n+1$, be a skew-symmetric $(2n+1)\mathrm{x}(2n+1)$-matrix of indeterminates and let I be the ideal in $k[X_{ij}]$ generated by all Pfaffians one gets from (X_{ij}) by deleting one column and the same row. Then $k[X_{ij}]/I$ is an extremal Gorenstein ring.

(3.2) Let (X_{ij}), $1 \leq i,j \leq n$, be an nxn-matrix of indeterminates and let I be the ideal in $k[X_{ij}]$ generated by all submaximal minors. Then $k[X_{ij}]/I$ is an extremal Gorenstein ring.

(3.3) Let (Q,m) be a d-dimensional Gorenstein ring of multiplicity m. Then the embedding dimension of Q is $\leq m+d-2$. A ring for which this inequality is an equality is called a Gorenstein ring of maximal embedding dimension. If Q is a Gorenstein ring of maximal embedding dimension, then $\mathrm{gr}_m Q$ is an extremal Gorenstein ring.

(3.4) If (Q,m) is the local ring of an elliptic surface singularity, then $\mathrm{gr}_m Q$ is an extremal Gorenstein ring.

(3.5) Let Δ be a triangulization of a sphere with n vertices and with a maximal number of faces in each dimension. Then the associated Stanley-Reisner ring $k[\Delta]$ is an extremal Gorenstein ring.

EXAMPLE 4. Compressed Gorenstein rings that are not extremal Gorenstein. In the previous example we saw that the Gorenstein compressed algebras with type of even degree are exactly the extremal Gorenstein rings. We shall here give two examples of Gorenstein compressed algebras with type of odd degree. These examples are not complete intersections. In example 5 we will also show that there are compressed complete intersections with type of any degree.

(4.1) Let (X_{ij}), $1 \leq i,j \leq 6$, be a 6x6 skew-symmetric matrix of indeterminates and let I be the ideal in $k[X_{ij}]$ generated by the Pfaffians one gets by deleting two columns and the same rows. Then it follows from [Jo-Pr, Thm 3.2] or [Kl-La, Prop. 21] that $\mathrm{Hilb}_{k[X_{ij}]/I}(z) = (1-z)^{-15}(1+6z+6z^2+z^3)$ and that $k[X_{ij}]/I$ is Gorenstein. A simple calculation shows that the ring is compressed of type z^3.

(4.2) $k[x_1,x_2,x_3]/(x_1x_2,x_1x_3,x_2x_3,x_1^3-x_2^3,x_1^3-x_3^3)$ is easily seen to be a Gorenstein ring with Hilbert series $1+3z+3z^2+z^3$ which is compressed of type z^3.

EXAMPLE 5. Compressed complete intersections.

Let A be a complete intersection, that is A is of the form $R/(f_1,f_2,\ldots,f_s)$ where $f_1,f_2,\ldots,f_s$ is a homogeneous R-regular sequence. The Koszul complex on $f_1,f_2,\ldots,f_s$ is then an R-resolution of A, see e.g. [Se, Ch IV,Prop. 2]. Hence it follows from (e) and (g) of section 4 that A is of type z^e where $e = \sum_{i=1}^{s} \deg f_i - s$. Moreover, from (b) of section 4 it follows that $\mathrm{Hilb}_A(z) = (1-z)^{-r} \prod_{i=1}^{s}(1-z^{\deg f_i})$. With this information it is easily checked that we in the following four cases obtain compressed algebras of the indicated type;

(1) $R/(f)$ is compressed of type $z^{\deg f - 1}$

(2) If $\deg f_1 = \deg f_2 = d$, then $R/(f_1,f_2)$ is compressed of type z^{2d-2}.

(3) If $\deg f_1 = \deg f_2 - 1 = d$, then $R/(f_1,f_2)$ is compressed of type z^{2d-1}.

(4) If $\deg f_1 = \deg f_2 = \deg f_3 = 2$, then $R/(f_1,f_2,f_3)$ is compressed of type z^3.

We shall prove in section 7 that these four cases are the only compressed complete intersections.

EXAMPLE 6. Lines in general position.

Let k be an infinite field and let A be the coordinate ring for t lines in $\mathbb{A}_k^{r+1}$ passing through the origin and in "sufficiently general" position. We will show that A is compressed. Let e be the integer defined by the inequalities

$$N(r+1,e-1) < t \leq N(r+1,e).$$

To avoid trivial cases we suppose that $e>1$, that is $t>r+1$. Then

$$\mathrm{Hilb}_A(z) = (1 + N(r,1)z + N(r,2)z^2 + \ldots + N(r,e-1)z^{e-1} + (t-N(r+1,e-1)z^e)/(1-z)$$

and A is a graded Cohen-Macaulay ring of dimension 1 and codimension r and is of type $S(A) = hz^{e-1} + (t-N(r+1,e-1))z^e$ for some integer h, see [Ge-Or, Thm 5 and Prop. 12]. The integer h is not known in general. According to Proposition 6(i) we see that in order to prove that A is compressed we have exactly to prove that $r_{e-1}(A) \leq 0$. However, $r_{e-1}(A) = N(r,e-1) - s_{e-1} - rs_e = N(r,e-1) - h - r(t - N(r+1,e-1))$. To see that the latter expression is not positive we rewrite $\mathrm{Hilb}_A(z)$ in the following way;

$\mathrm{Hilb}_A(z) = (1-z)^r(1 + N(r,1)z + \ldots + N(r,e-1)z^{e-1} + (t - N(r+1,e-1))z^e)/(1-z)^{r+1} =$
$=(1+b_1z+\ldots+b_{e+r-2}z^{e+r-2} + (-1)^{r-1}(r(t - N(r+1,e-1)) - N(r,e-1))z^{e+r-1} +$
$+(-1)^r(t - N(r+1,e-1))z^{e+r})/(1-z)^{r+1}$ for some integers $b_1,b_2,\ldots,b_{e+r-2}$.

By (g) of section 4 we have that $S(B) = \sum_{i=1}^{b_r} z^{n_{i,r}-r}$ for an artinian reduction B of A. However $S(A) = S(B)$ so that, comparing with the above expression for $S(A)$, we obtain that $n_{b_r,r} = r+e$. We conclude from (i) of section 4 that $n_{b_{r-1},r-1} \leq r+e-1$ and that $n_{b_s,s} < r+e-1$ for $s<r-1$. Let $\sum_{i=1}^{b_{r-1}} z^{n_{i,r-1}} = a_0 + a_1z + \ldots + a_{r+e-1}z^{r+e-1}$, where the a_i's are non-negative integers. Then comparing the coefficient of z^{r+e-1} in the numerator of the above expression for $\mathrm{Hilb}_A(z)$ with the same coefficient in the expression of section 4(j) we see that $a_{r+e-1} - h = r(t - N(r+1,e-1)) - N(r,e-1)$ and con-

sequently that $N(r,e-1)-h-r(t-N(r+1,e-1))=-a_{r+e-1} \leqq 0$ which shows that $r_{e-1}(A) \leqq 0$.

REMARK. In Example 6 we proved the inequality $h \geqq N(r,e-1)-r(t-N(r+1,e-1))$. This inequality also follows from the reasoning in [Ro].

7. CONSTRUCTION OF COMPRESSED ALGEBRAS.

In section 5 we have defined compressed algebras of a given type $p \in SP_r$. We have, however, not proved that such algebras exist for any p. The construction of compressed algebras of type p for every $p \in SP_r$ is an important part of this article. Parts (iv) of Propositions 4 and 6 show that to perform such a construction it is sufficient to construct a sequence of vector spaces $J_0, J_1, \ldots$ satisfying the properties of Proposition 4(iv). After a preliminary well known lemma we shall prove a rather innocent looking result about the behaviour of $(H:R)_c$ when H is a hyperplane in R_d, which turns out to be the crucial part of the construction and is the main mathematical contribution. Throughout this section we assume that k is an infinite field.

LEMMA 10. *Given a subspace* W *of* R_c. *Then for* m *points* $a_1, a_2, \ldots, a_m$ *in general position in* $\mathbb{A}^r$ *we have that*

$$\dim_k \{G \in W;\ G(a_i)=0 \text{ for } i=1,2,\ldots,m\} = \max(0, \dim_k W - m).$$

PROOF. We use induction with respect to m. If $n < \dim_k W$ and the dimension of the space $V = \{G \in W;\ G(a_i)=0 \text{ for } i=1,2,\ldots,n\}$ is $\dim_k W - n$, then we only have to choose a_{n+1} outside the variety of zeroes of the polynomials in V to obtain that $\dim_k \{G \in W;\ G(a_i)=0 \text{ for } i=1,2,\ldots,n+1\} = \dim_k W - n - 1$.

THEOREM 11. *Given integers* $c \leqq d$ *and a vector space* $W \subseteq R_d$. *For a hyperplane* H *in* R_d *in general position we have an equality*

$$\operatorname{codim}((H:R)_c, R_c) = \min(N(r,c), N(r,d-c))$$

and the space $(H:R)_c$ *intersects* W *properly.*

PROOF. By Lemma 2 and the reasoning of section 3 it suffices to show the existence of a single hyperplane H having the properties of the theorem. To this end we use Lemma 10 to choose points $a_1, a_2, \ldots, a_{N(r,d-c)}$ in $\mathbb{A}^r$

such that the following three conditions hold;

(i) $\dim_k\{E\varepsilon R_{d-c};\ E(a_i)=0 \text{ for } i=1,2,\dots,N(r,d-c)\}=0$

(ii) $\dim_k\{G\varepsilon R_c;\ G(a_i)=0 \text{ for } i=1,2,\dots,N(r,d-c)\}=\max(0,N(r,c)-N(r,d-c))$

(iii) $\dim_k\{G\varepsilon W;\ G(a_i)=0 \text{ for } 1=1,2,\dots,N(r,d-c)\}=\max(0,\dim_k W-N(r,d-c))$.

We choose H to be the space

$$\{F\varepsilon R_d;\ \sum_{i=1}^{N(r,d-c)} F(a_i)=0\}.$$

Then $(H:R)_c=\{G\varepsilon R_c;\ \sum_{i=1}^{N(r,d-c)} G(a_i)E(a_i)=0 \text{ for all } E\varepsilon R_{d-c}\}$ or

$(H:R)_c=\{G\varepsilon R_c;\ \sum_{i=1}^{N(r,d-c)} G(a_i)M(a_i)=0 \text{ for all monomials } M \text{ in } R_{d-c}\}$.

However, condition (i) above is clearly equivalent to the N(r,d-c) equations $\sum_M\lambda_M M(a_i)=0$ for $i=1,2,\dots,N(r,d-c)$ in the N(r,d-c) variables λ_M having only the trivial solution, where the sum is over all monomials in R_{d-c}. Hence we have that

$$(H:R)_c=\{G\varepsilon R_c;\ G(a_i)=0 \text{ for } i=1,2,\dots,N(r,d-c)\}.$$

Consequently it follows from condition (ii) that $\dim_k(H:R)_c=\max(0,N(r,c)-N(r,d-c))$. The first assertion of the theorem follows and the second part follows from assumption (iii) using similar arguments.

PROPOSITION 12. *Given a subspace* V *of* R_{d+1} *and an integer* t. *Then there exists an open dense subset* U *of* $\mathrm{Grass}_t((V:R)_d)$ *such that for each point* J *in* U *we have that*

$$\mathrm{codim}((J:R)_c,(V:R)_c)=\min(\dim_k(V:R)_c,tN(r,c),tN(r,d-c)).$$

PROOF. By Lemma 2 and the reasoning in section 3 it suffices to show the existence of a single J satisfying the codimension condition. Choose t hyperplanes $H_1,H_2,\dots,H_t$ in general position. We choose J to be the space $(V:R)_d\cap\bigcap_{i=1}^t H_i$. Then $\mathrm{codim}(J,(V:R)_d)=t$ and using Lemma 3 we obtain the following relations

$$(J:R)_c=\bigcap_{i=1}^t(H_i:R)_c\cap((V:R)_d:R)_c=\bigcap_{i=1}^t(H_i:R)_c\cap(V:R)_c.$$

Hence it follows from Theorem 11 that we have an equality

$$\mathrm{codim}((J:R)_c,(V:R)_c)=\min(\dim_k(V:R)_c,t\min(N(r,c),N(r,d-c)))$$

and the latter expression is clearly the same as the expression in the statement of the proposition.

COROLLARY 13. *Let* $p \in SP_r$. *Then there exist subspaces* J_d *of* R_d *for* $d=1,2,\ldots$ *such that for all* $0 \leq c \leq d$ *we have that* $J_d \subseteq (J_{d+1}:R)_d$ *and the equalities*

$$\dim_k (J_d:R)_c = \max\ (r_d(c),0)$$

hold.

PROOF. It follows from Proposition 11 with $d=e$, $t=s_e$ and $V=R_{e+1}$, that for a subspace J_e of R_e in general position we the dimension formula $\operatorname{codim}((J_e:R)_c,R_c) = \min(N(r,c),s_eN(r,c),s_eN(r,d-c))$, which shows that the formula of the corollary holds for J_e in general position. We shall reason by descending induction on d and assume that for J_{d+1} in general position in $(J_{d+2}:R)_{d+1}$ the dimension formula holds. Then it follows from Proposition 11 that for J_d in general position in $(J_{d+1}:R)_d$ we have equalities $\operatorname{codim}((J_d:R)_c,(J_{d+1}:R)_c) = \min(\dim_k(J_{d+1}:R)_c,s_dN(r,c),s_dN(r,d-c))$.
By the induction hypothesis we have that $\dim_k(J_{d+1}:R)_c = \max(r_{d+1}(c),0)$ so that we obtain $\dim_k((J_d:R)_c) = \max(0,r_{d+1}(c)-s_dN(r,c),r_{d+1}(c)-s_dN(r,d-c)) = \max(0,r_{d+1}(c)-s_dN(r,c),r_d(c))$. If $s_d = 0$ we have that $r_{d+1}(c) = r_d(c)$ so that we obtain the dimension formula of the corollary and if $s_d>0$ we have that $r_{d+1}(c)-s_dN(r,c) = N(r,c)-N(r,d+1-c)s_{d+1}-N(r,d+2-c)s_{d+2}-\ldots-N(r,e-c)s_e-N(r,c)s_d$ which is less than or equal to zero. Hence $\dim_k((J_d:R)_c) = \max(r_d(c),0)$ also when $s_d>0$ and we have proved the corollary.

The following result is an immediate consequence of Proposition 6(iv) and Corollary 13. Since it is a main result of this article we state it separately.

THEOREM 14. *Given* $p \in SP_r$ *and an integer* $t \geq 0$. *Then there exists an ideal* I *in* $R[y_1,y_2,\ldots,y_t]$, *where* $y_1,y_2,\ldots,y_t$ *are variables, such that* $R[y_1,y_2,\ldots,y_t]/I$ *is a Cohen-Macaulay graded ring of dimension* t *and type* p.

8. COMPRESSED COMPLETE INTERSECTIONS.

PROPOSITION 15. *Let* A *be a complete intersection of dimension* t *and codimension* s. *Then* A *is compressed if and only if it is one of the four types described in Example 4 of section 5 with* $r = s+t$ *and* $d = [e/2] - 1$ *where* z^e *is the type of* A.

PROOF. We saw in Example 4 that all the four types described are compressed.

Conversely assume that A is compressed. Since $\dim_k A_1 = s+t$ we can write A as a quotient $R/(f_1,f_2,\dots,f_s)$ where $f_1,f_2,\dots,f_s$ is a regular sequence of homogeneous elements in R. To check that the number s of forms and their degrees are as asserted in the proposition we can consider an artinian reduction $B = k[x_1,x_2,\dots,x_s]/(g_1,g_2,\dots,g_s)$ of A. We saw in Example 4 that $S(A) = z^e$ for some e, and we know that $S(A) = S(B)$. By the definition of a compressed algebra we have that

$$(5) \quad \mathrm{Hilb}_B(z) = (1-z)^t\mathrm{Hilb}_A(z) = \sum_{e=0}^{[e/2]} N(s,c)z^c + \sum_{c=[e/2]+1}^{e} N(s,e-c)z^c.$$

If $s = 1$ we have type (4.1) so we assume $s>1$. Assume first that $e = 2d+1$. Then it follows from (5) that $J_d=0$ where $J = (g_1,g_2,\dots,g_s)$ and that $\dim_k J_{d+1} = N(s,d+1) - N(s,d) = N(s-1,d+1)$. The latter expression equals 1 if s=2 and 3 if s=e=3. For all other values it is greater than s which is impossible because J has s generators. If s=2 we obtain from (5) that $\dim_k J_{d+2} = N(2,d+2) - N(2,d-1) = 3$. Since $\dim_k J_{d+1} = 1$ and J is an ideal in $k[x_1,x_2]$ we have that the number of generators for J of degree d+2 and not coming from elements in J_{d+1} is $\dim_k J_{d+2} - 2\dim_k J_{d+1} = 1$. We have thus proved that when s=2 the only possibility for A to be compressed is when $\deg f_1 = \deg f_2 - 1 = d+1$ and that when s=3 the only possibility is when $\deg f_1 = \deg f_2 = \deg f_3 = 2$.

Secondly we assume that $e = 2d$. Then it follows from (5) that $J_d = 0$ and that $\dim_k J_{d+1} = N(s,d+1) - N(s,d-1) = N(s-1,d+1) + N(s-1,d)$. The latter expression is 2 if s=2 and for all other values of s it is greater than s which is again impossible. Hence the only possibility is s=2 and $\deg f_1 = \deg f_2 = d+1$.

9. RESOLUTIONS OF COMPRESSED ALGEBRAS.

The purpose of this section is to show that the compressed algebras have minimal resolutions of a rather special kind.

We keep the notation of section 4. The minimal R-resolution F_* of that section is called *pure* if $n_{1,d}=n_{2,d}=\ldots=n_{b_d,d}=n(d)$ for each $d=1,2,\ldots,c$, that is, for each d, there exists an integer n(d) such that $\mathrm{Tor}_d^R(A,k)_j=0$ for $j\neq n(d)$. The resolution is called *linear* if it is pure and $n(d)=n(1)+d-1$ for $d=1,2,\ldots,c$.

PROPOSITION 16. *Let* $A=R/I$ *be a Cohen-Macaulay graded algebra of dimension* t *and codimension* $s=r-t$ *and let* $b=b(A)$. *Then the following six assertions hold;*
(i) If A *is compressed then* $\mathrm{Tor}_i^R(A,k)_j=0$ *for* $b-1+i\neq j\neq b+i$ *for all* $i=1,2,\ldots,s-1$. *In particular* I *is generated in degrees* b *and* b+1 *only.*
(ii) If A *is extremely compressed, then* $\mathrm{Tor}_i^R(A,k)_j=0$ *for* $j\neq b+i$ *for all* $i=1,2,\ldots,s-1$. *In particular* I *is generated in degree* b+1 *only.*
(iii) If A *is extremely compressed of type* cz^e *for some integers* c *and* e, *then* A *has a pure resolution.*
(iv) If A *is extremely compressed of type* $N(s,e)z^e$, *then* A *has a linear resolution.*
(v) If A *is extremely compressed, then the Betti numbers* $b_i(A)=\dim_k\mathrm{Tor}_i^R(A,k)$ *satisfy*

$$b_i(A)=\binom{b+i-1}{b}\binom{b+s}{s-i}-\sum_{c=b}^{e}s_c\binom{c-b-i+s-1}{c-b-1}\binom{c-b+s-1}{i-1}$$

for $i=1,\ldots,s-1$ *and*

$$b_s(A)=\sum_{c=b}^{e}s_c.$$

(vi) If $\mathrm{Tor}_i^R(A,k)_j=0$ *for* $j\neq b+i$ *for all* $i=1,\ldots,s-1$, *then* A *is extremely compressed.*

PROOF. We have seen in section 4 that the type, codimension and numerical characters of A are the same as those for an artinian reduction. Therefore we can assume that A is artinian of codimension $s=r$. Then we have that

$$\mathrm{Hilb}_A(z)=1+N(r,1)z+\ldots+N(r,b-1)z^{b-1}+(N(r,b)-r_b)z^b+\ldots+(N(r,e)-r_e)z^e$$

if A is compressed.

We substitute the expression $N(r,c)-r_c=\sum_{i=0}^{e-c}N(r,i)s_{c+i}$ for $c=b,b+1,\dots,e$ into the above expression for $Hilb_A(z)$ and obtain the formula

$$(1-z)^r Hilb_A(z)=(1-z)^r\Big(\sum_{c=0}^{b-1}N(r,c)z^c+z^b\sum_{i=0}^{e-b}N(r,i)s_{b+i}+\dots+z^e\sum_{i=0}^{e-e}N(r,i)s_{e+i}\Big).$$

Write $(1-z)^r Hilb_A(z)=1+c_1z+c_2z^2+\dots+c_{r+e}z^{r+e}$. We claim that the following four assertions hold;

(a) $c_i=0$ for $i=1,2,\dots,b-1$

(b) $(-1)^r c_{i+r}=s_i$ for $i=b,b+1,\dots,e$

(c) If A is extremely compressed then $c_b=0$

(d) If A is extremely compressed then

$$(-1)^i c_{b+i}=\binom{b+i-1}{b}\binom{b+r}{r-i}-\sum_{c=b}^{e}s_c\binom{c-b-i+r-1}{c-b-1}\binom{c-b+r-1}{i-1}$$

for $i=1,2,\dots,r-1$.

Claim (a) and (c) follow immediately from the formula $1=(1-z)^r(1-z)^{-r}=(1-z)^r(1+N(r,1)z+N(r,2)z^2+\dots)$.

To prove claim (b) we collect all terms containing s_i in the above expression for $(1-z)^r Hilb_A(z)$ and see that the "coefficient" of s_i is equal to $(1-z)^r\sum_{c=b}^{i}N(r,i-c)z^c$. In the latter expression all coefficients of z^j with $j\geq b+r$ coincide with the corresponding coefficient of $(1-z)^r\sum_{c=-\infty}^{i}N(r,i-c)z^c=(-z)^r(1-z^{-1})^r z^i\sum_{j=0}^{\infty}N(r,j)z^{-j}=(-1)^r z^{i+r}$ and consequently we have the formula $(-1)^r c_{i+r}=s_i$ for $i=b,b+1,\dots,e$ of claim (b).

To prove assertion (d) we note that the "coefficient" of s_i for $i=b,\dots,e$ becomes $(1-z)^r\sum_{c=b}^{i}N(r,i-c)z^c$. In this expression the coefficient of z^{j+b} for $j=1,2,\dots,r-1$ is $\sum_{l=0}^{j-1}(-1)^l N(r,i-b-j+l)\binom{r}{l}$ which can be written as $(-1)^{j-1}\binom{i-b-j+r-1}{r-j}\binom{r-1+i-b}{j-1}$. Indeed using induction on m one easily checks the formula

$$\sum_{l=0}^{m}(-1)^l N(r,n+l)\binom{r}{l}=(-1)^m\binom{n+r-1}{r-m-1}\binom{r+m+n}{n+r}. \tag{6}$$

The contribution to the coefficient of z^{j+b} for $j=1,2,\dots,r-1$ which does not contain the numbers s_i is coming from $(1-z)^r\sum_{i=0}^{b}N(r,c)z^c$ and is thus equal to $\sum_{l=j}^{r}(-1)^l N(r,b+j-l)\binom{r}{l}=(-1)^r\sum_{l=0}^{r-j}(-1)^l N(r,b+j-r-l)\binom{r}{l}$. By the formula (6) we see that the latter expression is equal to $(-1)^j\binom{b+j-1}{b}\binom{r+b}{r-j}$. Hence the total contribution to the coefficient of z^{j+b} is

$$(-1)^j\binom{b+j-1}{b}\binom{r+b}{r-j} + (-1)^{j-1}\sum_{i=b}^{e}\binom{i-b-j+r-1}{r-j}\binom{r-1+i-b}{j-1}s_i, \text{ for } j=1,2,\dots,r-1$$

which proves claim (d).

Comparing the above expression for $(1-z)^r\mathrm{Hilb}_A(z)$ with the expression

$$1+\sum_{i=1}^{r}(-1)^i\Big(\sum_{j=1}^{b_i}z^{n_{j,i}}\Big)$$

of section 4(j) and using the formula $\sum_{j=1}^{b_r}z^{n_{j,r}} = \sum_{c=b}^{e}s_c z^{c+r}$ of section 4(g) we see that

$$(7)\qquad 1+\sum_{i=1}^{r-1}(-1)^i\Big(\sum_{j=1}^{b_i}z^{n_{j,i}}\Big) = 1+c_b z^b + c_{b+1}z^{b+1}+\dots+c_{b+r-1}z^{b+r-1}.$$

From (7) it follows that $n_{1,1}=b$ and $n_{b_{r-1},r-1}=b+r-1$. Hence it follows from section 4(h) that $n_{1,i}\geqq b+i-1$ and from section 4(i) that $n_{b_i,i}\leqq b+i$ for $i=1,2,\dots,r-1$. Consequently we have that $b+i-1\leqq n_{j,i}\leqq b+i$ for $i=1,2,\dots,r-1$ and all $j=1,2,\dots,b_i$. The latter inequalities are clearly equivalent to assertion (i) of the proposition.

To prove assertion (ii) we note that when A is extremely compressed we have $c_b=0$ so that we obtain $n_{1,1}=b+1$. The same reasoning as we used to prove assertion (i) now gives that $n_{j,i}=b+i$ for $i=1,2,\dots,r-1$ and all $j=1,2,\dots,b_i$ and consequently proves assertion (ii).

When A is extremely compressed of type cz^e the formula of section 4(j) gives that $\sum_{j=1}^{b_r}z^{n_{j,r}}=cz^e$. Consequently we have equalities $n_{j,r}=e+r$ for $j=1,2,\dots,b_r$. These equalities together with part (ii) of the proposition proves assertion (iii). If $c=N(r,e)$ we have that $b=e$. Therefore, in this case, $n_{j,i}=e+i$ for $i=1,2,\dots,r$ and all j and part (iv) of the proposition is proved.

It follows from part (ii) of the proposition and formula (7) that $(-1)^i b_i = c_{b+i}$ which together with assertion (d) of the above claim proves part (v) of the proposition.

Finally, to prove (vi) we use section 4(j) to conclude that
$(1-z)^r\mathrm{Hilb}_A(z) = 1+c_{b+1}z^{b+1}+c_{b+2}z^{b+2}+\dots+c_{b+r-1}z^{b+r-1}+s_b z^b+\dots+s_e z^e$,
where $c_{b+i}=(-1)^i\dim_k\mathrm{Tor}_i^R(A,k)$ and $\mathrm{Hilb}_{\mathrm{Tor}_r^R(A,k)}(z)=\sum_{c=b}^{e}s_c z^{c+r}$. But $\mathrm{Hilb}_A(z)=1+h_1z+h_2z^2+\dots+h_e z^e$. Comparing these two expressions we get $h_i=N(r,i)$ for $i=1,2,\dots,b$. By definition $r_b\geqq 0$ and we know by Proposition 4(i) that $h_b\leqq N(r,b)-r_b$, so we can conclude that $r_b=0$, that is, A extremely compressed.

REMARK. The above results generalize results of Schenzel, who proves (iii) and (v) for extremal Gorenstein rings, [Sc, Thm B], and (iv) and (v) for extremal Cohen-Macaulay rings, [Sc, Thm A]. The artinian case of (ii) and (vi) is proved in [Bu-Ei]. An "almost p-linear" algebra in their terminology is in ours an artinian extremely compressed algebra A with $b(A) = p$.

EXAMPLE 7. We have proved in Example 3 that a compressed Gorenstein ring of type z^e is extremely compressed if and only if e is even. Proposition 16 shows that for e even, the Betti numbers $b_i(A) = \dim_k \mathrm{Tor}_i^R(A,k)$ are determined by the dimension, codimension and type. This contrasts to the case when e is odd when the Betti numbers are not unique. The simplest counterexample is given by the algebras $k[x_1,x_2,x_3]/(x_1x_2,x_1x_3,x_2x_3,x_1^3-x_2^3,x_1^3-x_3^3)$ and $k[x_1,x_2,x_3]/(x_1^2,x_2^2,x_3^2)$ which are both compressed of type z^3 (c.f. Example 4 and 5 of section 4) and with Betti numbers 1,5,5,1 and 1,3,3,1 respectively.

We now digress a moment from the theme of graded algebras.

PROPOSITION 17. *Suppose* (Q,m,k) *to be a local ring which is a factor ring of a regular ring* S. *Then, if* $\mathrm{gr}_m Q$ *is a compressed algebra,* Q *and* $\mathrm{gr}_m Q$ *have the same Betti numbers, that is* $\dim_k \mathrm{Tor}_i^S(Q,k) = \dim_k \mathrm{Tor}_i^{\mathrm{gr}S}(\mathrm{gr}Q,k)$ *for all* i.

PROOF. We use the spectral sequence $\mathrm{Tor}^{\mathrm{gr}S}(\mathrm{gr}Q,k) \Rightarrow \mathrm{Tor}^S(Q,k)$, see [Se, ch. II compl.]. It follows directly from the construction of the spectral sequence that if the E_1-term $\mathrm{Tor}^{\mathrm{gr}S}(\mathrm{gr}Q,k)$ satisfies

$$(8) \qquad \{\max j;\ \mathrm{Tor}_i^{\mathrm{gr}S}(\mathrm{gr}Q,k)_j \neq 0\} \leq \min\{j;\ \mathrm{Tor}_{i+1}^{\mathrm{gr}S}(\mathrm{gr}Q,k)_j \neq 0\}$$

for all i, then $E_1 = E_2 = \ldots = E_\infty$, which gives the equality of Betti numbers. (8) is satisfied according to Proposition 16(i).

REMARK. Proposition 17 confirms a conjecture in [Ge-Or, p. 56] about the Cohen-Macaulay type for a ring (Q,m,k) with grQ as in Example 6.

10. APPENDIX.

The purpose of the following section is to sketch the connection between our main result, Theorem 14, and the techniques used by Iarrobino. In particular we shall show that the results about powers of linear forms, due to Jordan, Grace and Young, that are central in Iarrobino's approach, follow from our result.

To make the results valid in arbitrary characteristic we shall work with divided power algebras of derivations, or as they are often called, the Hasse-Schmidt derivatives. We shall denote this algebra by Der and the graded piece of degree d elements by Der_d. Given a form F of degree d, we denote the corresponding element in Der_d by F(D). The algebra Der is built from the i'th partial derivatives D_i operating on R by the composition rules $D_iD_j = D_jD_i$ and the rule $D_iD_i^{(n)} = (n+1)D_i^{(n+1)}$ defining divided powers of D_i. The pairing $<\ ,\ >: Der_c x R_d \quad Der_{c-d}$ obtained by extending the operations $<D_i^{(n)}, x_i^m> = D_i^{(n-m)}$ and $<D_i^{(n)}, x_j^m> = 0$ if $i \neq j$, in the natural way, is easily seen to define a perfect pairing between Der_c and R_c. We record the following rule;

$$<l^{(d)}(D), E(x)> = E(a)l^{(c)}(D)$$

which holds for each form E in R_{d-c} and each linear form $l = \sum_{i=1}^{r} a_ix_i$ and where $a = (a_1, a_2, \ldots, a_r)$.

PROPOSITION 18. *Let* $a_i = (a_{i,1}, a_{i,2}, \ldots, a_{i,r})$ *for* $i=1,2,\ldots,m$ *be points in* A^r *and assume that* $m \leq N(r,d-c)$. *Moreover we let* $l_i = \sum_{j=1}^{r} a_{ij}x_j$ *be the corresponding linear forms and denote by* H *the hyperplane* $\{F \in R_d;\ \sum_{i=1}^{m} F(a_i) = 0\}$ *in* R_d. *If the points are chosen in general position, then the following three subspaces of* Der_c *are all equal to the dual space of* $(H:R)_c$ *under the above pairing between* Der_c *and* R_c;

(i) The space spanned by the forms $l_i^{(c)}(D)$ *for* $i=1,2,\ldots,m$.

(ii) The space $\{\sum_{i=1}^{m} E(a_i)l_i^{(c)}(D);\ E \in R_{d-c}\}$.

(iii) The space $<\sum_{i=1}^{m} l_i^{(c)}(D), R_{d-c}>$.

PROOF. The equality of the spaces in (i) and (ii) is easily seen to be a consequence of Lemma 10 and the equality of the spaces in (ii) and (iii) follows immediately from the derivation rule recorded above. The dual space of the space in (ii) is, by the same derivation rule equal to $\{G\in R_c;\ \sum_{i=1}^{m} E(a_i)G(a_i)=0\}$ for all $E\in R_{d-c}$ and the latter space is equal to $(H:R)_c$.

THEOREM 19. (Jordan, Grace-Young, Iarrobino). *Given a subspace* U *of* Der_c *and let* l_i *for* $i=1,2,\ldots,N(r,d-c)$ *be linear forms in general position. Then the subspace of* Der_c *spanned by the elements* $l_i^{(c)}(D)$ *for* $i=1,2,\ldots,N(r,d-c)$ *is of dimension equal to* $\min(N(r,c),N(r,d-c))$ *and is equal to the space* $<\sum_{i=1}^{m} l_i^{(d)}(D),R_{d-c}>$. *Moreover, this space intersects* U *properly. In characteristic* 0 *the latter space is isomorphic to the space of all derivatives of* $\sum_{i=1}^{m} l_i^{d}$ *of degree* $d-c$.

PROOF. In the proof of Theorem 11 we defined a space H satisfying the assertions of the theorem by $H=\{F\in R_d;\ \sum_{i=1}^{N(r,d-c)} F(a_i)=0\}$ where the a_i where points in $\mathbb{A}^r$ in general position. Choosing the points to be the coefficients of the linear forms l_i we see that the theorem follows from Theorem 11 and Proposition 18.

Exactly in the same way as we deduced Proposition 12 from Theorem 11 we obtain from Theorem 19 the following result;

PROPOSITION 20. (Iarrobino). *Given integers* $d(i)$ *for* $i=1,2,\ldots,n$. *Then if* F_i *for* $i=1,2,\ldots,n$ *are forms of degree* $d(i)$ *in general position, the subspace* $\sum_{i=1}^{n} <F_i(D),R_{d(i)-c}>$ *of* R_c *is of dimension* $\min(\sum_{i=1}^{m} N(r,d(i)-c),N(r,c))$.

REFERENCES.

[At-Ma] Atiyah-Macdonald, *Introduction to commutative algebra*, Addison-Wesley, 1969.

[Bu-Ei] D. Buchsbaum and D. Eisenbud, *Almost-linear resolutions in the Artinian case, appendix to* D. Eisenbud and S. Goto, *Linear Free Resolutions and Minimal Multiplicity*, Preprint, 1982.

[Ea] J.A. Eagon, *Examples of Cohen-Macaulay rings which are not Gorenstein*, Math. Z. 109 (1969), 109-111.

[Ge-Or] A.V. Geramita and F. Orecchia, *On the Cohen-Macaulay type of* s-*lines in* $\mathbb{A}^{n+1}$,J. Alg. 70 (1981), 116-140.

[Ge-Or'] A.V. Geramita and F. Orecchia, *Minimally generating ideals defining certain tangent cones*, J. Alg. 78 (1982), 36-57.

[Go-Ta] S. Goto and S. Tachibana, *A complex associated with a symmetric matrix*, J. Math. Kyoto Univ. 17 (1977), 51-54.

[Gu-Ne] T.H. Gulliksen and O.G. Negård, *Un complex résolvant pour certain idéaux déterminantiels*, C.R. Acad. Sci. Paris Sér. A 274 (1972), 16-19.

[Ho-Ra] M. Hochster and L. Ratliff, *Five theorems on Macaulay rings*, Pac. J. Math. 44 (1973), 147-172.

[Ia] A. Iarrobino, *Compressed Algebras: Artin Algebras having given Socle Degrees and Maximal Length*, to appear in Trans. A.M.S.

[Jo-Pr] T. Jozefiak and P. Pragacz, *Ideals generated by Pfaffians*, J. Alg. 61 (1979), 189-198.

[Kl-La] H. Kleppe and D. Laksov, *The algebraic structure and deformation of Pfaffian schemes*, J. Alg. 64 (1980), 167-189.

[Ma] Matsumura, *Commutative algebra 2:d ed.*, Benjamin, 1980.

[Na-v.O] Nastasescu and van Oystaeyen, *Graded ring theory*, North-Holland Math. Libr. vol. 28, 1982.

[Ro] L. G. Roberts, *A conjecture on Cohen-Macaulay type of* s-*lines in* $\mathbb{A}^{n+1}$, J. Alg. 70 (1981), 43-48.

[Sa] J. Sally, *Cohen-Macaulay rings of maximal embedding dimension*, J. Alg. 56 (1979), 168-183.

[Sc] P. Schenzel, *Über die freien Auflösungen extremaler Cohen-Macaulay-Ringe*, J. Alg. 64 (1980), 93-101.

[Se] J.-P. Serre, *Algèbre locale. Multiplicités*, Springer Lect. Notes 11 (1965).

[St] R. Stanley, *Hilbert functions of graded algebras*, Adv. Math. 28 (1978), 57-83.

[St'] R. Stanley, *The upper bound conjecture and Cohen-Macaulay rings*, Studies in Appl. Math. 54 (1975), 135-142.

[Wa] J.M. Wahl, *Equations defining rational singularities*, Ann. Sci. Ec. Norm. Sup. 4e sér. 10 (1977), 231-264.

SOME PROPERTIES OF SUBCANONICAL CURVES

Luca Chiantini and Paolo Valabrega (*)

Introduction

In 1942, G.Gherardelli obtained the following characterization of complete intersection curves in $\mathbb{P}^3$:

THEOREM ([G], n.1): A smooth irreducible curve $C \subseteq \mathbb{P}^3$ is complete intersection if and only if the following two conditions are fulfilled:

1) the canonical class ω_C of C is $O_C(a)$ for some $a \in \mathbb{Z}$;

2) C is projectively normal, i.e. the map $H^o(O_{\mathbb{P}^3}(t)) \longrightarrow$ $\rightarrow H^o(O_C(t))$ is surjective for all t (equivalently: $h^1(I_C(t))=0$ for all t, I_C being the ideal sheaf of C).

A smooth curve satisfying condition 1) is called "a-subcanonical"; more generally it is called "a-subcanonical" every locally complete intersection 1-dimensional subscheme Y of $\mathbb{P}^3$, such that $\omega_Y = O_Y(a)$ for some $a \in \mathbb{Z}$ (Y may also be reducible and not reduced). Therefore Gherardelli's theorem says that C is complete intersection if and only if it is a-subcanonical and projectively normal.

It has recently been remarked by several authors (see for instance [H]) that, if $C \subseteq \mathbb{P}^3$ is a-subcanonical, then the normal bundle $N_{C/\mathbb{P}^3}$ can be extended to a rank 2 vector bundle E over $\mathbb{P}^3$ and C is the zero locus of a section of E; conversely, if the zero locus of a section of a rank 2 vector bundle E over

(*) Paper written while P.Valabrega was member of C.N.R. (G.N.S.A.G.A.) and both authors were supported by M.P.I. funds.
Authors' address: Dipartimento di Matematica del Politecnico, Corso Duca degli Abruzzi, 24 10129 TORINO (Italy)

$\mathbb{P}^3$ is a 1-dimensional subscheme, then it is subcanonical. E is called the bundle "associated" to the curve C and has Chern classes $c_1 = a+4$, $c_2 = \deg C$. E and I_C are connected by the following exact sequence:

$$0 \longrightarrow O_{\mathbb{P}^3} \longrightarrow E \longrightarrow I_C(a+4) \longrightarrow 0 \qquad (1).$$

Using modern techniques, Griffiths and Harris (see [G-H]) gave two new proofs of Gherardelli's theorem.

Our starting point is the following remark: if a smooth curve C is only a-subcanonical, then this does not imply at all that C is complete intersection, but perhaps there is some condition weaker than projective normality, which forces a subcanonical curve to be complete intersection.

In the present talk we show how to find good conditions for an a-subcanonical smooth curve to be complete intersection in $\mathbb{P}^3$. Here we give only statements with sketches of proofs and produce some example; for more details we refer to a forthcoming paper.

n.1

Once for all, the schemes are defined over the complex field $\mathbb{C}$ and contained in the projective 3-space $\mathbb{P}^3 = \mathbb{P}^3_{\mathbb{C}}$; by "curve" we mean "smooth irreducible curve".

Our first (and strongest) refinement of Gherardelli's theorem is the following:

<u>THEOREM 1</u>: A curve $C \subseteq \mathbb{P}^3$ is complete intersection if and only if the following 2 conditions hold:

A) C is a-subcanonical for some $a \in \mathbb{Z}$;

B) $H^1(I_C(b))=0$, where b is one fixed integer, and precisely:

B1) if a is even, $b = 1+a/2$;

B2) if a is odd, we may take b to be either $(a+1)/2$ or $(a+3)/2$ or $(a+5)/2$.

<u>sketch of proof</u>: We use the bound on the genus of a space curve given in [G-P], from which we deduce that if C is a-subcanonical and there are numbers r,s, with $r \leq s$, such that $r+s=a+4$ and $rs \geq \deg C$, then C must lie on a surface of degree $< r$ or it is complete intersection.

With this in mind, a careful examination of the exact sequences

$$0 \longrightarrow H^0(I_C(n)) \longrightarrow H^0(O_{\mathbb{P}^3}(n)) \longrightarrow H^0(O_C(n)) \longrightarrow H^1(I_C(n))$$

for n in a "neighbourhood" of a/2 proves the statement.

Theorem 1 says that the condition 2) of Gherardelli's theorem ($H^0(O_{\mathbb{P}^3}(t)) \longrightarrow H^0(O_C(t))$ surjective for all t) can be replaced by the surjectivity at level b and nothing else. Such a number b cannot be replaced by any integer different from the ones stated in the theorem, as we see in the following:

<u>EXAMPLE 2</u>: First let us consider the case "a even". Let Y be the disjoint union of two lines; then Y is (-2)-subcanonical and moreover $h^1(I_Y(n))=0$ if $n \neq 0$, $h^1(I_Y) \neq 0$; for such a Y therefore $h^1(I_Y(1+a/2)) \neq 0$ and is the unique h^1 which does not vanish; let E be the bundle associated to Y; by [H], prop.1.4, E(t) gives a smooth irreducible curve C, provided that $t \gg 0$; such a C has the same behaviour of Y, i.e. $h^1(I_C(1+a'/2)) \neq 0$, $h^1(I_C(t))=0$ otherwise, a' being the new a; of course C is not complete intersection, because Y is not.

For the case "a odd", we can do similar computations starting with a disjoint couple of conics and twisting the associated bundle to produce an a-subcanonical, non complete intersection curve C for which $h^1(I_C(t)) \neq 0$ if and only if $t = (a+1)/2, (a+3)/2, (a+5)/2$.

n.2

Now we study the following situation: assume that $h^1(I_C(t_o))=0$ for some t_o different from b; then we can find bounds on the degree of C, functions of a and t_o, such that C is complete intersection if its degree does not exceed the proper bound.

If C is a-subcanonical but not complete intersection, then C cannot lie on a quadric surface (this follows from [H_1],IV, 6.4,6.4.1 by an easy trick); by Halphen's bound on the genus of curves not lying on quadric surfaces (see [G-P]) it is easy to deduce that deg C $\geqslant$ 3a+3; therefore a first rough bound for d= deg C is d_o= 3a+3; this means that if $d < d_o$ then C is complete intersection. The following example shows that, in the case $t_o \geqslant 2a$, this bound is really sharp.

<u>EXAMPLE 3</u>: Let C' be the disjoint union of two plane curves C_1 and C_2, both having degree q; then C' is (q-3)-subcanonical. Let I_1, I_2 be the homogeneous ideals of C_1 and C_2 respectively, in the homogeneous polynomial ring R of $\mathbb{P}^3$; then by [G-W], rem.3.1.6, I_1+I_2 contains all the homogeneous elements of R of degree $\geqslant$ 2q-1; this allows to see that $h^1(I_{C'}(t))=0$ for all t $\geqslant$ 2q-1. Furtherly, using sequence (1), it is easy to see that if E is the bundle associated to C', then E(1) gives a smooth irreducible curve C, (q-1)-subcanonical, whose degree is 3q+2=3(q-1)+5, such that $h^1(I_C(t))=0$ for all t $\geqslant$ 2q.

This shows that $h^1(I_C(t_o))=0$ for $t_o \geqslant 2a$ really does not produce any restriction on the degree of C.

On the other hand, let $h^1(I_C(t_o))=0$, with $(a/2)+1 < t_o < a$. To have an idea of the situation we put $h^1(I_C(ha))=0$, where $0.5 < h < 1$. Then with the same technique of theorem 1, we are able to prove the following:

PROPOSITION 4: Assume C is an a-subcanonical non complete intersection curve such that $h^1(I_C(a))=0$; then we have: $d = \deg C \geq (1-h)a(ha+4)$.

Therefore the lower bound for the degree of an a-subcanonical non complete intersection curve C, satisfying the above condition on the h^1 of the ideal sheaf, is quadratic in a.

EXAMPLE 5: Let Y be the union of two disjoint lines in $\mathbb{P}^3$ and let E be the associated bundle. Then $E(q), q \geq 1$, gives a smooth irreducible $(2q-2)$-subcanonical curve C such that, for all $n \neq q-1=b$, $h^1(I_C(n))=0$; C is not complete intersection and its degree is q^2+1, i.e. a quadratic function of a.

n.3

We are now led to consider what happens if we impose the condition $h^1(I_C(t_o))=0$ with t_o close to a. In fact the behaviour at $t_o=a$ seems to have deeper peculiarities than the one in $t_o < a$ and $t_o > a$, being between a linear and a quadratic bound with no apparent reason to fall into either of them. Moreover the following Sernesi's theorem makes it more interesting exactly $t_o=a$:

THEOREM (Sernesi; [S], 2.6): Let C be a-subcanonical in $\mathbb{P}^3$.
1) If $h^1(I_C(a+1))=0$, then $h^1(I_C(t))=0$ for all $t \geq a+1$.
2) If $h^1(I_C(a))=0$ and C is linearly normal, then $h^1(I_C(t))=0$ for all $t \geq a$ (linearly normal meaning that $H^o(O_{\mathbb{P}^3}(1)) \longrightarrow H^o(O_C(1))$ is onto).

So we investigate what happens when $h^1(I_C(a))=0$. Here the situation is quite complicated and we need to separate two different cases as follows:

<u>CASE A</u>. Suppose there exists a number $r \in \mathbb{Z}$ such that $d = r(a+4-r)$, where $d = \deg C$; in this case C has the same degree as an a-subcanonical complete intersection curve or, in other words, the bundle E associated to C has the same Chern classes of a sum of line bundles. The existence of such a curve which is not complete intersection can be proved by the following:

<u>EXAMPLE 6</u>: Let C_o be a smooth complete intersection in $\mathbb{P}^4$ of three hypersurfaces of degree 2,2,10. Let C be a generic projection of C_o to a $\mathbb{P}^3$, so that by construction C is a-subcanonical with $a=9$ and $d=40$; moreover C is not complete intersection since it is not linearly normal. If we put $r=5$, we see that $d=5.(9+4-5)=40$, so that C has the same characters of a complete intersection of surfaces of degree 5 and 8.

<u>THEOREM 7</u>: Let $C \subseteq \mathbb{P}^3$ be a curve, a-subcanonical of degree $d=r(a-r+4)$, with $r \in \mathbb{Z}$, and assume that $h^1(I_C(a))=0$. If C is not complete intersection then $d \geqslant r_o(a+4-r_o) = d_o$, where:

$$r_o = 3 + ((3/2)a^2-3a/2+\sqrt{q})^{1/3} + ((3/2)a^2-3a/2-\sqrt{q})^{1/3}$$

where $q = (9/4)a^4+(7/2)a^3-(7/4)a^2+2a/3-1/27$

So we have a lower bound for the degree in the non complete intersection case, which grows like $a^{5/3}$.

<u>CASE B</u>. If we drop the assumption that $d=r(a+4-r)$ with $r \in \mathbb{Z}$, then we have a different bound, as it is stated in the following:

<u>THEOREM 8</u>: Let $C \subseteq \mathbb{P}^3$ be a-subcanonical of degree d; if $h^1(I_C(a))=0$ and C is not complete intersection, then:

$$d \geqslant a^{3/2}+2a-2a^{1/2}+3 = d_1$$

Sketch of the proof of th.7 and th.8: The proofs follow by a numerical analysis of the dimensions of the vector spaces in the exact sequence:

$$0 \longrightarrow H^o(I_C(a)) \longrightarrow H^o(O_{\mathbb{P}^3}(a)) \longrightarrow H^o(O_C(a)) \longrightarrow H^1(I_C(a)) \longrightarrow 0$$

In this analysis we use once again the bound on the genus of a space curve given in [G-P] and we also gather information from a (possibly non-reduced) curve Y arising as the zero locus of a section of E(z), E being the bundle associated to C and z being the least integer such that $H^o(E(z)) \neq 0$.

Let us now give some example to show that the bounds are effective.

EXAMPLE 9: Let C be the curve of example 6 and let E be its associated bundle. By [G-L-P] we know that for all $t \geq 39$, $h^1(I_C(t))=0$ and $I_C(t)$ is generated by global sections; so by sequence (1), $h^1(E(t))=0$ and E(t) is generated by global sections for all $t \geq 26$; hence E(30) gives a smooth, irreducible 69-subcanonical curve C' with $h^1(I_{C'}(69))=0$ and deg C'= =1330=38.35; C' is not complete intersection but has the same degree and genus of a complete intersection curve.

EXAMPLE 10: Let C' be the curve we constructed in example 3, having associated bundle E. Since $h^1(E(s))=0$ for all $s \geq q-2$ and E is generated by global sections, we see that E(q+2) gives a (3q+1)-subcanonical curve L which is not complete intersection and further $h^1(I_L(3q+1))=0$. The degree of L is $2q^2+9q+6$, i.e. it grows quadratically with a=3q-2.

EXAMPLE 11: To find examples where the bound is almost sharp let us consider the case a=19; here the bound d_o of theorem 7 is 129; we can produce an example of degree 132 as follows.

In $\mathbb{P}^3$, with homogeneous coordinates x,y,z,t, let X be the line x=y=0 and let Y be the non reduced structure on X given by the homogeneous ideal $(x^2,xy,y^2,ex+fy)$, e,f homogeneous polynomials in z,t of the same degree q-1, without common zeroes along X. Using $[H_1]$,III,7.11, by an easy computation one can check that Y is (-q)-subcanonical; let E be the associated bundle.

Now put q=7, so that E has Chern classes $c_1=-3$, $c_2=2$; since $I_Y(7)$ is generated by global sections and $h^0(I_Y(6))=65$ by a direct computation, then by $0\to\mathcal{O}_{\mathbb{P}^3}\to E\to I_Y(-3)\to 0$ it follows that $h^0(E(9))=285$ and E(10) is generated by global sections; finally E(13) gives a smooth irreducible 19-subcanonical curve C of degree 132, such that $h^0(I_C(19))=285$ hence $h^1(I_C(19))=0$. C is not complete intersection, because Y is not, however it has the same degree and genus of a complete intersection of surfaces of degree 11 and 12.

REMARK 12: All the previous examples can be generalized, obtaining sharper examples, starting with more complicated non reduced structures on a line or with a disjoint union of complete intersections. The degree of these curves grows like $a^{5/3}$ in the situation of th.7 and like $a^{3/2}$ in the situation of th.8.

REFERENCES

[G-W] S.Goto-K.Watanabe On graded rings I, J.Math.Soc.Japan, 30, 1978

[G] G.Gherardelli Sulle curve sghembe algebriche intersezioni complete di due superficie, Rend.Reale Accad. Italia, vol.IV,1942

[G-H] P.Griffiths-J.Harris Two proofs of a Theorem concerning Algebraic Space Curves, Proc.8th Iranian Math.Conf.,1978

[G-P] L.Gruson-C.Peskine Genre des courbes de l'espace projectif, Proc.Tromsø Symp., Lect. Notes n.687,1978

[G-L-P] L.Gruson-R.Lazarsfeld-C.Peskine On a theorem of Castelnuovo and the Equations defining Space Curves, Inv.Math., 72 ,1983

[H] R.Hartshorne Stable Vector Bundles of rank 2 on $\mathbb{P}^3$, Math. Ann.,238,1978

[H_1] R.Hartshorne Algebraic Geometry, Springer 1977

[S] E.Sernesi L'unirazionalita' della varieta' dei moduli delle curve di genere dodici, Ann.Sc.Norm.Super. Pisa,IV,Ser.8, 1981

ABOUT THE CONORMAL SCHEME

by

Steven L. Kleiman[1]

Mathematics Department, M.I.T., 2-278

Cambridge, MA 02139; U.S.A.

CONTENTS

1. INTRODUCTION 161

2. THE CONTACT FORMULA 171

3. BASIC GENERAL THEORY 178

4. DUALITY AND REFLEXIVITY 188

REFERENCES 196

1. INTRODUCTION

Doubtless the conormal scheme of an embedded scheme -- the scheme of pointed tangent hyperplanes -- is a fundamental invariant. Its use in complex-analytic singularity theory was nicely summarized by Merle [10]. Its use in "algebraic analysis" was discussed by Pham [12] and by Oda [11]. Its use in the enumerative theory of contacts was developed by Fulton, MacPherson, and the author in [2]. In [2], however, some issues were treated only briefly. To introduce and amplify [2] are the aims of the present article.

[1] Supported in part by the Danish Natural Science Research Council, the Norwegian Research Council for Science and Humanities, il Consiglio Nazionalle delle Ricerche, and the National Science Foundation.

Section 2 is devoted to the contact formula. The formula gives the number of varieties X in a p-parameter family that touch, or are tangent to, p fixed varieties V in general position in an ambient projective space. Only the case $p = 1$ is treated here for the following reason: the case of arbitrary p is treated in [2]; the additional generality is obtained via a simple technical change in the initial setup of the proof, and no new property of the conormal scheme is involved. Likewise, relatively little will be said here about the enumerative significance of the formula. The formula, in fact, enumerates the contacts. Conceivably, a contact could appear with multiplicity greater than 1, an X could make 2 or more distinct contacts with the same V, every X could appear repeatedly in the family, etc. These matters are taken up in [2].

An introduction to the fomula is offered in (2.1). In (2.2), the formal general setup of the rest of Section 2 is presented. In (2.3), the formula is asserted as a theorem and established via a reduction to Lemma (2.4). Lemma (2.4) gives an expression for the fundamental class of a conormal scheme, modulo rational equivalence on the graph I of the point-hyperplane incidence correspondence. Lemma (2.4) is established using Theorem (2.5). Finally, Theorem (2.5) is an immediate consequence of certain results in Sections 3 and 4. Theorem (2.5) asserts that when a scheme is degenerated, then, at least in characteristic 0, its conormal scheme degenerates correspondingly into a union of conormal schemes. The theorem is illustrated in (2.6). The theorem may fail in positive characteristic; a counterexample is given in (4.11).

The preceding way of deriving the contact formula was given in essence in the case of plane curves by Schubert [14] pp. 13-14, to illustrate the use of the principle of conservation of number. A rigorous development of the method may be considered a direct contribution to the solution of Hilbert's 15th problem. Indeed, Hilbert [6] said, "The problem consists in this: To establish rigorously and with an exact determination of the limits of their validity those geometrical numbers which Schubert [14] especially has determined on the basis of the so-called principle of special position, or conservation of number, by means of the enumerative calculus developed by him."

Two more derivations of Lemma (2.4) are given in [2]. Both work in any characteristic and so establish the contact formula in any characteristic. The first of these involves a simple computation using the basis of Schubert varieties on I. This derivation too was given in essence by Schubert [14], pp. 50-51, 289-95. Schubert, moreover, obtained the basis in two related and clever ways. (See Grayson [3] for a lovely up-to-date version of this part of Schubert's work.) The other derivation involves the formula of special position of Schubert (1903) and Giambelli (1905), known today as Porteous's formula. This derivation yields a version of Lemma (2.4) on any smooth ambient space, see [2] Prop., p. 179.

Section 3 is devoted to the development of a basic general algebreo-geometric theory of conormal schemes. In (3.1), the conormal scheme is defined for any scheme V that, on a dense open subscheme, is smooth and immersed in a smooth ambient scheme Y of any dimension $N \geq 1$. This conormal scheme is denoted by CV or C(V/Y). The base scheme S is taken to be reasonably general.

Hence, the theory may be applied readily to families, including infinitesimal deformations.

In Section 3, the point of view is essentially that of a rudimentary part of the theory of Lagrangian geometry, or contact geometry, on the projectivization I of the cotangent bundle of Y, which in turn is part of the modern theory of partial differential equations. This theory is founded on the work of Lagrange, Monge, Plücker, Clebsch, Lie and others. The theory was revived around 1960 and has been actively pursued ever since. In particular, although few if any explicit references can be given, it seems that around 1960 many researchers independently discovered the characterization of conormal schemes as Lagrangians.

The idea behind this fundamental characterization is very simple. For the sake of discussion, assume for a moment that the base scheme S is the spectrum of an algebraically closed field and that V is a reduced closed subscheme of Y. In Articles (3.1)-(3.6), there is a coordinate free and substantially more general version of the discussion here.

The projectivized cotangent bundle I carries a canonical twisted differential 1-form ω. Locally this form is given as follows. Fix a rational point y of Y, and about it center a system of local coordinates $q = (q^1, \ldots, q^N)$. Then, correspondingly,

$$(q,p) = (q^1, \ldots, q^N ; p_1, \ldots, p_N) ,$$

where $p_i = \partial/\partial q_i$, is a system of local homogeneous coordinate for I/Y and the form is simply

$$\omega = p.dq \; (=p_1 \, dq^1 + \ldots + p_N \, dq^N).$$

Now, consider an "arc" $(q(t); p(t))$ on I emanating at $t = 0$ from y, and let $\dot{q}(t)$ denote the derivative with respect to t. Then, virtually by definition, the arc satisfies the equation $\omega = 0$ if and only if the dot-product vanishes,

$$p(o) \, . \, \dot{q}(0) = 0.$$

On the other hand, this vanishing is just the condition that the cotangent vector p(o) of Y be orthogonal to the tangent vector $\dot{q}(0)$.

The conormal scheme CV is defined as the closure in I of the subscheme ruled by the orthogonal complements of the tangent spaces at the smooth points of V. Hence, in view of the above "tautogology", it is evident that, if C is a reduced subscheme of I such that the structure map $I \to Y$ induces a generically smooth map $C \to V$, then C satisfies the differential equation $\omega = 0$ (such a C is called <u>isotropic</u>) if and only if C is contained in CV. Hence, in particular, every conormal scheme is <u>Lagrangian</u> (or <u>Legendre</u> or <u>holonomic</u>); that is, it is a solution (or integral scheme) of $\omega = 0$ of pure dimension N-1.

Conversely, let C be a Lagrangian reduced closed subscheme of I, and let V be its image in Y. Then, therefore, C = CV provided that the projection $C \to CV$ is generically smooth. In particular, C = CV in characteristic 0, because of a well-known and easily-proved version of Sard's lemma. This version asserts that, in characteristic 0, every map (locally of finite presentation) between reduced schemes is smooth over a dense open subset of the target. It holds because, in characteristic 0, every field extension is separable; hence, the open subset of the target over which the map is smooth will contain the generic point of each component.

The main result of Section 3 is Theorem (3.7). It asserts that, when S is a smooth curve over an auxiliary field of characteristic zero, then every reduced irreducible component D of every fiber $(CV)_s$ is itself a conormal scheme; namely, D = CW, where W is the image of D in the corresponding fiber V_s. (It is obviously equivalent to assert that the reduction of $(CV)_s$ is equal to the conormal scheme of its image.) Theorem (3.7) follows from the characterization of conormal schemes as Lagrangians and from Lemma (3.8). Lemma (3.8) says in effect that, at least in characteristic 0, if a scheme satisfying a (first order) differential equation $\eta = 0$ is degenerated, then each reduced irreducible component of the degeneration satisfies the equation too. The lemma is proved by reducing it to an explicit computation.

Related to Theorem (3.7) are Propositions (3.9) and (3.10), which hold for a general base scheme S. Proposition (3.9) implies notable that (i) each irreducible component W of V_s is the image of some component D of $(CV)_s$, and (ii) if V/S is smooth at the generic point of W, then CW is the unique such D and, as a component, CW appears with multiplicity 1, and (iii) if a component D of $(CV)_s$ is not of the form CW for some component W of V_s, then the image of D in V_s is entirely contained in the singular locus of V/S. Proposition (3.10) says that, if V dominates S and if S is reduced or, what is more general, V is smooth over a dense open subset of S, then there exists a dense open subset S' of S such that the formation of CV commutes with every base change $T \to S'$; in particular, of course, T may be a point of S.

C. Sabbah, after looking at [2], kindly pointed out to the author that, in the complex-analytic setting, a more refined version of (3.7), (3.9) and (3.10) can be proved easily using the theory developed by Hironaka [7] and by himself [13]. Namely, the hypothesis in (3.7) that S be a smooth curve may be replaced by the hypothesis that D be (N-1)-dimensional. Moreover and more importantly, the fundamental cycle of the fiber

$$[(CV)_s] = \Sigma\, n_W [CW]$$

is determined by the following relation, which holds for every point x of V_s:

$$(-1)^{\dim V_s}\, \Sigma_W\, (-1)^{\dim W}\, n_W\, Eu_W(x) = \chi(F_x, Eu_x)$$

where $Eu_W(x)$ denotes the Euler obstruction of W at x, where F_x denotes the Milnor fiber of V/S at x (that is, $F_x = B_\varepsilon(x) \cap V_t$ for t near s), and where $\chi(F_x, Eu_x)$ denotes the Euler characteristic of F_x with weight Eu_x (in a stratification (U_i) of F_x such that Eu_x is constant on each stratum, this Euler characteristic is given by

$$\chi(F_x, Eu_x) = \Sigma(Eu_x|U_\alpha).\chi(U_\alpha) \quad .)$$

It would be nice to have an abstract algebreo-geometric treatment of these issues.

Section 4 is devoted to the particular case in which Y is a bundle of projective spaces $\mathbb{P}(E)$, where E is a locally free sheaf on S. In this case, as is well known and reproved in a more precise form in (4.2, i), the projectivization I of the tangent bundle of Y is equal to the graph of the point-hyperplane incidence correspondence. Hence I is also equal to the projectivization of the tangent bundle of the dual bundle of projective spaces,

$$Y^\vee = \mathbb{P}(E^\vee).$$

Therefore, I carries two contact forms ω and $\omega^\vee$, and it is easy to check that their sum is 0. Indeed, let

$$q = (q^0, \ldots, q^N) \text{ and } p = (p_0, \ldots, p_N)$$

be systems of local homogeneous coordinates for Y/S and $Y^\vee/S$ respectively. Then I is defined by the equation

$$I : p.q(=p_0q^0 + \ldots + p_Nq^N) = 0.$$

Differentiating this equation yields

$$p.dq + q.dp = 0;$$

whence, $\omega + \omega^\vee = 0$. A coordinate-free version of this proof is given in (4.2, ii).

For simplicity, assume now that S is the spectrum of an algebraically closed field and that V is a reduced closed subscheme of Y. Then CV is defined in I. Denote its image in $Y^\vee$ by V*. Intuitively, V* is the locus of hyperplanes tangent to V. It has been studied intensively since the turn of the 19th century, and it is known as the _dual_, or _reciprocal_, of V. The name is justified by a celebrated theorem, sometimes called the _biduality theorem_. It asserts that the double dual of V, the dual V** of V*, is equal to V always in characteristic 0 and ordinarily in positive characteristic. Intuitively, the equation V = V** says that V is the scheme enveloped by the family of hyperplanes tangent to V. (The scheme enveloped by a family of hyperplanes is the locus of points of intersection of pairs of infinitely near hyperplanes.)

The biduality theorem holds because (i) CV is Lagrangian for ω, hence (ii) CV is Lagrangian for $\omega^\vee$, hence (iii) CV = CV* provided that the characteristic is 0, more generally, the projection CV → V* is smooth on a dense open subscheme of CV; finally, it is evident that the image of CV in Y is V and that the image of CV* in Y is V**. This proof is essentially the one that C. Segre [15], no. 16, gave over the complex numbers and that A. Wallace [16] gave over a field of any characteristic (see also Kleiman [9]). Here the proof is broken into three substantive pieces, (i)-(iii), of independent interest. Furthermore, this version works over a general base scheme S, see Theorem (4.4).

There is a more refined and more important notion than biduality. It is the notion that CV* = CV, and it will be called _reflexivity_. The Segre-Wallace proof does, in fact, establish reflexivity. In positive characteristic, biduality and reflexivity are distinct and nonautomatic notions. Indeed, it is easy to give examples of plane curves that are not bidual and of ones that are bidual but not reflexive; see (4.11) and Wallace [16], Section 7. In fact, Wallace indicates how to construct a plane curve V such that the separable degree d of the projection is arbitrarily large; these V are not only not bidual but have the curious property that every tangent line touches V at at least d distinct points.

At the end of Section 4 (which is also the end of the article), there are three discussions. The first, in (4.8), suggests a possible theory of reflexivity for normal nonembedded schemes. The second, in (4.10), amplifies the discussion in [2] about the way in which the contact formula is self-dual. The third, in (4.11), provides a counterexample in positive characteristic to Theorem (2.5), and so one to Theorem (3.7) too. The three discussions also illustrate the use of various of the results of the present article.

In an article [5] under preparation by Hefez and the author, there will be some additional results in the theory of the conormal scheme of a reduced projective scheme V over a field.

Two main results, which were quoted in [2], are the following:

(1) The ith rank r_i of V is nonzero exactly over a certain interval; more precisely,

$$r_i \neq 0 \text{ if and only if } (N-1-\dim(V^*)) \leq i \leq \dim(V).$$

(2) If V is reflexive, then so is a general hyperplane section, except if V* is a hypersurface in characteristic 2, where it is not.

Finally, it is a pleasure for the author to thank Abramo Hefez for numerous useful discussions about the material in this article. In addition, the author would thank Anne Clee for her swift fine typing. Last but not least, the author would thank the mathematics departments of the universities of Copenhagen, Denmark, Oslo, Norway, Turin, Italy, and Ferrara, Italy, for their kind hospitality during early stages of the present work.

2. THE CONTACT FORMULA

(2.1) A quick introduction to the formula. In the plane, consider a 1-parameter family of curves X and a fixed curve V in general position. Let r_0 denote the class of V -- the number of lines that pass through a point and that touch V -- and let r_1 denote the degree of V -- the number of points that lie on a line and that lie in V. Let λ_0 denote the number of X passing through a point, and λ_1 the number of X touching a line. Collectively, r_0 and r_1 are called the ranks of V, and λ_0 and λ_1 are called the characteristics of the family. Now, the number n of X touching V is given by the formula,

(2.1.1) $$n = r_0\lambda_0 + r_1\lambda_1 \; .$$

This is the contact formula. It was discovered by Cremona (1862) and independently by Charles (1864).

For example, consider a family of concentric circles X. (Over the complex numbers, it is simply the family of conics with given tangents at two given points; the points are the two "circular" points at infinity, $(0, \sqrt{-1}, 1)$ and $(0, -\sqrt{-1}, 1)$, and the given tangents meet at the common center. Another real form of the family is the family of coasymptotic hyperbolas.) The characteristics of the family are obviously $\lambda_0 = 1$ and $\lambda_1 = 1$. Hence the number of concentric circles X touching a curve V of class r_0 and degree r_1 in general position is simply

$$n = r_0 + r_1 \; .$$

For instance, if V is a conchoid of Nicomedes or a folium of Descartes, both nodal cubics, then $r_1 = 3$ and, by one of the

Plücker formulas, $r_0 = 4$; hence, V is touched by $n = 7$ concentric circles (and by $n = 7$ coasymptotic hyperbolas). For any V, it is evident that n may also be interpreted as the number of lines through a point that are normal to V, or what is the same, as the class of the evolute of V (the _evolute_ is the curve enveloped by the normals).

Schubert ([14], Biespiel 4, pp. 13-14) suggests deriving Formula (2.1.1) as follows. By the principle of conservation of number, the number n of X touching V remains constant when V is varied. Degenerate V into an r_1-fold line. Correspondingly, the set of tangent lines degenerates into the union of r_0 pencils -- that is, the set of lines through r_0 points. The formula is now evident: through each of the r_0 points, there are λ_1 curves X, and touching the r_1-fold line there are λ_1 curves X; or in all, $r_0\lambda_0 + r_1\lambda_1$ curves X.

(2.2) _The general setup_. Fix an ambient projective space of any dimension $N \geq 2$ over an algebraically closed ground field of any characteristic. In this space, consider a 1-parameter family of reduced closed subschemes X and a fixed reduced closed subscheme V; these subschemes need not be irreducible nor equidimensional, but assume, of course, that the parameter scheme is reduced and of finite type over the ground field. For $i = 0, \ldots, N-1$, the ith characteristic λ_i of the family is defined as the number of X touching a general i-plane. (The precise meaning of the word "touch" is reviewed in the course of the proof of (2.3).) The ith _rank_ r_i of V is defined as the number of (N-i-1)-planes that lie in a general linear pencil and that touch V. (A _linear pencil_ of j-planes consists of those nested between a (j-1)-plane and a (j+1)-plane.) That λ_i and r_i are finite and independent of the choice of i-plane and of pencil is a corollary of the following theorem; for λ_i apply it with an i-plane as V, and for r_i apply it with a linear pencil of (N-i-1)-planes as the family of X. (More information about r_i is given in [2], pp. 4-5, 15-16, 22.)

(2.3) <u>Theorem</u> (the Contact Theorem). In the setup of (2.2), the number n of X touching V is finite and given by the formula,

$$n = r_0\lambda_0 + \dots + r_{N-1}\lambda_{N-1} ,$$

provided that V is in general position; moreover, any finite number of X may be discarded without loss. More precisely, there exists a dense open subset of the linear group consisting of transformations g such that the (weighted) number n of X touching gV is finite, is independent of g, excludes any given finite number of X, and is given by the stated formula.

<u>Proof</u>. Consider the graph of the point-hyperplane incidence correspondence,

$$I = \{(P,H) | P \in H\}.$$

Recall (or see (4.3)) that the conormal scheme CV is equal to the closure in I of the set of pairs (P,H) such that P is a smooth point of V and H is a hyperplane touching V at P (that is, H contains the embedded tangent space).

Let S be any compactification of the parameter space of the family. Extend the family over S, and let F denote the conormal scheme of the total space in $\mathbb{P}^N \times S$. It is not hard to show (see (3.10)) that there is a dense open subset S_0 of the original parameter space such that the fiber of F over a point of S_0, representing a scheme X, is just CX.

By definition, an X <u>touches</u> a translate gV if and only if CX meets CgV. Hence the number n of X touching gV is finite and is given as an intersection number by the expression,

$$(2.3.1) \qquad n = \int_{I\times T} [F] \cdot pr_I{}^*[C(gV)] ,$$

provided that g is such that C(gV) meets F in a finite number of points lying over S_0. Now, obviously

$$C(gV) = g(CV) .$$

Moreover, obviously, the linear group acts transitively on I. Hence, by the theorem of (dimensional) transversality of the general translate (see for example [8], (2,i), p. 290), there exists a dense open subset of the general linear group consisting of such g. Finally, since the various translates g(CV) are all rationally equivalent, the number n is independent of g, and n is given by the asserted formula because of the following lemma.

(2.4) Lemma. In the setup of (2.2), as continued in the first paragraph of the proof of (2.3), let A_i be an i-plane for i = 0, ..., N-1. Then, modulo rational equivalence on I,

$$[CV] = r_0[CA_0] + \dots + r_{N-1}[CA_{N-1}] .$$

<u>Proof</u>. In (2.7) the lemma will be derived from the following theorem. (Two other proofs of the lemma are given in [2]. One involves showing that the $[CA_i]$ form a basis for the (N-1)-cycles on I. The other involves applying Porteous's formula. Both proofs are valid in any characteristic, unlike the present proof.)

(2.5) <u>Theorem</u>. In the setup of (2.2), as continued in the first paragraph of the proof of (2.3), if V is degenerated to a scheme

V_0 in a flat family, then, at least in characteristic 0, the conormal variety CV degenerates correspondingly and within I into a scheme C_0 whose reduced and irreducible components are each of the form CW for an appropriate closed, reduced and irreducible subscheme W of V_0. Moreover, each component of V_0 appears as a W, and if a W is not a component, then it lies in the singular locus of V_0.

Proof. The parameter space of both degenerations is the same smooth curve S. The total space of the degeneration of CV is, by (3.10), just the conormal scheme of the total space $\underline{V}$ of the degeneration of V. Hence, the first assertion is an immediate consequence of (3.7), and the second is one of (3.9), because the formation of the smooth locus of $\underline{V}/S$ commutes with base change, since $\underline{V}/S$ is flat, as S is a smooth curve.

(2.6) Example. Under the conditions of (2.5), suppose that V is a curve. Then C_0 is set-theoretically the union of the conormal varieties of the irreducible components of V_0 and the conormal varieties of certain points lying in the singular locus of V_0. For instance, if V is a smooth plane cubic and V_0 is a nodal cubic, then C_0 is the union of CV_0 and CW, where W is the node; in fact, it is not hard to show using the reasoning at the end of (2.7) that CV_0 appears with multiplicity 1 and that CW appears with multiplicity 2 (= $r_0(V) - r_0(V_0)$). (That CV_0 appears with multiplicity 1 is also a consequence of (3.9).)

If V_0 has a multiple component, then the points W lying in it are no longer determined by V_0 alone but depend on the degeneration. For instance, an explicit computation shows that, if V is an ellipse degenerating in a confocal family to a double line V_0, then the points W in question are the two foci, and they may be placed at will anywhere along V_0 by changing the family.

(2.7) Proof of (2.4). Fix a point P off V, fix a hyperplane H, and consider the corresponding homolography (or homography) -- that is, the family of all homologies with center P and axis H. The homologies degenerate into the projection from P to H as their cross ratios tend to 0. In terms of a coordinate "tetrahedron" with vertex P and base H, the family is given by multiplying the vertical coordinate by a number, which tends to 0. (Schubert uses a homolography to effect the degeneration in his derivation of the contact formula (see (2.1)), a procedure he ([14], Lit. 24, p. 338) says he learned from Zeuthen.)

Under the homolography, CV degenerates, according to Theorem (2.5), within I into a scheme whose reduced, irreducible components are of the form CW for certain subvarieties W of H. If V is a hypersurface, then one of the W will be H itself. In any event, proceed treating all the W properly contained in H with a homolography within H. Repeat and repeat, concluding that modulo rational equivalence on I

(2.7.1) $$[CV] = s_0[CA_0] + \dots + s_{N-1}[CA_{N-1}]$$

for suitable integers s_i.

Finally, $s_i = r_i$ for each i. Indeed, the rank $r_i = r_i(V)$ is defined in (2.2) as the number of (N-i-1)-planes in a general linear pencil that touch V. So (2.7.1) and (2.3.1) yield

$$r_i = s_0 r_0(A_0) + \dots + s_{N-1} r(A_{N-1}) .$$

Thus it remains to prove

(2.7.2) $\quad r_i(A_j) = \delta_{ij}$ (the Kroneker function).

Now, it is evident that there is an (N-i-1)-plane in a given general linear pencil, that touches A_j if and only if j=i, and that if j=1, there is exactly one, and it makes a single contact. In characteristic 0 (which must be assumed anyway to apply (2.5)), this contact automatically counts with multiplicity 1 by the theorem of (differentiable) transversality, [8], (2,ii), p. 290. In any characteristic, (2.7.2) may be established by using the projection formula to evaluate (2.3.1) with $V = A_j$, see [2], pp. 15-16.

3. BASIC GENERAL THEORY

(3.1) Setup. Work in the category of noetherian schemes and maps of finite type. Fix a base scheme S and a smooth ambient scheme Y of constant relative dimension $N \geq 1$. Form the projectivization of the cotangent bundle

$$I = \mathbb{P}(\Omega^{1\ \vee}_{Y})$$

and denote its structure map by

$$p : I \to Y.$$

The bundle I carries two canonical maps, whose composition

$$\omega : O_I(-1) \to p^*\Omega^1_Y \xrightarrow{\partial p} \Omega^1_I$$

is called the contact form.

Let $g : C \to I$ be an S-map. For any map of O_I-modules

$$\eta : L \to \Omega^1_I ,$$

let η/C stand for the composition $\partial g \circ g^*\eta$,

$$\eta/C : g^*L \to g^*\Omega^1_I \to \Omega^1_C .$$

Then C or C/I will be said to satisfy the (twisted first-order partial differential) equation $\eta = 0$ if η/C vanishes on an S-smooth, dense open subscheme C^0 of C. If C satisfies the equation $\omega = 0$ and if C^0/S has pure relative dimension N-1, then C will be called Lagrangian.

Let $f : V \to Y$ be an S-map. Let V^0 denote the largest open subscheme of V on which V/S is smooth and on which f is an immersion; the latter condition means that on V^0 the Jacobian map

$$\partial f : f^*\Omega^1_Y \to \Omega^1_V$$

is surjective. Assume that V^0 is scheme-theoretically dense in V. Then the scheme CV defined by

$$CV = \text{the closure of } \mathbb{P}((\ker(\partial f|V^0))^\vee) \text{ in } V\times_Y I \text{ is called}$$

the conormal scheme of V or V/Y. It will also be denoted by C(V/Y).

(3.2) <u>Proposition</u>. In the setup of (3.1), CV is Lagrangian. In fact, the scheme

$$C^0 = CV^0$$

is a dense open subscheme of CV, it is S-smooth of pure relative dimension N-1, and on it $\omega|C$ vanishes. Moreover, $p : I \to Y$ induces a proper and surjective map,

$$q : CV \to V,$$

and the inverse image of V^0 is exactly C^0,

$$q^{-1}V^0 = C^0 .$$

<u>Proof</u>. The map q is just the restriction of the first projection of $V\times_Y I$. Now, in view of the definition of CV, it is obvious that CV is just the closure of C^0 in $V\times_Y I$ and that C^0 is closed in $V^0\times_Y I$. Hence, by general topology,

$$C^0 = (CV) \cap (V^0\times_Y I) = q^{-1}V^0 .$$

Since V^0 is open in V, therefore C^0 is open in CV. Obviously, C^0 is S-smooth of pure relative dimension N-1. Now, since p is proper, so is q. Hence, qCV is closed in V. Since qcv contains V^0 and since V^0 is dense in V, therefore qCV is all of V. Thus q is surjective. Finally, $\omega|C$ vanishes on C^0 by (iii) ⇒ (ii) of the next proposition.

(3.3) Proposition. In the setup of (3.1), suppose that there is a map $q : C \to V$ that is smooth on a dense open subscheme C' of C and that the following diagram is commutative:

$$\begin{array}{ccc} C & \xrightarrow{g} & I \\ {\scriptstyle q}\downarrow & & \downarrow{\scriptstyle p} \\ V & \xrightarrow{f} & Y \end{array}$$

Then the following three statements are equivalent:

(i) $\omega|C$ vanishes on $q^{-1}V^0$.

(ii) C satisfies the equation $\omega = 0$.

(iii) $(q,g) : C \to V \times_Y I$ factors through CV.

Moreover, the following open subscheme of C is also dense and S-smooth:

$$C_1^0 = C' \cap q^{-1} V^0 .$$

Proof. Since $(q|C')$ is smooth, it is open. Hence, since V^0 is open and dense in V, by general topology C_1^0 is open and dense in C'. By hypothesis, C' is open and dense in C; hence, C_1^0 is also. Now, C_1^0/S is smooth, because C_1^0/V^0 and V^0/S are smooth. Thus the "moreover" assertion holds, and the implication (i) ⇒ (ii) follows immediately.

Consider the following diagram of natural maps:

$$\begin{array}{ccccc} g^*O_I(-1) & \to & g^*p^*\Omega^1_Y & \to & g^*\Omega^1_I \\ & \searrow^{u} & \downarrow{\scriptstyle q_1^*\partial f} & & \downarrow \\ & & q^*\,\Omega^1_V & \xrightarrow{\partial q} & \Omega^1_C \end{array}$$

It is obviously commutative. So

$$\omega | C = \partial q \circ u.$$

Hence, if u vanishes on $q^{-1}V^0$, then so does $\omega|C$; that is, then (i) holds.

Suppose on the other hand that (ii) holds; that is, suppose that $\omega|C$ vanishes on a dense open subscheme C^0 of C. Now, by hypothesis, q is smooth on the dense, open subscheme C' of C; hence, ∂q is injective on C'. Therefore, u vanishes on $C' \cap C^0$. So, u vanishes on $C' \cap C^0 \cap q^{-1}V^0$, which is open and dense in $q^{-1}V^0$. Now, since V^0 is S-smooth, $q^*\Omega^1_V$ is locally free on $q^{-1}V^0$. Therefore, u vanishes on $q^{-1}V^0$.

Finally, $q^{-1}V^0$ is open in C, and it is dense in C because its subset C^0_1 is. Hence (q,f) factors through CV (that is, (iii) holds) if and only if the restriction $(q,f)|q^{-1}V^0$ factors through CV^0. However, the latter obtains if and only if u vanishes on $q^{-1}V^0$, by virtue of the following lemma applied with $v = (\partial f|V^0)$. (Note that a map factors through a closed subscheme of its target if and only if the pullback of the ideal of the subscheme is equal to 0.)

(3.4) Lemma. For any exact sequence of quasi-coherent sheaves

$$F \xrightarrow{v} G \to H \to 0$$

on any scheme, the image of the composition of natural maps

$$F_{\mathbb{P}(G)}(-1) \to G_{\mathbb{P}(G)}(-1) \to O_{\mathbb{P}(G)}$$

is equal to the ideal of $\mathbb{P}(H)$ in $\mathbb{P}(G)$.

Proof. In short, the homogeneous ideal is $F[-1] \otimes \mathrm{Sym}(G)$. For more details, see [1], (2,6,i), p. 17.

(3.5) Corollary. In the setup of (3.1), suppose that C is a closed subscheme of Vx_YI. Then (i) C = CV if and only if (ii) C is Lagrangian and the projection $q : C \to V$ is smooth on a dense open subscheme C' of C such that qC' is a dense open subscheme of V.

Proof: (i) ⇒ (ii) by (3.2). Conversely, (ii) implies by (3.3), (ii) ⇒ (iii) and the "moreover" assertion, that C is a closed subscheme of CV and that

$$C_1^0 = C' \cap q^{-1}V^0 = C' \cap CV^0$$

is open and dense in C and is S-smooth of relative dimension N-1. Now, C_1^0 is open in CV^0, because CV^0 too is S-smooth of relative dimension N-1. Hence, C_1^0 is open and dense in CV^0; indeed, qC_1^0 is open and dense in V^0, and CV^0 is of the form $\mathbb{P}(K)$ where K is locally free on V^0. Therefore, C = CV, as (i) asserts.

(3.6) Corollary. In the setup of (3.1), suppose that the characteristic is 0; that is, S is $\mathbb{Q}$-scheme. Suppose that C is a reduced closed subscheme of I and that V is the image of C in Y, equipped with the induced reduced closed subscheme structure. Then C = CV if and only if C is Lagrangian.

Proof. The assertion is an immediate consequence of (3.5), because the condition on $q : C \to V$ holds automatically by Sard's lemma (see the introduction), since C and V are reduced and the characteristic is 0.

(3.7) Theorem. In the setup of (3.1), suppose that S is a smooth curve over an auxiliary field of characteristic 0. Let $s \in S$. Let D be a reduced irreducible component of the fiber $(CV)_s$.

Then

$$D = C(W/Y_s)$$

for a unique reduced and irreducible closed subscheme W of V_s; in fact, W is just the image of D.

Proof: By (3.6), it suffices to prove that D/I_s is Lagrangian. By (3.2), CV is Lagrangian. Obviously, the formation of ω commutes with base change. Hence, the assertion follows from the following lemma.

(3.8) Lemma. In the setup of (3.1), suppose that S is a smooth curve over an auxiliary field k of characteristic 0. Suppose also that C satisfies the equation $\eta = 0$. Let $s \in S$. Let D be a reduced, irreducible component of the fiber C_s. Then D is of codimension 1 in C, and D satisfies $\eta | I_s = 0$.

Proof. By hypothesis, $\eta | C$ vanishes on an S-smooth, dense open subscheme C^0 of C. Since C^0 is smooth, open and dense, D would dominate a component of S if D were a component of C; hence, D is of codimension 1 in C. Again, since C^0 is smooth, open and dense, and since S is reduced, then C is reduced.

Let C' be the normalization of C. Let D' be a closed, reduced and irreducible subscheme of C' mapping onto D. Since the characteristic is 0, the map from D' to D is separable; hence, it is generically étale. It follows that D will satisfy $\eta | I_s = 0$ if D' does.

Since C' is normal and k is of characteristic zero, C'/k is smooth in codimension 1. By the same token, D'/k is smooth in codimension 0. (This is a less important use of the hypothesis that the characteristic is zero; here it would suffice that k be perfect.) Hence, there exists a dense open subscheme C" of C' that is k-smooth and meets D' in a k-smooth, dense, open subscheme D". Replacing C" with a smaller subset if necessary, we may assume that there exist regular functions $t_1, \ldots, t_m$ on C" such that t_1 generates the ideal of D" and $dt_1, \ldots, dt_m$ form a basis of $\Omega^1_{C''/k}$. Moreover, we may assume that D" is set-theoretically the entire fiber of C" over s and that scheme-theoretically the fiber is defined by the vanishing of a function v which is the pullback of a function defined on a neighborhood of s in S. Then

$$v = ut_1{}^n$$

for some invertible function u and some integer $n \geq 1$. Hence

$$\Omega^1_{C''/S} = (O_{C''}/t_1{}^{n-1})dt_1 \oplus \Omega$$

where Ω is generated by $dt_2, \ldots, dt_m$ and is free. Moreover, the differential of the inclusion map of D" into C" factors through the projection onto Ω,

$$(3.8.1) \qquad \Omega^1_{C''/S} \to \Omega \to \Omega^1_{D''/k(s)},$$

because dt_1 vanishes on D".

Finally, by hypothesis, C satisfies $\eta = 0$. It follows that C" satisfies $\eta = 0$; that is, $\eta|C"$ with the projection onto Ω,

$$(3.8.2) \qquad L|C'' \xrightarrow{\eta|C''} \Omega^1_{C''/S} \to \Omega,$$

vanishes on C"'. Hence (3.8.2) vanishes everywhere on C", because Ω is free and C" is reduced. So the restriction of (3.8.2) to D" vanishes too. In view of (3.8.1), therefore $\eta|D"$ vanishes. Thus D satisfies $\eta|I_s = 0$, as asserted.

(3.9) Proposition. In the setup of (3.1), consider an arbitrary base-change map $T \to S$.

(i) The fiber of surjective map $q : CV \to V$ of (3.2) is also a surjective map,

$$q_T : (CV)_T \to V_T .$$

(ii) Let W' be an arbitrary open subscheme of the fiber $(V^0)_T$, and let W denote the closure of W' in the fiber V_T. Then (a) the conormal schemes

$$CW' = C(W'/I_T) \text{ and } CW = C(W/I_T)$$

are defined; (b) CW' is a dense open subscheme of CW; (c) CW' is an open subscheme of $(CV)_T$; and (d) set-theoretically,

$$((CV)_T - CW') = q_T^{-1}(V_T - W') .$$

Proof: (i) It is well known and easy to prove that, in general, surjectivity is conserved whenever the base is changed.

(ii) Consider W^0, the largest open subscheme of W on which W/T is smooth and on which $f_T : V_T \to I_T$ is an immersion. Obviously W' is open and dense in W^0. Hence (a) and (b) hold. Now, obviously CW' is open in $C((V^0)_T)$. Obviously, scheme-theoretically

$$C((V^0)_T) = (CV^0)_T . \tag{3.9.1}$$

Hence (c) holds. Finally, obviously, set-theoretically

$$(C((V^0)_T) - CW') = q_T^{-1}((V^0)_T - W') ,$$

$$((CV)_T - (CV^0)_T) = q_T^{-1}(V_T - (V^0)_T) .$$

Hence, (d) holds.

(3.10) Proposition. In the setup of (3.1), assume that V dominates S and assume that S is reduced or (what is more general) that both V and CV are flat over a topologically-dense open subscheme of S. Then there exists a topologically-dense open subscheme S' of S such that, for every base-change map $T \to S'$, the conormal scheme $C(V_T)$ (that is, $C(V_T/I_T)$) is defined and satisfies the relation (of commuting with base-change),

$$C(V_T) = (CV)_T .$$

Proof. It suffices to find a topologically-dense open subscheme S' of S such that, for every map $T \to S'$,
(i) $(V^0)_T$ is scheme-theoretically dense in V_T, and
(ii) $(CV^0)_T$ is scheme-theoretically dense in $(CV)_T$.
Indeed, (i) implies by (3.9, ii, a,b) that $C((V^0)_T)$ and $C(V_T)$ are both defined and that the former is scheme-theoretically dense in the latter. Hence (ii) and (3.9.1) imply the assertion. Finally, such an S' exists by virtue of (3.11,ii) below.

(3.11) Lemma. In the setup of (3.1), let X be an S-scheme and X^o an open subscheme.
(i) Consider the subset S" of S of all points s such that the fiber X_s is geometrically reduced and its open subscheme X^0_s is dense. Then S" is constructible.
(ii) Assume that X^0 is smooth over S and scheme-theoretically dense in X, that X dominates S, and that S is reduced or X is flat over a topologically-dense open subscheme of S. Then there exists a topologically-dense open subscheme S' of S such that, for every base-change map $T \to S'$, the fiber X^0_T is scheme-theoretically dense in the fiber X_T.

Proof. (i) The set of s such that X_s is geometrically reduced is constructible by [4], (9.9.5, iv), p. 94. The set of s such that X_s^0 is topologically dense in X_s is constructible by [4], (9.5.3), p. 67.

(ii) Replacing S by a suitable open subscheme if necessary, we may assume that S is free of embedded components and that X is flat over all of S. Indeed, if S is reduced, then X is flat over a topologically-dense open subscheme of S by the theorem of generic flatness [4], (6.9.3), p. 154.

Let s be a generic point of a component of S. Let x be an associated point of the fiber X_s. Then x is also an associated point of X because X/S is flat. So x lies in X^0 because X^0 is open and scheme-theoretically dense in X. Hence x lies in X_s^0, which is a smooth open subset of X_s. It follows that X_s^0 is dense and that X_s is geometrically reduced; that is, s lies in the subset S" of (i). Since S" is constructible, by [4], (9.2.3), p. 58, therefore it contains a dense open subset S'.

Consider a map $T \to S'$. Let x be an associated point of X_T. Let t denote the image of x in T. Then x is also an associated point of X_t because X_T/T is flat. So x lies in X_t^0 because t lies over a point of S". Therefore, X_T^0 is scheme-theoretically dense in X_T, as asserted.

4. DUALITY AND REFLEXIVITY

(4.1) Setup. Let S be a locally noetherian base scheme, E a locally free sheaf of constant rank N+1. Let $E^\vee$ denote the dual sheaf, and set

$$Y = \mathbb{P}(E) \quad \text{and} \quad Y^\vee = \mathbb{P}(E^\vee) \ .$$

Let p and $p^\vee$ denote the projections of $Y \times Y^\vee$ onto Y and $Y^\vee$. Finally, let I denote the divisor of zeros of the composition of canonical maps,

$$p^{\vee *}O_{Y^\vee}(-1) \to E_{Y\times Y^\vee} \to p^*O_Y(1).$$

Then (as is well known and easily proved), I is the graph of the point-hyperplane incidence correspondence.

(4.2) Proposition. (i) There is a canonical isomorphism from I/Y (resp. $I/Y^\vee$) to the projectivization of the cotangent bundle of Y (resp. of $Y^\vee$). This isomorphism carries the invertible sheaf

$$(p^*O(1) \otimes p^{\vee *}O(1))|I$$

onto the tautological sheaf O(1).

(ii) Let ω and $\omega^\vee$ denote the contact forms of I/Y and $I/Y^\vee$ (see (3.1)). Then

$$\omega + \omega^\vee = 0 \ .$$

Proof. (i) Both assertions are immediate consequences of Lemma (3.4), applied to the dual of the standard exact sequence,

$$0 \to \Omega^1_Y(1) \to E_Y \to O_Y(1) \to 0 \ ,$$

(resp. to the corresponding sequence on $Y^\vee$).

(ii) The assertion follows immediately from the definition (3.1) of ω and $\omega^\vee$, from (i), and from the exactness (more precisely, the central nullity) of the conormal sheaf-cotangent sheaf exact sequence,

$$0 \to (p^*O(-1) \otimes p^{\vee *}O(1))|I \to (p^*\Omega^1_Y \otimes p^{\vee *}\Omega^1_{Y^\vee})|I \to \Omega^1_I \to 0 .$$

(4.3) <u>Setup continued and some observations</u>. Use the notation and hypotheses of both (4.1) and (3.1). The two setups are compatible by (4.2, i).

Consider a point-hyperplane pair (P,H) and say that H is defined by an equation $F = 0$. Obviously, the isomorphism of (4.2, i) carries (P,H) onto the cotangent vector (dF)(P). Moreover, it identifies the conormal scheme CV with the closure in $V\times_Y I$ of the locus of (P,H) such that P is in V^0 and

$$(\partial f)(dF)(P) = 0 ,$$

that is, such that P is in V^0 and the tangent space of H at fP contains the image of the tangent space to V at P.

Assume that there exists a commutative diagram

$$\begin{array}{ccc} CV & \xrightarrow{g} & I \\ \downarrow q^\vee & & \downarrow p^\vee \\ V^* & \xrightarrow{f^\vee} & Y^\vee \end{array}$$

such that (i) $f^\vee$ is an immersion on a dense open subscheme of V^* that is contained in the smooth locus of V^*/S and (ii) the following map is a closed embedding:

$$(q^\vee, g) : CV \to V^*\times_{Y^\vee} I .$$

Note that (i) implies that the conormal scheme CV^* is defined in $V^* x_{Y^\vee} I$.

(4.4) <u>Theorem</u> (A generalized Segre-Wallace theorem). In the setup of (4.3), the following two conditions are equivalent:
(i) The embedding $(q^\vee, g)$ induces an isomorphism,

$$CV \tilde{\to} CV^* .$$

(ii) The map $q^\vee : CV \to V^*$ is smooth on a dense open subscheme of CV, whose image is a dense open subscheme of V^*.

<u>Proof</u>. The scheme CV is Lagrangian for ω by (3.2). Hence, it is Lagrangian for $\omega^\vee$ by (4.2, ii). Therefore, the assertion holds by (3.5).

(4.5) <u>Definition</u>. In the setup of (4.3), suppose that $f : V \to Y$ is a closed embedding and that $f^\vee : V^* \to Y$ is the scheme-theoretic image of $p^\vee g : CV \to Y$. Then V^* is called the <u>dual</u>, or <u>reciprocal</u>, of V. If in addition

$$CV = CV^* ,$$

then call V <u>reflexive</u>.

(4.6) <u>Corollary</u>. In the setup of (4.1), suppose that S is reduced and defined over a field of characteristic 0. Then every reduced closed subscheme V of Y is <u>reflexive</u>.

<u>Proof</u>. The assertion is an immediate consequence of (4.4), because of Sard's lemma (see the introduction).

(4.7) <u>Corollary</u> (The second derivative test). Suppose that S is the spectrum of a field and that V is a reduced, irreducible plane curve, not a line, which is given in affine coordinates by an equation $P(x,y) = 0$ such that the partial derivative P_y is nonzero. Then V is reflexive if and only if the second derivative $y''(x)$ is not identically 0.

Proof. Let K and K* denote the function fields of V and V* respectively. By (4.4), V is reflexive if and only if K/K* is separable. Now, x is a separating transcendental for K because $P_y \neq 0$. Moreover, K* is obviously generated by $y'(x)$ and $b = y - y(x).x$. Hence K/K*, if finite, is inseparable if and only if $y''(x) = 0$ and $b'(x) = 0$. However, $b'(x) = y''(x).x$.

(4.8) Remark. The greater generality of (4.4) suggests the intriguing possibility of developing a theory of reflexivity for nonembedded schemes. Such a theory might run as follows. In the setup of (4.3), assume that S is reduced and universally Japanese (that is, for every reduced and irreducible S-scheme X of finite type and every extension K of the function field of X, the normalization of X in K is a finite X-scheme). Assume that V is normal and that $f : V \to Y$ is finite. Drop the hypothesis (ii) of (4.3). Let C^nV denote the normalization of CV, consider the composition,

$$h^\vee : C^nV \to CV \to I \to Y^\vee ,$$

and form its Stein factorization,

$$C^nV \xrightarrow{q^\vee} V^* \xrightarrow{f^\vee} Y^\vee .$$

Then V* is normal, and it might be called the nonembedded dual, or nonembedded reciprocal, of V. Suppose now that $h^\vee$ is separable. Then, by (3.2), (4.2, ii) and (3.3), there is a natural map,

$$C^nV \to CV^* .$$

The crucial question is this: Is this map finite and birational? If so, then

$$C^nV = C^nV^*$$

and it would be reasonable to call V (something like) <u>nonembeddedly reflexive</u>.

(4.9) <u>Proposition</u>. In the setup of (4.3), assume that V and V* are free of embedded components, that V and V* dominate S, and that S is reduced or (what is more general) all of V, V*, CV and CV* are flat over a topologically-dense open subscheme of S.

(1) There exists a topologically-dense open subscheme S' of S such that, for every map $T \to S'$, (a) the maps $q^{\vee}_T$ and $f^{\vee}_T$ satisfy the corresponding form of the conditions (i) and (ii) of (4.3), and (b) the conormal schemes $C(V_T)$ and $C(V^*_T)$ are defined and equal to $(CV)_T$ and $(CV^*)_T$ respectively.

(2) The two equivalent conditions (i) and (ii) of (4.4) are also equivalent to each of the following four conditions. In the first two of these, S' is as in (1).

(i_T) For every map $T \to S'$, the embedding $(q^{\vee}_T, g_T)$ induces an isomorphism, $CV_T \xrightarrow{\sim} CV^*_T$.

(ii_T) For every map $T \to S'$, the map $q^{\vee}_T : CV_T \to V^*_T$ is smooth on a dense open subscheme of CV_T whose image is a dense open subscheme of V^*_T.

(i_G) For the generic point G of every component of S, the embedding $(q^{\vee}_G, g^{\vee}_G)$ induces an isomorphism, $CV_G \xrightarrow{\sim} CV^*_G$.

(ii_G) For the generic point G of every component of S, the map $q^{\vee}_G : CV_G \to V^*_G$ is smooth on a dense open subscheme of CV_G whose image is a dense open subscheme of V^*_G.

(3) Assume that S is a Jacobson scheme. Then the equivalent conditions of (2) are also equivalent to each of the following two conditions. Again S' is as in (1).

(i_F) There exists a topologically-dense open subscheme S" of S' such that for every closed point F of S", the embedding $(q^{\vee}{}_F, g^{\vee}{}_F)$ induces an isomorphism, $CV_F \xrightarrow{\sim} CV^*_F$.

(ii_F) There exists a topologically-dense open subscheme S" of S' such that for every closed point F of S", the map $q_F : CV_F \to V^*_F$ is smooth on a dense open subscheme of CV_F whose image is a dense open subscheme of V^*_F.

Proof. (1) Assertion (b) is a special case of (3.10). Assertion (a) is obvious, except for the requirement that the base extension of the open subscheme of (4.3.i) still be dense, which follows from (3.11,ii).

(2), (3). That (i) implies (i_T) is obvious. The equivalence of (i_T) and (ii_T) is a special case of (4.3). Obviously, (i_F) and (i_A) are special cases of (i_T), and (ii_G) and (ii_F) are special cases of (ii_T).

Finally, note that CV and CV* are free of embedded components because V and V* are. Hence, to prove CV and CV* are equal (that is, (i) holds), we may replace S by any topologically-dense open subscheme. So we may assume that CV and CV* are flat over S. Hence, if for any s in S, the fibers $(CV)_s$ and $(CV^*)_s$ are equal, then CV and CV* are equal on a neighborhood of $(V^* x_Y I)_s$. It follows that (i_G) implies (i) and that (i_F) implies (i).

(4.10) Discussion (Reflexivity and the contact formula). Return to the setup of (2.2) and assume that V and almost all X are reflexive. Since finitely many X may be discarded without altering the enumeration so long as V remains in general position by virtue of (2.3), there is no loss in assuming that in fact all the X are reflexive. By virtue of (2.3) and of (4.9), there is no loss in assuming that the duals X* form a family, whose parameter space is the same as the parameter space of the X and whose total space is the dual of the total space of the X. Similarly, there is no loss in assuming that the conormal schemes CX (resp., CX*) form a family whose total space is the conormal scheme of the total space of the X (resp. of the X*).

It is evident from the first part of the proof of (2.3) that the number n of X touching V is equal to the number n* of X* touching V*; in fact, X touches V if and only if X* touches V*, and X and X* are counted with the same multiplicity. In particular, since obviously an i-plane of Y is reflexive and its dual is an (N-1-i)-plane of $Y^{\vee}$, the i-th characteristic λ_i of the family of X is equal to the (N-1-i)-th characteristic $\lambda^*_{(N-1-i)}$ of the family of X*, and the i-th rank r_i of V is equal to the (N-1-i)-th rank of V*. (This identity of the ranks is the content of the Piene-Urabe Theorem, see [9].) Thus, when the X and V are replaced by their duals, the contact formula

$$n = r_0\lambda_0 + \dots + r_{(N-1-i)}\lambda_{(N-1-i)}$$

is left virtually unchanged, only the order of the summands is reversed. In this sense, the contact formula is self-dual.

(4.11) Counterexample. Theorem (2.5) (and so also Theorem (3.7)) may fail in characteristic $p \geq 3$. For example, let S be the affine line and consider a smooth family V/S of plane curves V_s such that V_s is reflexive if and only if $s \neq 0$. (A specific example, given in affine coordinates, is this:

$$V_s : y^p + y + x^{p+1} + sx^{p-1} = 0 .$$

It is easy to check on the smoothness of V_s by using the Jacobian criterion, and on the reflexivity of V_s by using the second derivative test, Corollary (4.7). Note moreover in passing that, although V_0 is not reflexive, nevertheless V_0^{**} is equal to V_0.)

Since V/S is smooth, Proposition (3.9) yields

$$(CV)_s = C(V_s)$$

for all s in S. Now, consider the dual family V*/S. Proposition (4.9) yields

$$(CV^*)_s = C((V^*)_s)$$

for almost all s in S and because, by hypothesis, V_s is reflexive for $s \neq 0$, it yields

$$CV^* = CV .$$

For each s in S, the first and third of these equations show that $(CV^*)_s$ is smooth and irreducible, because V_s is so. Nevertheless, the second equation fails for s = 0; in fact,

$$(CV^*)_0 \neq CW$$

for any reduced, irreducible subscheme W of $Y^\vee$. Indeed, suppose that equality held for some W. Then, by virtue of the first and third equations, W would be the image of $C(V_0)$ in $Y^\vee$, or $V_0{}^*$. Hence, V_0 would be reflexive, contrary to hypothesis. Finally, V* is irreducible because V is so; hence, V*/S is flat, because S is a smooth curve. Thus, Theorem (2.5) fails for V*/S.

REFERENCES

[1] Altman, A. and Kleiman, S.: "Foundations of the Theory of Fano Schemes", Compositio Math., 34(1) (1977), 3-47.

[2] Fulton, W., Kleiman, S. and MacPherson, R.: "About the enumeration of contacts", Algebraic Geometry -- Open Problems (Proceedings, Ravello 1982), Ciliberto, C., Ghione, F. and Orecchia, Lecture Notes in Math., 997. Springer-Verlag (1983), 156-196.

[3] Grayson, D.: "Coincidence formulas in enumerative geometry", Communications in Algebra 7(16) (1979), 1685-1711.

[4] Grothendieck, A. and Dieudonné, J.: Eléments de Geometrie Algébrique, Publ. Math. IHES N°. 24, N°. 28. Bures-sur-Yvette (S. et O.), (1965, 1966).

[5] Hefez, A. and Kleiman, S.: "Notes on duality for projective varieties", in preparation.

[6] Hilbert, D.: "Mathematical Problems", translated for the Bull. Amer. Math. Soc., with the author's permission, by M.W. Newson, Bull. Amer. Math. Soc.; v. 8 (1902), 437-479. = Proceedings of Symposia in Pure Math.; v. 28. Browder, F., ed., Amer. Math. Soc. (1976), QA1.S897, pp. 1-34.

[7] Hironaka, H.: "Stratifications and flatness", Real and complex singularities, Oslo 1976, Holm, P., ed., Sijihoff & Noordhoff (1977), 199-265.

[8] Kleiman, S.: "The transversality of a generic translate", Compositio Math. 28 (1974), 287-297.

[9] Kleiman, S.: "Concerning the dual variety", 18th Scandinavian Congress of Mathematicians; proceedings, 1980, Balslev, E., ed., Progress in Math., 11. Birkhäuser Boston (1981), 386-396.

[10] Merle, M.: "Variétés polaires, stratifications de Whitney et classes de Chern des espaces analytiques complexes [d'après Lê-Teissier], Sém. Bourbaki, Nov. 1982, exp. 600.

[11] Oda, T.: "Introduction to Algebraic Analysis on Complex Manifolds", Algebraic Varieties and Analytic Varieties, Proc. Symposium in Tokyo, 13-24 July 1981, Iitaka, S., ed., North Holland (1982), pp. 29-48.

[121 Pham, F.: Singularités de Systèmes Différentiels de Gauss-Manin, Progress in Math., 2, Birkhäuser Boston (1979).

[13] Sabbah, C.: Quelques Remarques sur la Géometrie des Espaces Conormaux, Prepublication Ecole Polytechnique, Palaiseau 91128, France (Fall 1983).

[14] Schubert, H.: Kalkül der Abzählenden Geometrie, Teubner, Leipzig (1879), reprinted with an introduction by S. Kleiman and a list of publications prepared by W. Burau, Springer-Verlag (1979).

[15] Segre, C.: "Preliminari de una teoria delle varietà luoghi di spazi", Rendiconti Circolo Mat. Palermo XXX (1910), 87-121 = Opere vol. II, Cremonese, Roma (1958), 71-114.

[16] Wallace, A.: "Tangency and duality over arbitrary fields", Proc. Lond. Math. Soc. (3) 6 (1956), 321-342.

ON THE UNIQUENESS OF CERTAIN LINEAR SERIES ON SOME CLASSES OF CURVES

Ciro Ciliberto

Istituto di Matematica

"R. Caccioppoli"

Università di Napoli (Italia)

Robert Lazarsfeld

Department of Mathematics UCLA

Los Angeles (California)

U.S.A.

Introduction.

This note is a report of work in progress about some problems of uniqueness for certain linear series on some curves in a projective space. The problem we started with was to show that on any smooth,non degenerate, complete intersection curve in $\mathbb{P}^3$, of degree $n > 4$ the linear series cut out by the planes is the unique simple g_n^3. Once we proved this (for the proof see § 2), we got aware that the same method of proof could be applied to obtain other uniqueness statements. The results we have achieved in this direction are exposed in § 3, and concern, for instance, projectively normal curves in $\mathbb{P}^3$, some determinantal curves in $\mathbb{P}^r$, etc. Our feeling at the present state of affairs, is that similar results should hold for several and large classes of curves in a projective space.

The key ingredient in many of the proofs of this paper turns out to be a classical method of Castelnuovo, which we briefly recall in § 1. In fact this method, and some applications of it to uniqueness questions for curves of high genus in a projective space, was the main topic of the talk given by the first named author at the CIME Conference (Acireale, June 1983). At that time most of the results exposed here were only conjectured; their proofs were achieved also thanks to the opportunity, which the Conference gave the authors, of meeting and discussing.

Notation.

We work on an algebraically closed field k of characteristic zero. If X is a k-scheme, we denote by O_X its structure sheaf. If D is any Cartier divisor on X, we denote by $O_X(D)$ the sheaf of sections of the corresponding line bundle, with $H^i(x, O_X(D))$, $h^i(X,O_X(D))$ its cohomology spaces and their dimensions over k. If X is integral and projective, $|D|$ will denote the complete linear system determined by D on X. If X is smooth, K_X will be any canonical divisor on X. If $X \subset \mathbb{P}^r$ is a variety, X will be said to be <u>non degenerate</u> if no hyperplane of $\mathbb{P}^r$ contains X. If $X \subset \mathbb{P}^r$ is smooth, irreducible, it is said to be projectively normal if for any $t \in \mathbb{N}$, the linear system of hypersurfaces of degree t cuts out on X a complete linear system.

1. Preliminaries: Castelnuovo's lemma.

Let C be a smooth irreducible, projective curve and let E be a divisor of degree $n > 0$ on C. Any $(r + 1)$-dimensional vector subspace V of $H^0(C,O_C(E))$ corresponds to a linear series g^r_n on C, contained in the complete series $|E|$. If D is an effective divisor of degree d on

C, we set $V(-D) = \{s \in V : (s) \geq D\}$, (s) denoting the zeroes divisor of the section $s \in V$. $V(-D)$ corresponds to a linear series having D as a fixed divisor: the linear series we get from this removing D from its divisors will be denoted by $g_n^r(-D)$.

Let D be a divisor and g_n^r a linear series on C; the non negative integer

$$c(D, g_n^r) = r - \dim g_n^r(-D)$$

is the so-called _number of conditions which_ D _imposes to the linear series_ g_n^r. Clearly it is

$$c(D, g_n^r) \leq \min \{d, r+1\} \tag{1.1}$$

d being the degree of D. If g_m^s is another linear series on C, we say that g_m^s is _contained_ in g_n^r, writing $g_m^s \leq g_n^r$, if there is an effective divisor D', of degree n - m, such that $g_n^r(-D')$ contains g_m^s as a linear subseries. If $g_m^s \leq g_n^r$ and D and D' have no point in common, then

$$c(D, g_m^s) \leq c(D, g_n^r)$$

If $d \leq r + 1$ and in (1.1) the equality holds, D is said to _impose independent conditions to_ g_n^r. We shall assume, from now on, D formed by d distinct points. Given any integer $t \geq 0$, D is said to be in t-_uniform_ _position_ with respect to the g_n^r if any divisor of degree t contained in D imposes independent conditions to g_n^r. Then it is

$t \leq c(D, g^r_n)$. If $t = c(D, g^r_n)$, D is said to be in <u>uniform position</u> with respect to g^r_n.

Before stating Castelnuovo's lemma, we recall the definition of minimal sum of some linear series $g^{r_i}_{n_i}$, $i = 1, \ldots, k$, on C. Let $V^{(i)}$ be the vector sub-space of $H^o(C, O_c(E_i))$ corresponding to $g^{r_i}_{n_i}$, $i = 1, \ldots, k$, and consider the linear map

$$f : \bigotimes_{i=1}^{k} V^{(i)} \to H^o(C, O_c(\sum_1^K {}_i E_i))$$

such that

$$f(\bigotimes_{i=1}^{k} s^{(i)}) = \prod_{i=1}^{k} s^{(i)}$$

The linear series corresponding to Im f is the so-called <u>minimal sum</u> of the given series $g^{r_i}_{n_i}$, and will be denoted by $\bigoplus_{i=1}^{k} g^{r_i}_{n_i}$. If $g^{r_i}_{n_i} = g^r_n$ for any $i = 1, \ldots, k$, we simply set $k\, g^r_n = \bigoplus_{i=1}^{k} g^{r_i}_{n_i}$. The basic fact about minimal sums is the following:

(1.2) <u>Lemma</u> (Castelnuovo, cfr. [C]). <u>Let</u> $g^{r_i}_{n_i}$, $i = 1, \ldots, k$ be <u>linear series on</u> C <u>and</u> D <u>an effective divisor of degree</u> d <u>on</u> C <u>formed by distinct points</u>. <u>If</u> D <u>is in</u> t_i-<u>uniform position with respect to</u> $g^{r_i}_{n_i}$, $i = 1, \ldots, k$, <u>then</u>:

(i) <u>if</u> $t = \sum_1^k {}_i t_i - k + 1 < d$, D <u>is in</u> t-<u>uniform position with respect to</u> $\bigotimes_{i=1}^{k} g^{r_i}_{n_i}$;

(ii) *if* $t \geq d$, D *imposes independent conditions to* $\bigoplus_{i=1}^{k} g_{n_i}^{r_i}$.

We omit the proof, which can be found, for example, in [CI]. The problem, in order to apply the above lemma, is to verify, for a given divisor D, the conditions of uniform position. For this reason it is useful to have a few "uniformity criteria"; the following two will be enough for our purposes.

(1.3) *Proposition* (Bertini, cfr. [B]; Harris, cfr [H1]). *Let* g_n^r *be a linear series, without base points, not composed of an involution on* C, *and let* D *be its generic divisor. If* g_m^s *is any linear series on* C, *then* D *is in uniform position with respect to* g_m^s. *In particular* D *is in r-uniform position with respect to* g_n^r.

(1.4) *Proposition* (Accola, cfr [A]). *Let* g_n^r, g_m^s *be distinct linear series, without base points on* C, *not composed of the same involution, with* $s \geq r$. *Then the generic divisor of* g_n^s *is in (r + 1) -uniform position with respect to* g_n^r.

For the proofs of the above propositions we refer to the quoted references. A proof of (1.3) will also be found in [CI], where Castelnuovo's lemma is applied to study linear series on curves of the following types:

(i) curves in a projective space with high genus with respect to the degree;

(ii) in particular, smooth plane curves;

(iii) subcanonical curves in a projective space, namely curves whose canonical divisors are linearly equivalent to a multiple of a hyperplane section; in particular complete intersection curves in a projective space.

In what follows we shall focus our attention on curves of the third kind above and, taking the point of view of [CI], we shall prove a number of uniqueness theorems for some linear series on sub-canonical curves and, in particular, for complete intersections in $\mathbb{P}^3$. Later on we shall show how to extend these results to more general classes of curves.

2. Subcanonical curves.

Let $\Gamma \subset \mathbb{P}^r$, $r \geq 2$, be an irreducible, non degenerate, complete curve of degree n and let

$$p : C \to \Gamma \subset \mathbb{P}^r$$

be its normalization. The morphism p of C to $\mathbb{P}^r$ corresponds to a linear series g^r_n on C, pull-back, via p, of the linear series cut out on Γ by the hyperplanes in $\mathbb{P}^r$. We denote by H the generic divisor of this g^r_n. For each integer k, the linear series kg^r_n is pull-back, via p, of the linear series cut out Γ by the hypersurfaces of degree k in $\mathbb{P}^r$.

Let us put now

$$l(\Gamma) = \max \{t \in \mathbb{Z} : h^0(C, O_C(k_C - tH)) \neq 0\}$$

We call $l(\Gamma)$ the <u>level</u> of Γ. If K_C is linearly equivalent to $l(\Gamma)H$, Γ is said to be subcanonical of level $l(\Gamma)$.

(2.1) <u>Proposition</u>. <u>If</u> $l = l(\Gamma) \geq 1$, $n \leq r(l + 1)$ <u>and</u> g^s_m <u>is a linear series on</u> C <u>with</u> $m \leq n$, $s \geq r$, <u>then</u> $g^s_m = g^r_n$.

Proof. Of course we can reduce to the case g^s_m base points free. Let D be the generic divisor in g^s_m. If $g^r_n \neq g^s_m$, by proposition (1.4) and Castelnuovo's lemma, we would get

$$c(D, |K_C|) \geq c(D, l\, g^r_n) \geq \min \{m, l\, r + 1\}$$

By Riemann-Roch theorem, it is

$$c(D, |K_C|) = m - \dim |D| \leq m - s < m$$

and therefore we should have

$$c(D, |K_C|) \geq l\, r + 1$$

Since

$$c(D, |K_C|) \leq m - s \leq n - r$$

we should also get

$$n - r \geq c(D, K_C) \geq l\, r + 1$$

contradicting the hypotheses.

With the further assumption that Γ is projectively normal and subcanonical, and some more arithmetical conditions on n and $l(\Gamma)$ one has more information.

(2.2) Theorem. Let Γ be a subcanonical, projectively normal curve of level $l \geq 2$, with $r(l + 1) < n \leq r(l + 1) + l - 1$. If g^s_m is a linear series on C, not composed of an involution, with $m \leq n$, $r \leq s$, then $g^s_m = g^r_n$.

Proof. We again can assume g^s_m base points free. If $g^r_n \neq g^s_m$, the generic divisor D in g^s_m should be in (r + 1)-uniform position with respect to g^r_n, so that, by proposition (1.3)

$$c(D, 2g^r_n) \geq \min\{m, 2r+1\} = 2r+1 \tag{2.3}$$

the last equality holding, because

$$m > m - s \geq c(D, |K_c|) \geq c(D, 2g^r_n)$$

Moreover, D is in uniform position with respect to $2g^r_n$. Let $l = 2k$. Applying Castelnuovo's lemma we then get

$$n - r \geq m - s \geq c(D, |K_c|) \geq c(D, k(2g^r_n)) \geq$$

$$\geq k(c(D, 2g^r_n) - 1) + 1$$

whence, by the hypotheses

$$c(D, 2g^r_n) \leq 2r + 3 - \frac{2}{l}$$

Similarly, if l is odd, we have

$$c(D, 2g^r_n) \leq 2r + 3 - \frac{2}{l-1}$$

so that, in any case, it is

$$c(D, 2g^r_n) \leq 2r + 2 \tag{2.4}$$

Assume now that Γ has not maximal genus in $\mathbb{P}^r$. If in (2.3) the equality held, since

$$n > r(l+1) \geq 3r \geq 2r + 2$$

the points of D would lie on a rational normal curve $\Gamma_0 \subset \mathbb{P}^r$ (see [H2]).

On the other hand, the ideal of Γ in $\mathbb{P}^r$ is generated by forms of degree $\leq l + 1$ (see [AS], thms (4.3), (4.7)). Since $n > r(l + 1)$, any such a form would contain Γ_0 as well as Γ, which is impossible. Let us discuss the case in which the equality holds in (2.4). If this happened the points of D would lie either on a rational normal curve, or on an irreducible curve $\Gamma_1 \subset \mathbb{P}^r$ of degree $r + 1$ and arithmetic genus 1 (see [H3]). It is also easy to see that no point of D would be singular on Γ_1. Then it would be

$$c(D,|K_C|) = c(D, l\, g^r_n) = \begin{cases} n, \text{ if } n < l(r + 1) \\ \text{either } n \text{ or } n - 1 \text{ if } n = l(r + 1) \\ l(r + 1), \text{ if } n > l(r + 1) \end{cases}$$

But this leads to contradictions. In the first two cases, for instance, we would get

$$m - s \geq c(D,|K_C|) \geq n - 1 \geq m - 1$$

whence $r \leq s \leq 1$. In the last case it would be

$$n - r \geq m - s \geq c(D,|K_C|) \geq l(r + 1)$$

whence $n \geq r(l + 1) + l$. Finally, if Γ has maximal genus the theorem follows by theorem (2.11) of [CI] or by the results of Accola (see [A]).

The hypothesis of projective normality on Γ in (2.2) is too strong (see [AS], remark (4.6)). Moreover also the hypothesis Γ subcanonical is too strong. It could be easily replaced by the hypothesis that the ideal of Γ is generated by forms of degree $\leq l + 1$. Theorem (2.2) and proposition (2.1), already proved in [CI], readily apply to smooth complete intersection curves. One has the:

(2.5) Corollary. *Let* $C \subset \mathbb{P}^3$ *be a smooth complete intersection of two surfaces of degrees* h, k, *with* $h \leq k$. If $h = 2$, $k > 2$ *and if* $h = 3, 4$, *the linear series cut out on* C *by planes of* $\mathbb{P}^3$ *is the unique simple* g^3_{hk} *on* C. *A similar results holds for any smooth complete intersection of two quadrics and a hypersurface of degree* $h \geq 2$ in $\mathbb{P}^4$.

The above corollary, which has been independently proved by P. Maroscia (see [M]), suggests the problem of seeing if an analogous results holds, more generally, for any complete intersection curve of positive level. To this question we can give an affirmative answer,at least for curves in $\mathbb{P}^3$.

(2.6) Theorem. *Let* $C \subset \mathbb{P}^3$ *be a smooth, complete intersection of two surfaces of degrees* h,k *with* $4 < h \leq k$. *If* g^s_m *is a linear series on* C, *without base points, not composed of a pencil, with* $s \geq 2$, $m \leq hk$ *then* $g^s_m \leq |H|$.

Proof. Let D be the generic divisor in g^s_m, and let x_0 be the least integer for which the following property holds: there exists a surface G of degree x_0 in $\mathbb{P}^3$ such that

(i) $G \not\supset C$;

(ii) G cuts out on C a divisor containing D.

We denote by D + D' the divisor cut out by G on C , and consider the linear system Σ formed by all surfaces of degree x_0 which either contain C or cut out on C a divisor containing D'. It is not difficult, using the minimality of x_0 and the projective normality of C, to see that

Σ has no fixed components. Thus, if F', F" are generic elements in Σ, the 1-dimensional subscheme Γ of $\mathbb{P}^3$, of degree x_0^2, complete intersection of F', F" can be considered. Let F now be the generic surface of degree k containing C. Since the ideal of C is generated in degree at most k, F can be chosen to be smooth, and to contain no component of Γ, so that the cycle of intersection $F \cdot \Gamma$ is well defined, its degree being $x_0^2 k$. Now D' has degree at least $(x_0 - 1)h\,k$, and simple arguments of local algebra show that

$$x_0^2 k - (x_0 - 1)h\,k = k\,(x_0^2 - hx_0 + h) \geq 0$$

Since $h > 4$, this can only happen if either $x_0 = 1$, in which case the theorem is proved, or $x_0 > h - 2$. We shall now prove that only the first case can occur. Let us put

$$h + k - 4 = a(h - 2) + b\ , \quad 0 \leq b < h - 2$$

Since C is subcanonical of level $h + k - 4$, applying proposition (1.3) and Castelnuovo's lemma, one gets

$$h\,k - 2 \geq m - s \geq c(D, |K_C|) = c(D, a((h - 2)H) + b\,H)$$
$$\geq a(c(D, |(h - 2)H|) - 1) + 3b + 1$$

Hence, by easy computations, we have

$$c(D, |(h - 2)H|) \leq h(h - 2)$$

and, since

$$h(h - 2) \leq \binom{h + 1}{3} - 1 = \dim\ |(h - 2)H|$$

it is $x_0 \leq h - 2$ (compare [R]).

By virtue of theorem (2.6), corollary (2.5) can be extended

to any value of $h > 4$. Anyhow the disappointing feature in theorem (2.6) is the hypothesis of simplicity on g^s_m. It is likely that it may be removed, but it seems that further assumptions are necessary. For example, we are able to prove the following:

(2.7) Proposition. The conclusions of theorem (2.6) still hold without assuming g^s_m simple, as soon as $h \geq 12$.

The idea of the proof is as follows. The linear series $g^s_m \oplus |H|$ is certainly simple. Thus everything amounts to prove that if $h \geq 12$ and $g^{s'}_{m'}$ is a simple linear series, without base points on C, with $s' \geq 2$, $m' \leq 2\,hk$, then $g^{s'}_{m'} \leq |2H|$. This can be done by the same reasoning of the proof of theorem (2.7).

3. Extension of the above results. Final remarks.

Once theorem (2.6) has been proved, a natural problem is to look for other, or larger, classes of curves in $\mathbb{P}^3$, or preferably in $\mathbb{P}^r$, $r \geq 2$, for which an analogous result holds. A first extension can be made to smooth projectively normal curves in $\mathbb{P}^3$. Let C be such a curve. Then it is well known that the homogeneous ideal of C can be minimally generated by the minors of maximal order of a homogeneous matrix of forms of the type $u \times (u+1)$, which we write in the form

$$A = \begin{pmatrix} f_{11} & \cdots\cdots & f_{1,u+1} \\ & & \\ f_{u1} & \cdots\cdots & f_{u,u+1} \end{pmatrix}$$

We put $m_{ij} = \deg f_{ij}$. It is known that A can be taken such that

$$m_{u,1} \leq m_{u-1,1} \leq \dots \leq m_{1,1} \leq \dots \leq m_{1,u+1}$$

In this case $u = 1$ if and only if C is a complete intersection: therefore we shall assume $u \geq 2$. We have the:

(3.1) Theorem. *Let $m_{11} \geq 9$ and let n be the degree of C. If g_m^s is a linear series on C, without base points, not composed of a pencil, with $s \geq 2$, $m \leq n$, then $g_m^s \leq |H|$.*

Proof. We put $h = m_{1,1}$, and call k the highest degree of a minor of A. Simple computations show that

(3.2) $$n > hk$$

Moreover it is

(3.3) $$l(\mathbb{C}) \geq h + k - 4$$

(see [G], pg 36). Now the proof goes like that of theorem (2.6), the role of F being played by the generic surface of degree k through C. One has

$$x_0^2 - \frac{n}{k} x_0 + \frac{n}{k} \geq 0$$

By (3.2) and by the hypotheses, it is $n > 4k$, implying that either $x_0 = 1$, or $x_0 > \frac{n}{k} - 2$. But the second case cannot happen: in fact, we set $q = [\frac{n}{k}]$ and show that $x_0 \leq q - 2$. If

$$h + k - 4 = a(q - 2) + b, \quad 0 \leq b < q - 2$$

one can easily see that $a \geq 1$. Then applying Castelnuovo's lemma and (3.3) we get

$$n - 2 \geq a(c(D, |(q - 2)H|) - 1) + 3b + 1$$

whence

$$c(D, |(q-2)H|) \leq 2(q-1)(q-2)$$

Now it is not difficult to check that

$$\dim |(q-2)H| = \binom{q+1}{3} - 1$$

Thus $c(D, |(q-2)H)|) \leq \dim |(q-2)H|$ if $q \geq 9$, and this is certainly the case by virtue of (3.2), since $h \geq 9$. So the theorem is proved.

The hypothesis $m_{1,1} \geq 9$ in (3.1) is probably too strong. It is likely that the theorem can be substantially improved. By contrast the statement is not true for any $m_{1,1}$: there are in fact projectively normal, but not special, curves. But even for projectively normal, special curves the uniqueness can fail: a counterexample is the projection in $\mathbb{P}^3$ of a canonical curve of degree 8 in $\mathbb{P}^4$ from a generic point on it.

It seems quite difficult to extend rather precise statements like theorems (2.6) and (3.1), to curves in $\mathbb{P}^r$, $r > 3$. Anyhow, if one seeks for sort of asymptotic results, the same strategy of the proofs of these theorems appears to be very fruitful. Moreover the use of Castelnuovo's lemma can be avoided, so that the hypothesis of simplicity can be removed. As an example we state here a rather general asymptotic uniqueness theorem for determinantal curves, which can be proved in the same vein as theorems (2.6) and (3.1), but with some more technical devices:

(3.4) Theorem. Let E, F be vector bundles of ranks n, $n + r - 2$ respectively on $\mathbb{P}^r$, $r \geq 2$. Then there exists an integer $c(E,F)$, such that for any $a \geq c(E,F)$ and for any homomorphism $u : E \to F(a)$ which *drops rank on a smooth irreducible curve* C *of degree* n, *the following happens*:

if g^s_m *is a linear series on* C *without base points, with* $m \leq n$, *then* $g^s_m \leq |H|$, H *being a hyperplane section of* C. *In particular* C *has no* g^r_m *for* $m < n$ *and* $|H|$ *is the only* g^r_n *on* C.

REFERENCES

[A] R.D.M. Accola, on Castelnuovo's inequality for algebraic curves, I, Trans. Ann. Math. Soc., 251 (1979), 357-373.

[AS] E. Arbarello, E. Sernesi, Petri's approach to the study of the ideal associated to a special divisor, Inventiones Math. 49 (1978), 99-119.

[B] E. Bertini, Intorno ad alcuni problemi della geometria sopra una curva algebrica, Atti Accad. Sci. Torino, 26 (1890).

[C] G. Castelnuovo, Sui multipli di una serie lineare di gruppi di punti appartenenti ad una curva algebrica, Memorie Scelte, Zanichelli, (Bologna), 1937.

[CI] C. Ciliberto, Alcune applicazioni di un classico procedimento di Castelnuovo, Pubbl. Ist. Mat. "R. Caccioppoli" Univ. Napoli, 39 (1983) (to appear in Seminario di variabili complesse 1982, Università di Bologna).

[G] F. Gaeta, Nuove ricerche sulle curve sghembe algebriche di resi duale finito e sui gruppi di punti del piano, Ann. Mat. Pura e Appl., 31 (1950), 1 - 64.

[H1] J. Harris, The genus of spaces curves, Math. Ann., 249 (1980) 191-204.

[H2] J. Harris, A bound on the geometric genus of projective varieties, Ann. S.N.S. Pisa, serie IV, 8 (1981), 35-68.

[H3] J. Harris, Curves in projective space, Les presses de L'Université de Montréal (Montréal), 1982.

[M] P. Maroscia, Some problems and results on finite sets of points in $\mathbb{P}^n$, Proceedings of Ravello Conference ,Springer L.N. in M., vol. 997 (1983), 290-314.

[R] Z. Ran, On projective varieties of codimension 2, Inventiones Math. 73 (1983), 333-336.

ON THE LOCAL COHOMOLOGY MODULES $H^i_{\mathfrak{a}}(R)$ FOR IDEALS $\mathfrak{a}$ GENERATED BY MONOMIALS IN AN R-SEQUENCE

by Gennady Lyubeznik

Let R be a local commutative ring containing a field k and $\mathfrak{a}$ an ideal in R generated by monomials in an R-sequence. In this paper we describe the local cohomology modules $H^i_{\mathfrak{a}}(R)$, determine when they are zero and compute the cohomological dimension of R with respect to $\mathfrak{a}$ (denoted $cd(R, \mathfrak{a})$). It is usually difficult to compute $cd(R, \mathfrak{a})$ for a given $\mathfrak{a}$. For monomial ideals the method used until now was to represent $\mathfrak{a}$ as the intersection of two simpler monomial ideals $\mathfrak{a}_1$ and $\mathfrak{a}_2$ and then apply the Mayer-Vietoris sequence (cf., for example [3], [9]). An ingenious choice of $\mathfrak{a}_1$ and $\mathfrak{a}_2$ can indeed solve the problem for some $\mathfrak{a}$. In this paper we give a straightforward way to compute $cd(R, \mathfrak{a})$, which works for ALL monomial ideals independently of anyone's ingenuity (Section 1).

In Section 2 we discuss applications to computing the arithmetical rank of $\mathfrak{a}$ (denoted $ara(\mathfrak{a})$), i.e. the minimum number of elements generating $\mathfrak{a}$ up to radical. In particular, we answer the question from [10], p.250, by proving that Reisner's subvariety of P^5_k is not a set-theoretic complete intersection if char k = 2.

1

For basic definitions and results on local cohomology modules $H^i_{\mathfrak{a}}(R)$ and the integer $cd(R, \mathfrak{a})$ the reader is referred to [4], [5]. Let $x_1, \ldots, x_n$ be an R-sequence of elements of R. Denote by S the subring of R generated by k and $x_1, \ldots, x_n$. It is isomorphic to the polynomial ring in n variables over k. Put $S' = S_{(x_1, \ldots, x_n)}$. We will frequently use the fact that R is flat over S (cf. [7], p. 150). It also follows from [7], p. 28, Theorem 3, that R is faithfully flat over S'.

Let $y_i = x_{j_1} \ldots x_{j_{t_i}}$ $(1 \leq i \leq m)$ be m square-free monomials in the x_i's. For any positive integer q and $X = S$, S' or R denote by $\mathfrak{a}^{(q)}_X$ the ideal generated by $y_1^q, \ldots, y_m^q$ in X. We occasionally denote these ideals by $\mathfrak{a}^{(q)}$ if the ring in question is clear from the context and we also write $\mathfrak{a}$ for $\mathfrak{a}^{(1)}$. Then for any v

$$H^v_{\mathfrak{a}_R}(R) \cong \lim_{q\to\infty} \mathrm{Ext}^v_R(R/\mathfrak{a}^{(q)}_R, R) \qquad (1)$$

Recall that $\mathrm{cd}(R, \mathfrak{a})$ equals the biggest integer v such that $H^v_{\mathfrak{a}}(R) \not\cong 0$.

Theorem 1. Let X and $\mathfrak{a}^{(q)}_X$ be as above. Then

(i) The natural maps $\varphi^v_X(q)\colon \mathrm{Ext}^v_X(X/\mathfrak{a}^{(q)}_X, X) \longrightarrow \mathrm{Ext}^v_X(X/\mathfrak{a}^{(q+1)}_X, X)$ are injective for all q, v and X.

(ii) For any q and v $\mathrm{Ext}^v_R(R/\mathfrak{a}^{(q)}_R, R) \cong 0$ if and only if $\mathrm{Ext}^v_{S'}(S'/\mathfrak{a}_{S'}, S') \cong 0$.

(iii) $H^v_{\mathfrak{a}}(R) \cong 0$ if and only if $\mathrm{Ext}^v_{S'}(S'/\mathfrak{a}_{S'}, S') \cong 0$

(iv) $\mathrm{cd}(R,\mathfrak{a}) = \mathrm{cd}(S',\mathfrak{a}) = \mathrm{proj.dim.}_{S'}(S'/\mathfrak{a}) = n - \mathrm{depth}_{S'}(S'/\mathfrak{a})$

Proof: (i) Let F be the free X-module of rank m. We denote its free generatores by $Z_1, \ldots, Z_m$. Consider the well-known Taylor resolution of $X/\mathfrak{a}^{(q)}_X$ which we denote by T^q_X and which remains valid for any monomial ideal, not necessarily generated by square-free monomials (cf. [11], see [2] for a simpler proof):

$$0 \xrightarrow{d} \Lambda^m F \xrightarrow{d} \Lambda^{m-1} F \xrightarrow{d} \cdots\cdots \xrightarrow{d} \Lambda^1 F \xrightarrow{d} \Lambda^0 F \longrightarrow X/\mathfrak{a}^{(q)}_X$$

where $\Lambda^i F$ is the i-th exterior power of F and the differentials are given by

$$d(Z_{i_1}\wedge \ldots \wedge Z_{i_v}) = \sum_j (-1)^j \frac{\mathrm{lcm}(y^q_{i_1}, \ldots, y^q_{i_v})}{\mathrm{lcm}(y^q_{i_1}, \ldots, \widehat{y^q_{i_j}}, \ldots, y^q_{i_v})} \; Z_{i_1}\wedge \ldots \wedge \widehat{Z_{i_j}} \wedge \ldots \wedge Z_{i_v}$$

Here lcm denotes least common multiple and ^ means that the corresponding element is omitted. We note that

$$\frac{\mathrm{lcm}(y^q_{i_1}, \ldots, y^q_{i_v})}{\mathrm{lcm}(y^q_{i_1}, \ldots, \widehat{y^q_{i_j}}, \ldots, y^q_{i_v})} = \left(\frac{\mathrm{lcm}(y_{i_1}, \ldots, y_{i_v})}{\mathrm{lcm}(y_{i_1}, \ldots, \widehat{y}_{i_j}, \ldots, y_{i_v})}\right)^q$$

Since $\mathrm{Hom}(\Lambda^v F, X) \cong \Lambda^v F$, the modules $\mathrm{Ext}^v_X(X/\mathfrak{a}^{(q)}_X, X)$ can be computed from the following complex which we denote by K^q_X:

$$0 \longrightarrow \Lambda^0 F \xrightarrow{\sigma^{(q)}_{X,0}} \Lambda^1 F \xrightarrow{\sigma^{(q)}_{X,1}} \cdots\cdots \xrightarrow{\sigma^{(q)}_{X,m-1}} \Lambda^m F \xrightarrow{\sigma^{(q)}_{X,m}} 0$$

where the differentials are given by

$$\sigma^{(q)}_{X,v}(Z_{i_1}\wedge \ldots \wedge Z_{i_v}) = \sum_{t=1}^{t=m} \left(\frac{\operatorname{lcm}(y_{i_1},\ldots, y_{i_v}, y_t)}{\operatorname{lcm}(y_{i_1},\ldots, y_{i_v})} \right)^q Z_{i_1}\wedge \ldots \wedge Z_{i_v} \wedge Z_t$$

We have $\operatorname{Ext}^v_X(X/\mathfrak{a}^{(q)}_X, X) = \operatorname{Ker}\sigma^{(q)}_{X,v}/\operatorname{Im}\sigma^{(q)}_{X,v-1}$. Clearly $T^q_X = T^q_S \otimes_S X$ and $K^q_X = K^q_S \otimes_S X$ for $X = S'$ and R. The flatness of $X = S'$, R over S implies that $\operatorname{Im}\sigma^{(q)}_{X,v-1} = \operatorname{Im}\sigma^{(q)}_{S,v-1} \otimes_S X$ and $\operatorname{Ker}\sigma^{(q)}_{X,v} = \operatorname{Ker}\sigma^{(q)}_{S,v} \otimes_S X$. Therefore

$$\operatorname{Ext}^v_X(X/\mathfrak{a}^{(q)}_X, X) = \operatorname{Ext}^v_S(S/\mathfrak{a}^{(q)}_S, S) \otimes_S X \tag{2}$$

which means that it is sufficient to prove (i) only only for $X = S$.

Put $X = S$ and consider the S-linear map $f_v\colon \Lambda^v F \longrightarrow \Lambda^v F$ defined by $f_v(Z_{i_1}\wedge \ldots \wedge Z_{i_v}) = \operatorname{lcm}(y_{i_1},\ldots, y_{i_v})\cdot Z_{i_1}\wedge \ldots \wedge Z_{i_v}$. Clearly, f_v is an injection and satisfies the following identity: $\sigma^{(q+1)}_{v-1}\circ f_{v-1} = f_v \circ \sigma^{(q)}_{v-1}$. Therefore $\varphi^v_S(q)$ is induced by f_v. Take an element $b = \sum b_{i_1\ldots i_v} Z_{i_1}\wedge \ldots \wedge Z_{i_v} \in \operatorname{Ker}\sigma^{(q)}_v$. Assume that $f_v(b) \in \operatorname{Im}\sigma^{(q+1)}_{v-1}$, i.e. that there exists an element $b' = \quad b'_{i_1\ldots i_{v-1}} Z_{i_1}\wedge \ldots \wedge Z_{i_{v-1}} \in \Lambda^{v-1}F$ such that $\sigma^{(q+1)}_{v-1}(b') = f_v(b)$. Denote by $b'''_{i_1\ldots i_{v-1}}$ the sum of all monomials with respective coefficients which take part in $b'_{i_1\ldots i_{v-1}}$ and are not divisible by $\operatorname{lcm}(y_{i_1},\ldots, y_{i_{v-1}})$. Then

$$b'_{i_1\ldots i_{v-1}} = b'''_{i_1\ldots i_{v-1}} + \operatorname{lcm}(y_{i_1},\ldots, y_{i_{v-1}}) b''_{i_1\ldots i_{v-1}}$$

for some polynomial $b''_{i_1\ldots i_{v-1}}$. Put

$$b'' = \sum b''_{i_1\ldots i_{v-1}} Z_{i_1}\wedge \ldots \wedge Z_{i_{v-1}} \quad\text{and}\quad b''' = \sum b'''_{i_1\ldots i_{v-1}} Z_{i_1}\wedge \ldots \wedge Z_{i_{v-1}}$$

Then $b' = b''' + f_{v-1}(b'')$ and

$$f_v(b) = \sigma^{(q+1)}_{v-1}(b') = \sigma^{(q+1)}_{v-1}(b''') + \sigma^{(q+1)}_{v-1}(f_{v-1}(b'')) = \sigma^{(q+1)}_{v-1}(b''') + f_v(\sigma^{(q)}_{v-1}(b'')). \tag{3}$$

The coefficients of $f_v(b)$ and $f_v(\sigma^{(q)}_{v-1}(b''))$ at every basis element $Z_{i_1}\wedge \ldots \wedge Z_{i_v}$ are divisible by $\operatorname{lcm}(y_{i_1},\ldots, y_{i_v})$. However, not a single monomial taking part in the coefficient of $\sigma^{(q+1)}_{v-1}(b''')$ at every basis element $Z_{i_1}\wedge \ldots \wedge Z_{i_v}$

is divisible by $\mathrm{lcm}(y_{i_1}, \ldots, y_{i_v})$. (Here we are using the fact that the y_i's are square-free). Thus (3) is possible only if $\sigma_{v-1}^{(q+1)}(b''') = 0$ and $f_v(b) = f_v(\sigma_{v-1}^{(q)}(b''))$. Since f_v is injective, $b = \sigma_{v-1}^{(q)}(b'')$, i.e. $b \in \mathrm{Im}\,\sigma_{v-1}^{(q)}$ and it follows that $\varphi_S^v(q)$ is injective.

(ii) In view of (2) and the fact that R is faithfully flat over S' it is sufficient to prove that $\mathrm{Ext}_{S'}^v(S'/\mathfrak{a}_{S'}^{(q)}, S') \cong 0$ if and only if $\mathrm{Ext}_{S'}^v(S'/\mathfrak{a}_{S'}, S') \cong 0$. Put $w_i = x_i^q$ and consider the subring $W = k[w_1, \ldots, w_n]_{(w_1, \ldots, w_n)}$ of S'. Put $\tilde{y}_i = y_i^q \in W$ and denote by $\mathfrak{a}_W$ the ideal of W generated by $y_1, \ldots, y_m$. Since $w_1, \ldots, w_n$ form an S'-regular sequence, S' is faithfully flat over W and similarly to (2) we have $\mathrm{Ext}_{S'}^v(S'/\mathfrak{a}_{S'}^{(q)}, S') \cong \mathrm{Ext}_W^v(W/\mathfrak{a}_W, W) \otimes_W S'$. Thus it is sufficient to prove that $\mathrm{Ext}_{S'}^v(S'/\mathfrak{a}_{S'}, S') \cong 0$ if and only if $\mathrm{Ext}_W^v(W/\mathfrak{a}_W, W) \cong 0$ But these two modules are isomorphic as abelian groups under the isomorphism induced by the ring isomorphism $\psi : W \longrightarrow S'$ which sends every w_i to x_i.

(iii) This is immediate from (1), (ii) and (i).

(iv) Let $0 \longrightarrow B_p \xrightarrow{\phi_p} B_{p-1} \longrightarrow \;\; \cdots\cdots \;\; \longrightarrow B_0 \longrightarrow S'/\mathfrak{a}_{S'}$ be the minimal free resolution of $S'/\mathfrak{a}_{S'}$. Then $\mathrm{Ext}_{S'}^v(S'/\mathfrak{a}_{S'}, S') \cong 0$ if $v > p$ and therefore by (iii) $H_{\mathfrak{a}}^v(R) \cong 0$ if $v > p$. Thus $\mathrm{cd}(R, \mathfrak{a}) \leq p$.

Since the resolution is minimal, the matrix representing ϕ_p has all its entries in the maximal ideal of S' and so does the matrix representing the induced map ψ_p: $\mathrm{Hom}(B_{p-1}, S') \longrightarrow \mathrm{Hom}(B_p, S')$. By Nakayama's Lemma $\mathrm{Ext}_{S'}^p(S'/\mathfrak{a}_{S'}, S') \cong \mathrm{Hom}(B_p, S')/\mathrm{Im}\,\psi_p \not\cong 0$ and therefore by (iii) $H_{\mathfrak{a}}^p(R) \not\cong 0$.

Clearly $p = \mathrm{proj.dim.}_{S'} S'/\mathfrak{a}_{S'} = n - \mathrm{depth}_{S'} S'/\mathfrak{a}_{S'}$, where the second eqality follows from the Auslander-Buchsbaum theorem. Q.E.D.

2

Remark 1. For some straightforward ways to compute $\mathrm{proj.dim.} S'/\mathfrak{a}_{S'}$ and depth $S'/\mathfrak{a}_{S'}$ see for example [1] , Theorem 4.6 or [6], Theorem 5.2.

Remark 2. If R is not assumed to contain a field, then for a pathologically chosen R-sequence $x_1, \ldots, x_n$ our theorem may not be true. For example, put $R = \mathbb{Z}[x_0, x_1, x_2, x_3, x_4, x_5]_{(2, x_0, x_1, x_2, x_3, x_4, x_5)}$ and consider the ideal $\beta = 2\cdot\mathfrak{a}$, where $\mathfrak{a}$ is the ideal of Reisner defined below in Example 1. The reader may compute the modules $\mathrm{Ext}^4_R(R/\beta^{(q)}, R)$explicitly (using for example the Taylor resolution) and convince himself that $\mathrm{Ext}^4_R(R/\beta^{(q)}, R) \cong R/(2, x_0^q, x_1^q, x_2^q, x_3^q, x_4^q, x_5^q) \neq 0$ while all the maps $\varphi^v_R(q)$ are zero.

Our theorem has obvious applications to computing $\mathrm{ara}(\mathfrak{a})$, which in the case when $R = k[x_0,\ldots, x_n]_{(x_0,\ldots, x_n)}$ equals the minimum number of hypersurfaces needed to define a monomial subvariety of P^n_k set-theoretically (k is assumed to be algebraically closed and of arbitrary characteristic). As is well-known $\mathrm{ara}(\mathfrak{a}) \geq \mathrm{cd}(R,\mathfrak{a})$.

Example 1. Consider Reisner's subvariety of P^5_k defined by the ideal $\mathfrak{a} = (x_0x_1x_2, x_0x_1x_3, x_0x_2x_4, x_0x_3x_5, x_1x_2x_5, x_1x_3x_4, x_1x_4x_5, x_2x_3x_4, x_2x_3x_5, x_0x_4x_5)$. Reisner [8], p. 35 proved that $\mathrm{depth} R/\mathfrak{a}$ equals 3 if char $k \neq 2$ and 2 if char $k = 2$. Schmitt and Vogel [10], p. 250 constructed 4 equations with integer coefficients defining Reisner's variety set-theoretically for an algebraically closed field of any characteristic. As they pointed out, it was an open question to compute $\mathrm{ara}(\mathfrak{a})$. Our theorem implies that if char $k = 2$, then $\mathrm{cd}(R,\mathfrak{a}) = 6 - 2 = 4$ and therefore $\mathrm{ara}(\mathfrak{a}) = 4$, i.e. Resner's variety is not a set-theoretic complete intersection.

Our theorem gives a convenient source of examples of subvarieties of P^n_k whose local cohomological dimensions are easily computable while the minimum numbers of hypersurfaces needed to define them set-theoretically are not. Thus it sheds some light on the limitations of the cohomological dimension technique in computing $\mathrm{ara}(\mathfrak{a})$.

Example 2. Let $R = k[x_0, x_1,\ldots, x_{2n}]_{(x_0,\ldots, x_{2n})}$ and put $y_i = x_{i-1}x_i\ldots x_{i+n-1}$ if $1 \leq i \leq n+1$ and $y_{n+2} = x_{n+1}x_{n+2}\ldots x_{2n}x_0$. Put $\mathfrak{a} = (y_1,\ldots, y_{n+2})$. I claim that $\mathrm{proj.dim.} R/\mathfrak{a} = 2$. For the proof consider the Taylor resolution again and denote by K the submodule of $\wedge^2 F$ spanned by all elements of the form $Z_j \wedge Z_{j+1}$ $(1 < j < n+1)$. It is easy to check that

$$d(Z_i \wedge Z_j) = \sum_{t=i}^{t=j-1} \frac{\operatorname{lcm}(y_i, y_j)}{\operatorname{lcm}(y_t, y_{t+1})} d(Z_t \wedge Z_{t+1})$$

which means that the second differential sends K onto the kernel of the first differential. Considering the exact sequence

$$0 \longrightarrow \operatorname{Ker} d \longrightarrow K \longrightarrow \wedge^1 F \longrightarrow R \longrightarrow R/\mathfrak{a} \longrightarrow 0$$

and localizing it at the zero ideal of R (i.e. inverting every element of R) we will get an exact sequence of vector spaces over T, the field of quotients of R. Clearly, $\dim_T(\operatorname{Ker} d)_0 = \dim_T(K_0) - \dim_T((\wedge^1 F)_0) + \dim_T T - \dim_T((R/\mathfrak{a})_0) = 0$, i.e. $(\operatorname{Ker} d)_0 \cong 0$. Since Ker d is a submodule of a free R-module and R is a domain, this implies that $\operatorname{Ker} d \cong 0$, i.e. $0 \longrightarrow K \longrightarrow \wedge^1 F \longrightarrow R \longrightarrow R/\mathfrak{a}$ is a free resolution of $R/\mathfrak{a}$, i.e. $\operatorname{proj.dim.} R/\mathfrak{a} = 2$. Therfore by our theorem $\operatorname{cd}(R, \mathfrak{a}) = 2$ and it turns out that $\operatorname{cd}(R, \mathfrak{a})$ gives no non-trivial information about $\operatorname{ara}(\mathfrak{a})$ at all.

For simplicity's sake we now assume that $n = 2^h - 3$, where $h \geq 3$ is an integer (a similar construction may be carried out for all n, but the formulas for the corresponding indices will look rather tedious). Put

$$g_t = \sum_{s=1}^{s=2^{t-1}} y_{2^{h-t}(2s-1)} \qquad (1 \leq t \leq h).$$

By the main lemma from [10] $\mathfrak{a} = \operatorname{rad}(g_1, \dots, g_h)$. Therefore in this case we have $2 \leq \operatorname{ara}(\mathfrak{a}) \leq \log_2(n+3)$. It is an open problem (for me) to compute $\operatorname{ara}(\mathfrak{a})$ and even to improve the above estimate.

I would like to thank Professor David Eisenbud for a helpful discussion of the results of this paper.

References

1. C. De Concini, D. Eisenbud, C. Procesi. Hodge Algebras, Asterisque 91 (1982), Societe Mathematique de France.

2. D. Gmaeda, Multiplicative structure of finite free resolutions of ideals generated by monomials in an R-sequence. Thesis, Brandeis University, 1978.

3. H-G. Gräbe. Über den arithmetischen Rank quadratfreier Potenzproductideale. Preprint, 1982.

4. A. Grothendieck. (notes by R. Hartshorne). Local Cohomology, Lecture Notes in Mathematics 41, Springer, Berlin, 1967.

5. R. Hartschorne. Cohomological dimension of algebraic varieties. Ann. of Math., 88 (1968), 403 - 450.

6. M. Hochster. Cohen-Macaulay rings, combinatorics and simplicial complexes. in Ring Theory II, Ed B. R. MacDonald and R. Morris, Lecture Notes in pure and applied math. 26, Marcel Dekker, New York (1975).

7. H. Matsumura. Commutative Algebra. Math. Lect. Notes Series, 56 Benjamin/Cummings, Massachusetts, (1980).

8. G. A. Reisner. Cohen-Macaulay quotients of polynomial rings. Adv. in Math. 21 (1975) 30 - 49.

9. P. Schenzel , W. Vogel. On set-theoretic intersections. J. of Algebra 48 (1977) 401 - 408.

10. T. Schmitt, W. Vogel. Note on set-theoretic intersections of subvarieties of projective space. Math. Ann. 245 (1979), 247 - 253.

11. D. Taylor. Ideals generated by monomials in an R-sequence. Thesis, University of Chicago (1960).

Gennady Lyubeznik
Math. Dept., Columbia Univ.
New York, N.Y. 10027

Home address:
Gennady Lyubeznik
6402 23rd Avenue
Brooklyn, New York, N.Y. 11204

IN CHARACTERISTIC p=2 THE VERONESE VARIETY $V^m \subset \mathbb{P}^{m(m+3)/2}$ AND EACH OF ITS GENERIC PROJECTION IS SET-THEORETIC COMPLETE INTERSECTION.

Remo Gattazzo(°)

Istituto di Matematica Applicata
via Belzoni 7 - Padova (Italia)

INTRODUCTION.

Since many years ago they have been trying to know if every non singular variety $V^r \subset \mathbb{P}^N$, of dimension r, $r > 1$, k an algebraically closed field, is complete intersection (or at least set-theoretic complete intersection) in the projective space $\mathbb{P}^N$. The answers to these questions are not exahustive yet. As regards the non singular varieties $V^r \subset \mathbb{P}^N$, $r > \frac{2}{3}N$, each of them was conjectured in /1/ to be complete intersection. In the case $1 \leqslant r \leqslant \frac{2}{3}N$ one knows some criteria (see /1/ and also /4/ for a large discussion in this subject) which permit to find varieties $V^r \subset \mathbb{P}^N$ which are not set-theoretic complete intersections. However such criteria depend on the characteristic of the field k and become weaker in characteristic p, $p > o$ (see /2/).

In this paper we show some results over a field of characteristic p, $p > o$ having infinite many elements; more precisely: if $n=p^s$, $s > o$, the Veronese varieties $V_n^m \subset \mathbb{P}^N$, $N=\binom{m+n}{n}-1$ and $m,n > 1$ are set-theoretic complete intersections in $\mathbb{P}^N$. Furthermore the same result holds when $n=p=2$ for each (non singular) projection of V_2^m from a point in an open set in $\mathbb{P}^N$.

1. THE VERONESE VARIETY V_n^m.

Let us assume $m > o$, $n > 1$, $N=\binom{m+n}{n}-1$. Let I_n^m be a set, with lexicographic order, of the m+1-ples

$$(i)=(i_o,\dots,i_m) \qquad \text{with } o \leqslant i_o,\dots,i_m \leqslant n \text{ and } i_o+\dots+i_m=n.$$

The Veronese variety (or m-embedding) is the image in $\mathbb{P}^N$ of the morphism:

$$\{(T_o,\dots,T_m)\} \longrightarrow \{(T_o^n,\dots,T_o^{i_o}\dots T_m^{i_m},\dots,T_m^n)\}$$

(°) Lavoro svolto nell'ambito del gruppo G.N.S.A.G.A. del C.N.R..

for each $(i) \in I_n^m$ and $T_o,\dots,T_m$ indeterminates over the (infinite) field k. Let us denote $Y_{(i)}$, for each $(i) \in I_n^m$, indeterminates over k and $k[Y_{(i)}]$ the polynomial ring in $Y_{(i)}$.

The properties of the ideal $I(V_n^m) \subset k[Y_{(i)}]$ are well known (e.g./3/); in particular $I(V_n^m)$ is generated by the 2x2-minors of a symmetric matrix whose enters are $Y_{(i)}$.

PROPOSITION 1.Let k an infinite field of characteristic p, $p > o$. For each $m > o$, $n=p^s$, $s > o$, the Veronese variety $V_n^m \subset \mathbb{P}^N$ is set-theoretic complete intersection of the following N-m hypersurfaces:

$$(1) \qquad Y_{(i)}^n - Y_{(n,o,o,\dots,o)}^{i_o} Y_{(o,n,o,\dots,o)}^{i_1} \cdots Y_{(o,o,\dots,n)}^{i_m} = o$$

for each $(i) \in I_n^m$ having at least two indices different from zero.

PROOF. It is easy to verify that each point in V_n^m satisfies (1). We have to show the converse. We use induction on m. If m=1 then we have $N=n=p^s$, $s > o$. Hence V_n^1 is the curve

$$V_n^1=\{(T_o^n, T_o^{n-1}T_1,\dots,T_1^n)\} \subset \mathbb{P}^n$$

and (1) becomes $Y_{(n-i,i)}^n - Y_{(n,o)}^{n-i} Y_{(o,n)}^i = o$ for $o \leqslant i \leqslant n-1$. These, as it is easy to see, define in the hyperplane $\{Y_{(n,o)}=o\}$ precisely a point which is just $V_n^1 \cap \{Y_{(n,o)} = o\}$. Now we consider in the affine open set $A_{(n,o)} = \mathbb{P}^n - \{Y_{(n,o)}=o\}$ the coordinate functions

$$t_j = Y_{(n-j,j)}/Y_{(n,o)} \qquad 1 \leqslant j \leqslant n.$$

In $A_{(n,o)}$ the equations (1) become $t_i^n - t_n^i = o$ for $1 \leqslant i \leqslant n-1$. For i=1 we have $t_n = t_1^n$ and, after replacement of this in the others, we can use the formula $a^n+b^n=(a+b)^n$ being $n=p^s$, $s > o$, and the field k of characteristic p. We then get:

$$t_i^n - (t_1^n)^i = (t_i - t_1^i)^n = o \qquad 2 \leqslant i \leqslant n-1;$$

that means (1) defines in $A_{(n,o)}$ the curve $\{(t_1,t_1^2,\dots,t_1^n)\} = V_n^1 \cap A_{(n,o)}$. Thus proposition 1 holds for m=1. Let us suppose that the proposition is true for V_n^{m-1} whenever $m > 1$. First we calculate $V_n^m \cap \{Y_{(n,o,\dots,o)}=o\}$. It must be $T_o^n=o$, hence our set is:

$$\{(o,\dots,o,T_1^n,\dots,T_1^{i_1}\cdots T_m^{i_m},\dots,T_m^n)\} \qquad \text{for each } (i) \in I_n^m \text{ with } i_o=o$$

and it can be identified with V_n^{m-1}. The equations (1) define in the hyperplane $\{Y_{(n,o,\dots,o)}=o\}$ the set defined by:

$$\begin{cases} Y^n_{(i)}=o & \text{for each } (i)\in I^m_n \text{ with } i_o\neq o \\ Y^n_{(i)}-Y^{i_o}_{(o,n,\dots,o)}\cdots Y^{i_m}_{(o,o,\dots,n)}=o & \text{for each } (i)\in I^m_n \text{ with } i_o=o. \end{cases}$$

This set coincides thus with $V_n^m\cap\{Y_{(n,o,\dots,o)}=o\}$ owing to its identification with V_n^{m-1} and to induction on m.

Let us denote the affine open set $A_{(n,o,\dots,o)}=\mathbb{P}^N-\{Y_{(n,o,\dots,o)}=o\}$ and the coordinate functions:

$$y_{(i)}=Y_{(i)}/Y_{(n,o,\dots,o)} \qquad \text{for each } (i)\in I^m_n-\{(n,o,\dots,o)\}.$$

Let us consider I^m_n as disjoint union of the subsets:

$$A=\{(o,\dots,o,n_i,o,\dots,o) : n_i=n \quad \text{for } o\leqslant i\leqslant m\}$$

$$B=\{(n-1,o,\dots,1_i,\dots,o) : 1_i=1 \quad \text{for } o<i\leqslant m\}$$

$$C=I^m_n-A\cup B.$$

Therefore the equations (1) are parametrized by $(i)\in B\cup C$. Let us denote:

$$(°)\qquad t_i=y_{(i)} \qquad \text{for each } (i)\in B.$$

After dividing each (1) by $Y^n_{(n,o,\dots,o)}$ we get first:

$$(')\qquad t_i^n=y_{(o,\dots,n_i,\dots,o)} \qquad \text{for } 1\leqslant i\leqslant m$$

and replacing these in the others with $(i)\in C$, we obtain:

$$y^n_{(i)}-(t_1^n)^{i_1}\cdots(t_m^n)^{i_m}=o \qquad \text{for each } (i)\in C.$$

Furthermore because $n=p^s$, $s>o$, and the characteristic of the field k is p, the last equations can be written under the form:

$$('')\qquad (y_{(i)}-t_1^{i_1}\cdots t_m^{i_m})^n=o \quad \text{hence} \quad y_{(i)}=t_1^{i_1}\cdots t_m^{i_m} \text{ for each } (i)\in C.$$

Thus the assumption (°),(') and ('') mean that (1) defines in $A_{(n,o,\dots,o)}$ precisely $V^m_n\cap A_{(n,o,\dots,o)}$. Q.E.D.

NOTE 1. When n=2 we have $I^m_n = A \cup D$, where A is the set as above and

$$D = \{(o,\dots,o,1_i,o,\dots,o,1_j,o,\dots,o) \quad \text{for } o \leqslant i < j \leqslant m\}.$$

For this reason one introduces instead of $Y_{(i)}$:

Y_{ii} if $(i) \in A$; Y_{ij} if $(i) \in D$ respectively.

Furthermore one uses to identify, to within a not zero factor, a point of $\mathbb{P}^N$ with a not zero symmetric matrix of order m+1 with enters in k. By this the ideal $I(V^m_n) \subset k[Y_{ij}]$ is generated by 2x2-minors of the symmetric matrix $W=(Y_{ij})$, $o \leqslant i \leqslant j \leqslant m$.

The proposition 1 asserts then that, whenever the field k is infinite and of characteristic 2, V^m_2 is set-theoretic complete intersection of the quadrics:

$$Y^2_{ij}+Y_{ii}Y_{jj} = o \qquad\qquad \text{for } o \leqslant i < j \leqslant m$$

which are precisely defined by <u>principal</u> 2x2-minors of the matrix W.

This result is more or less known as E.STAGNARO kindly told me.

2. PROJECTIONS OF V^m_n FROM A POINT OF $\mathbb{P}^N$.

From proposition 1 one can see easily that all projections of V^m_n from a vertex of coordinate reference in $\mathbb{P}^N$ are again set-theoretic complete intersections. For example if we leave out only one equation of (1) corrisponding to a fixed (i'), the other remained equations in (1) define N-m-1 hypersuperfaces whose intersection is the cone projecting V^m_n from the point P whose coordinates are zero except the one of index (i'). This cone (and its sections by hyperplanes not passing through P) is of course set-theoretic complete intersection. These projections however can be singular. On the other hand if the point P is not a vertex of the coordinate reference in $\mathbb{P}^N$, it is very cumbersome to calculate the projection of V^m_n from P. This calculation is easy enough in the case n=p=2.

From now on will be $V^m = V^m_2$, $M=m(m+3)/2$ instead of N; $Sec(V^m)$ denotes the set of the points in the straigh lines which are tangent or meeting V^m in at least two points. Let M_3 be the set in $\mathbb{P}^M$ vanishing all 3x3-minors of matrix W.

LEMMA. Let k be an infinite field of arbitrary characteristic. Then:

a) $Sec(V^m) \subseteq M_3$.

b) If k is of characteristic $p \neq 2$ then $Sec(V^m) = M_3$.

c) The projection of V^m from every $P \in \mathbb{P}^M - M_3$ on every hyperplane not passing through P is not singular.

PROOF. a) Let $A=(a_{ij})$, $B=(b_{ij})$ points in V^m and $C=(c_{ij})$ an arbitrary point in the line joining A,B or in the tangent line to V^m at A, if A=B. Then we can find $a,b \in k$ such that $c_{ij}=aa_{ij}+bb_{ij}$ for $o \leq i \leq j \leq m$. Let us consider for each pair of (i_1,i_2,i_3) and (j_1,j_2,j_3) with distinct elements belonging to $\{o,1,\dots,m\}$ the 3x3-matrix $(c_{i_r j_s})$. The determinant of this matrix is a linear combination, with coefficients $a^u b^v$, for $o \leq u,v \leq 3$ and $u+v=3$, of determinants of 3x3-matrices whose u columns have enters in the coordinates of A and v columns in the coordinates of B. Such matrices have at least two proportional columns because $A,B \in V^m$ and we can consider these as matrices of rank 1. Thus $(c_{i_r j_s})$ has determinant zero. Every point of $Sec(V^m)$ is then in M_3.

b) Let us suppose the characteristic of k is $p \neq 2$. If $R=(r_{ij}) \in M_3$, R as symmetric matrix has rank 1 or 2. In the former case $R \in V^m$ and then $R \in Sec(V^m)$; in the second case we can find an invertible matrix H with enters in k such that:

$$HR\tilde{H} = aE_{11}+bE_{22} \qquad a,b \in k \text{ not zero, } \sim = \text{tranposition}$$

where E_{ii} denotes the matrix with 1 in the place (i,i) and zero everywhere, for i=1,2. We have also:

$$R = a(H^{-1}E_{11}\tilde{H}^{-1})+b(H^{-1}E_{22}\tilde{H}^{-1}).$$

Therefore $A=H^{-1}E_{11}\tilde{H}^{-1}$, $B=H^{-1}E_{22}\tilde{H}^{-1}$ are symmetric matrices of rank 1 which we can see as distinct points of V^m and A,B,R belong to the same line. Thus $R \in Sec(V^m)$.

c) Evident. Q.E.D.

NOTE 2. If p=2 the proof of b) is incorrect. For example let m=2 ; $R=\begin{vmatrix} o & 1 & o \\ 1 & o & o \\ o & o & o \end{vmatrix}$ is a symmetric matrix of rank 2 <u>not congruent</u> to a diagonal matrix of rank 2! It easy to verify that R is a point in $M_3 - Sec(V^2)$.

Furthermore we note that all points $P=(p_{ij})$ satisfying $p_{oo}=\dots=p_{mm}=o$ belong to M_3. Indeed (p_{ij}) as symmetric matrix is, because p=2, skew-symmetric to, and so all principal minors of (p_{ij}) of odd order are zero.

PROPOSITION 2. Let k be an infinite field of characteristic p=2. Let P be an arbitrary point in $\mathbb{P}^M$ which has at least a principal 3x3-minor not zero. The cone projecting V^m from P is set-theoretic complete intersection of m hypersurfaces of degree 4 and of m(m-1)/2 -1 quadrics.

PROOF. From linear algebra we know that it is possible to suppose, to within a permutation of raws and columns of (p_{ij}), that the principal 3x3-minor

$$\det \begin{vmatrix} p_{oo} & p_{o1} & p_{o2} \\ p_{o1} & p_{11} & p_{12} \\ p_{o2} & p_{12} & p_{22} \end{vmatrix} \neq o$$

and, for example, $p_{oo} \neq o$ (see note 2).

Let us change the coordinates in $\mathbb{P}^M$ by

$$Y_{ij}=X_{ij}+a_{ij}X_{oo} \qquad \text{where } a_{oo}=o,\ a_{ij}=p_{ij}/p_{oo} \qquad \text{for } o \leqslant i \leqslant j \leqslant m,\ i+j\neq o.$$

Therefore P is, to within a factor, the point $\{(1,o,\dots,o)\}$ while

$$V^m=\{(T_o^2,T_oT_1+a_{o1}T_o^2,\dots,T_1^2+a_{11}T_o^2,T_1T_2+a_{12}T_o^2,\dots,T_m^2+a_{mm}T_o^2)\}.$$

We know (note 1) that V^m in $\mathbb{P}^M$ is set-theoretic complete intersection of the quadrics given by the principal 2x2-minors of the matrix $W'=(X_{ij}+a_{ij}X_{oo})$.

Let us denote by G_{oi} and G_{ij} the principal minors of W' involving the pairs of distinct columns of indices o,i and i,j respectively, and J the ideal in $k[X_{ij}]$ generated by G_{oi} and G_{ij}. We have:

$$G_{oi} = A_{oi}X_{oo}^2 +X_{oo}X_{ii}+X_{oi}^2 \qquad G_{ij} = A_{ij}X_{oo}^2 +X_{oo}(a_{ii}X_{jj}+a_{jj}X_{ii})+X_{ij}^2+X_{ii}X_{jj}$$

where

$$A_{oi}= \det\begin{vmatrix} 1 & a_{oi} \\ a_{oi} & a_{ii}\end{vmatrix} \qquad A_{ij}= \det\begin{vmatrix} a_{ii} & a_{ij} \\ a_{ij} & a_{jj}\end{vmatrix} \qquad 1 \leqslant i < j \leqslant m.$$

Let us denote furthermore by

$$A_{oij}=\det\begin{vmatrix} 1 & a_{oi} & a_{oj} \\ a_{oi} & a_{ii} & a_{ij} \\ a_{oj} & a_{ij} & a_{jj}\end{vmatrix} \qquad Z_{ij}= X_{ij}^2+X_{ii}X_{jj}+a_{ii}X_{oj}^2+a_{jj}X_{oi}^2$$

and consider $S_{ij}= G_{ij}+a_{ii}G_{oj}+a_{jj}G_{oi} = A_{oij}X_{oo}^2+Z_{ij} \qquad 1 \leqslant i < j \leqslant m.$

It is easy to see that $S_{ij} \in J$ and it is indeed

$$J = (G_{o1},\dots,G_{om},S_{12},S_{13},\dots,S_{m-1\,m}).$$

If we calculate now the resultants, with respect X_{oo}, of all the pairs of distinct generators of J, they give hypersurfaces whose intersection is the cone projecting V^m from P. In order to prove our proposition it is enough to find only $M-m-1 = m + +m(m-1)/2 -1$ such resultants and verify that these define the cone itself. For this we take only the resultants of the pairs (G_{oi},S_{12}) and (S_{ij},S_{12}) and denote them by R_i and R_{ij}. We have:

$$R_i = A^2_{oi}Z^2_{12}+A^2_{o12}X^4_{oi}+A_{o12}Z_{12}X^2_{ii} \qquad 1 \leqslant i \leqslant m$$

$$R_{ij} = F^2_{ij} \qquad \text{where } F_{ij} = A_{oij}Z_{12}+A_{o12}Z_{ij} \qquad 1 \leqslant i < j \leqslant m,\ i+j > 3.$$

Let us denote by $J°$ the ideal in $k[X_{ij}]$ generated by

$$R_1,\dots,R_m,F_{13},\dots,F_{1m},F_{23},\dots,F_{2m},F_{34},\dots,F_{m-1\,m}$$

and Δ the set which $J°$ defines in $\mathbb{P}^M$. We note that $R_1,\dots,R_m$ define m hypersurfaces of degree 4 while $F_{13},\dots,F_{m-1\,m}$ define $m(m-1)/2 -1$ quadrics and Δ is their intersection. Now we prove that Δ is the cone projecting V^m from P.

First it is evident that Δ is a cone with vertex P. Later we consider

$$\Delta = (\Delta \cap \{Z_{12} = o\}) \cup (\Delta \cap \{Z_{12} \neq o\}) \ ; \ V^m = (V^m \cap \{X_{oo} = o\}) \cup (V^m \cap \{X_{oo} \neq o\})$$

and we prove that each piece of Δ is a cone projecting from P just one piece of V^m.

<u>Calculation of</u> $\Delta \cap \{Z_{12}=o\}$. For assumption on $P=(p_{ij})$ we have $A_{o12} \neq o$. Hence

$$\Delta \cap \{Z_{12} = o\} = \left[\begin{array}{l} Z_{ij} = o \\ X^4_{oi} = o \end{array}\right. = \left[\begin{array}{l} X^2_{ij}+X_{ii}X_{jj} = o \\ X_{oi} = o \end{array}\right. \qquad 1 \leqslant i < j \leqslant m$$

and such set intersects the hyperplane $\{X_{oo}=o\}$ exactly in

$$\{X_{oo} = G_{oi} = G_{ij} = o\} = V^m \cap \{X_{oo} = o\}.$$

Thus $\Delta \cap \{Z_{12}=o\}$ is the cone projecting $V^m \cap \{X_{oo}=o\}$ from P.

<u>Calculation of</u> $\Delta \cap \{Z_{12} \neq o\}$. Let $Q \in \Delta \cap \{Z_{12} \neq o\}$ be a fixed point. We choose a suitable factor of proportionality for the coordinates of Q such that $Q=(x_{ij})$, for $o \leqslant i \leqslant j \leqslant m$, $i+j \neq o$, and furthermore it results $Z_{12}(Q) = A_{o12}$.

From $F_{ij}(Q) = o$ follows $Z_{ij}(Q) = A_{oij}$ $\qquad 1 \leqslant i < j \leqslant m.$

From $F_{ij}(Q)=o$ follows $Z_{ij}(Q)=A_{oij}$ $\qquad 1 \leq i < j \leq m.$

For $1 \leq i \leq m$ from $R_i(Q)=o$ follows:

$$A_{oi}^2 A_{o12}^2 + A_{o12}^2 x_{oi}^4 + A_{o12}^2 x_{ii}^2 = A_{o12}^2 (A_{oi} + x_{oi}^2 + x_{ii})^2 = o$$

and remembering that $A_{oi} = a_{oi}^2 + a_{ii}$, $A_{o12} \neq o$, we get

$$(') \qquad x_{ii} = (x_{oi} + a_{oi})^2 + a_{ii}.$$

By replacing (') in $Z_{ij}(Q)=A_{oij}$ and remembering what A_{oij} denotes, we obtain:

$$('') \qquad x_{ij} = (x_{oi} + a_{oi})(x_{oj} + a_{oj}) + a_{ij}.$$

(') and ('') mean that the point Q is in the line joining P and the point

$$Q' = \{(T_o^2, T_o T_1 + a_{o1} T_o^2, \ldots, T_1^2 + a_{11} T_o^2, T_1 T_2 + a_{12} T_o^2, \ldots, T_m^2 + a_{mm} T_o^2)\}$$

as we can easily verify fixing $1/T_o^2$ as factor of proportionality of the coordinates of Q'. Because $Q' \in V^m \cap \{X_{oo} \neq o\}$, we have then stated that $\Delta \cap \{Z_{12} \neq o\}$ is just the cone projecting $V^m \cap \{X_{oo} \neq o\}$ from P. Q.E.D.

COROLLARY. Every section of the cone projecting V^m from a point taken as above by hyperplane not passing through P is a non singular variety set-theoretic complete intersection.

REFERENCES

/1/ R.HARTSHORNE, Varieties of small codimension in projective space. Bull.A.M.S. 80 (1974) 1017-1032.

/2/ R.HARTSHORNE, R.SPEISER, Local cohomological dimension in characteristic p. Annals of Math. 105 (1977) 45-79.

/3/ W.GROEBNER, Ueber Veronesesche Varietäten und deren Projectionen. Archiv d. Math. 16 (1965) 257-264.

/4/ G.VALLA, Curve algebriche insiemisticamente intersezioni complete. Atti del Convegno di Geometria Algebrica - Firenze 1981.(Ist.Analisi Globale).

Idéaux de définition des courbes monomiales*

Shalom ELIAHOU

Introduction. On appelle courbe monomiale une courbe $\Gamma \subset \mathbb{A}_k^n$ paramétrée par $x_i = t^{d_i}$ $(i = 1,\ldots,n)$ où $d_1,\ldots,d_n$ sont des entiers positifs sans facteurs communs. Le corps de base, k, est supposé être algébriquement clos et de caractéristique nulle.

On s'intéresse au problème suivant: les courbes monomiales dans $\mathbb{A}^n$ sont-elles intersection complète ensembliste (I.C.E.), i.e. sont-elles le lieu des zéros de n-1 polynômes à n variables?

Sauf erreur, les seuls résultats connus pour l'instant sont les suivants:

-Les courbes monomiales dans $\mathbb{A}^3$ sont I.C.E. ([B_1], [H], [V]).

-Les courbes $(t^{d_1},t^{d_2},t^{d_3},t^{d_4})$ dans $\mathbb{A}^4$ dont le semi-groupe associé $\sum_{i=1}^{4} \mathbb{N}d_i$ est symétrique, sont I.C.E. ([B_2]).

Dans cet article, on donne tout d'abord une caractérisation effective des systèmes de binômes qui

---

*Cet article est une partie de la thèse de doctorat présentée par l'auteur à l'Université de Genève (juin 1983).

définissent une courbe monomiale $\Gamma \subset \mathbb{A}^n$ donnée (proposition 1). On montre ensuite que Γ peut être définie par n binômes (comparer avec [E-E]), et que Γ réunion un axe de coordonnées de $\mathbb{A}^n$ est I.C.E. (proposition 5). Enfin, on donne une condition suffisante pour les courbes monomiales dans $\mathbb{A}^4$ d'être I.C.E., qui étend le résultat de Bresinsky dans $[B_2]$ (proposition 6 et applications).

Ce travail est né d'une question de M. Kervaire, et des fructueuses discussions que nous avons eues ensemble. Je l'en remercie sincèrement, ainsi que T. Vust pour ses nombreux conseils.

Soit $\Gamma \subset \mathbb{A}^n_k$ $(k = \bar{k})$ la courbe monomiale paramétrée par $x_i = t^{d_i}$, où $d_1,\dots,d_n$ sont des entiers positifs sans facteurs communs.

On note P l'idéal de Γ. C'est par définition le noyau de l'homomorphisme

$$\Phi : k[X_1,\dots,X_n] \to k[T]$$
$$X_i \mapsto T^{d_i}$$

On gradue $k[X_1,\dots,X_n]$ en posant $\deg X_i = d_i$; Φ est alors un homomorphisme gradué, et P un idéal <u>homogène</u> de $k[X_1,\dots,X_n]$.

On appelle <u>augmentation</u> d'un polynôme la somme de

ses coefficients. Il est clair que les éléments de P sont d'augmentation nulle.

On appelle <u>binôme</u> un polynôme de la forme $s-t$, où s et t sont des monômes de $k[X_1,\ldots,X_n]$. Un binôme appartient à P si et seulement si il est homogène.

On note M_+ l'ensemble des monômes de $k[X_1,\ldots,X_n]$, M le groupe multiplicatif des monômes rationnels de $k(X_1,\ldots,X_n)$, et M_0 le sous-groupe des monômes rationnels de degré 0.

A un binôme $g = s-t$, on associe l'élément $\hat{g} := s/t$ de M. Si $g \in P$ (i.e. si g est homogène), $\hat{g} \in M_0$.

Si $u_1,\ldots,u_r \in M$, on note $\langle u_1,\ldots,u_r\rangle$ le sous-groupe de M qu'ils engendrent.

Enfin, si $f_1,\ldots,f_s \in k[X_1,\ldots,X_n]$, on note $V(f_1,\ldots,f_s)$ l'ensemble de leurs zéros communs dans $\mathbb{A}^n_k$.

<u>Proposition 1</u>.* Soient $g_1,\ldots,g_r \in P$ des binômes homogènes. Alors $V(g_1,\ldots,g_r) = \Gamma$ si et seulement si

a) $\langle\hat{g}_1,\ldots,\hat{g}_r\rangle = M_0$

b) $V(g_1,\ldots,g_r,X_i) = \{0\} \quad \forall\, i = 1,\ldots,n$.

La preuve requiert quelques préliminaires.

I. A la donnée des binômes $g_1,\ldots,g_r$, on associe la relation d'équivalence suivante sur M_+:

* Note: ce résultat est faux en caractéristique $p>o$. Par exemple pour $\Gamma = (t,t)$, prendre $g = x_1{}^p - x_2{}^p = (x_1 - x_2)^p$. Clairement $V(g) = \Gamma$, mais $\langle\hat{g}\rangle \neq M_0$. Cette remarque est de D. Eisenbud.

Def: $s \sim t \Leftrightarrow s/t \in \langle \hat{g}_1,\dots,\hat{g}_r \rangle$ $(s,t \in M_+)$.

On appelle simple un polynôme dont tous les monômes sont équivalents (pour $\sim$).

Clairement, tout polynôme f s'écrit de façon unique à l'ordre près comme somme de polynômes simples, $f = f_1 + \dots + f_m$. On appelle $f_1,\dots,f_m$ les composantes simples de f.

Remarque 2. Si $f \in (g_1,\dots,g_r)$, les composantes simples de f appartiennent aussi à $(g_1,\dots,g_r)$.

En effet, la propriété ci-dessus est vraie pour $g_1,\dots,g_r$, qui sont simples. D'autre part, si elle est vraie pour f et g, elle l'est aussi pour $f+g$, et pour sf $\forall s \in M_+$. Donc elle est vraie pour tout $f \subset (g_1,\dots,g_r)$.

II. Lemme 3. Soient $g_1,\dots,g_r \in P$ des binômes homogènes t.q. $\langle \hat{g}_1,\dots,\hat{g}_r \rangle = M_0$. Alors $V(g_1,\dots,g_r) \subset \Gamma \cup V(X_1 \dots X_n)$.

preuve: Soit $x \in V(g_1,\dots,g_r) \setminus V(X_1 \dots X_n)$. A voir: $x \in \Gamma$. Comme P est engendré par des binômes homogènes, il suffit de voir: $g(x) = 0$ pour tout binôme homogène g. Par hypothèse, $g_i(x) = 0$ $\forall i = 1,\dots,r$, ce qu'on peut aussi écrire sous la forme $\hat{g}_i(x) = 1$, puisque les coordonnées de x sont toutes non-nulles. Or $\hat{g} \in M_0 = \langle \hat{g}_1,\dots,\hat{g}_r \rangle$, donc on a aussi $\hat{g}(x) = 1$, i.e. $g(x) = 0$. $\square$

Preuve de la proposition 1

$\Rightarrow$) l'hypothèse $V(g_1,\ldots,g_r) = \Gamma$ et le théorème des zéros de Hilbert donnent: $\mathrm{rad}(g_1,\ldots,g_r) = P$.

Preuve de a). L'inclusion $<\hat{g}_1,\ldots,\hat{g}_r> \subset M_0$ est banale. Réciproquement, soit s/t un élément quelconque de M_0, où s et t sont des monômes dans M_+ de même degré. A montrer: $s/t \in <\hat{g}_1,\ldots,\hat{g}_r>$, i.e. $s \sim t$, où $\sim$ est la relation d'équivalence introduite dans les préliminaires. Le binôme $s - t$ étant homogène, on a $s - t \in P = \mathrm{rad}(g_1,\ldots,g_r)$.
Soit p un nombre premier t.g. $(s-t)^p \in (g_1,\ldots,g_r)$.

Assertion: $s^p \sim t^p$. En effet, notons f la composante simple de $(s-t)^p$ qui "contient" s^p. Par la remarque 2, $f \in (g_1,\ldots,g_r) \subset P$, en particulier f est d'augmentation nulle. Or les coefficients des monômes de $(s-t)^p$ sont nuls mod p, sauf ceux de s^p et de t^p. Donc f "contient" aussi t^p, sinon il serait d'augmentation 1 mod p, ce qui est absurde. Cela signifie $s^p \sim t^p$, i.e. $(s/t)^p \in <\hat{g}_1,\ldots,\hat{g}_r>$.

Si p' est un nombre premier $> p$, le même argument donne $(s/t)^{p'} \in <\hat{g}_1,\ldots,\hat{g}_r>$. Donc, par Bezout, $s/t \in <\hat{g}_1,\ldots,\hat{g}_r>$.

Preuve de b). Découle directement de $\Gamma \cap V(X_i) = \{0\}$ $\forall i$.

$\Leftarrow$) L'inclusion $\Gamma \subset V(g_1,\ldots,g_r)$ est banale. Réciproquement, $V(g_1,\ldots,g_r) \subset \Gamma \cap V(X_1\ldots X_n)$ d'après le lemme 3. Mais $V(g_1,\dots,g_r) \cap V(X_1\dots X_n) \subset \bigcup_i V(g_1,\dots,g_r,X_i)$, qui est $\{0\}$ par b). Donc $V(g_1,\dots,g_r) \subset \Gamma$. □

Lemme 4. Le groupe M_0 admet une base $\hat{g}_2,\dots,\hat{g}_n$, où $g_i = X_i^{a_{ii}} - X_{i-1}^{a_{i,i-1}} X_{i+1}^{a_{i,i+1}} \dots X_n^{a_{in}}$, avec $a_{ij} > 0 \ \forall j \geq i-1$.

preuve: par récurrence sur n.

Si $n = 2$, c'est banal. Si $n > 2$, considérons l'application $\lambda: M_0 \to \mathbb{Z}$ qui envoie $X_1^{a_1}\dots X_n^{a_n}$ sur a_1. L'image $\lambda(M_0)$ est un sous-groupe non-nul de $\mathbb{Z}$, donc engendré par un entier positif m. Soit $\hat{g} = X_1^{-m} X_2^{a_2} \dots X_n^{a_n} \in M_0$. Quitte à multiplier $\hat{g}$ par $X_2^{d_i} X_i^{-d_2}$ autant que nécessaire, on peut supposer $a_2 > 0$ et $a_i < 0 \ \forall i \geq 3$.

L'élément $g_2 := g = X_2^{a_2} - X_1^m X_3^{|a_3|} \dots X_n^{|a_n|}$ est de la forme voulue.

Si $\hat{h}$ est un élément quelconque de M_0, $\lambda(\hat{h})$ est divisible par m. On peut donc supposer, modulo $\hat{g}_2$, que $\lambda(\hat{h}) = 0$, i.e. que $\hat{h}$ ne contient plus X_1. On est ainsi ramené au cas $n-1$. □

Proposition 5. Soient $g_2,\dots,g_n$ les binômes du lemme 4, et $g_1 = X_1^{d_n} - X_n^{d_1}$. Alors

1) $V(g_2,\dots,g_n) = \Gamma \cup$ l'axe des X_1.

2) $V(g_1,g_2,\dots,g_n) = \Gamma$.

preuve:

1) L'inclusion $\Gamma \cup V(X_2,\ldots,X_n) \subset V(g_2,\ldots,g_n)$ est banale. Réciproquement: la forme particulière des g_i entraîne $V(g_2,\ldots,g_n) \cap V(X_1 \ldots X_n) \subset V(X_2,\ldots,X_n)$. Mais par le lemme 3, $V(g_2,\ldots,g_n) \subset \Gamma \cup V(X_1 \ldots X_n)$, donc $V(g_2,\ldots,g_n) \subset \Gamma \cup V(X_2,\ldots,X_n)$.

2) La condition a) de la proposition 1 est vraie par construction, et la condition b) découle directement de la forme particulière des g_i. □

Voici maintenant une condition suffisante pour les courbes monomiales dans $\mathbb{A}^4$ d'être I.C.E., qui étend celle donnée par Bresinsky dans [B_2].

Proposition 6. Soit $\Gamma \subset \mathbb{A}^4$ une courbe monomiale d'idéal P. Supposons que P contienne 4 binômes homogènes $g_1,\ldots,g_4$ de la forme $g_i = X_i^{a_{ii}} - \prod_{j \neq i} X_j^{a_{ij}}$ t.q.

a) $V(g_1,\ldots,g_4) = \Gamma$

b) $\prod_i \hat{g}_i = 1$

c) l'un des g_i ne contient que 3 variables.

Alors Γ est I.C.E.

preuve (inspirée de [H]): La condition b) équivaut à l'existence de 4 monômes $s_1,\ldots,s_4$ t.q.

$$s_1 g_1 + s_2 g_2 + s_3 g_3 + s_4 g_4 = 0 \qquad (*)$$

(appliquer 3 fois l'identité $xy-1 = (x-1)+x(y-1)$ à la formule $\Pi\hat{g}_i - 1 = 0$, puis chasser les dénominateurs; réciproquement: laissé au lecteur).

En utilisant le point c), on peut supposer que $g_2 = X_2^{a_{22}} - X_3^{a_{23}}X_4^{a_{24}}$. Posons $B = k[X_1,\ldots,X_4]/(g_1,g_2)$, et $^{-}$ la projection de $K[X_1,\ldots,X_4]$ sur B. Il est clair que B est isomorphe à l'algèbre

$$\bigoplus_{\substack{0\leq\nu_1<a_{11}\\ 0\leq\nu_2<a_{22}}} k[X_3,X_4]y_1^{\nu_1}y_2^{\nu_2}, \quad \text{où} \quad \begin{cases} y_1^{a_{11}} := y_2^{a_{12}}X_3^{a_{13}}X_4^{a_{14}} \\ y_2^{a_{22}} := X_3^{a_{23}}X_4^{a_{24}} \end{cases} \qquad (**)$$

Dans B, on a $\overline{s_3g_3} = -\overline{s_4g_4}$. En élevant ceci à la puissance $a_{11}a_{22}$, et en utilisant (**), on trouve des monômes t_3, t_4 en X_3, X_4 seulement t.q.

$$t_3\bar{g}_3^{a_{11}a_{22}} = \pm t_4\bar{g}_4^{a_{11}a_{22}} \qquad (***)$$

On peut supposer t_3, t_4 sans facteurs communs, puis résoudre (***) composante par composante dans B, pour trouver $\bar{h} \in B$ t.q. $\bar{g}_3 = t_4\bar{h}$, $\bar{g}_4 = \pm t_3\bar{h}$. Si $h \in k[X_1,\ldots,X_4]$ relève $\bar{h}$, alors $g_3, g_4 \in \mathrm{rad}(g_1,g_2,h)$, donc $P = \mathrm{rad}(g_1,\ldots,g_4) \subset \mathrm{rad}(g_1,g_2,h) \subset P$, i.e. $P = \mathrm{rad}(g_1,g_2,h)$. □

Applications

I. La proposition 6 s'applique au cas des courbes dans

$\mathbb{A}^4$ dont le semi-groupe associé est symétrique. Par définition, un semi-groupe $\Delta \subset \mathbb{N}$ est dit symétrique si

1) $\mathbb{N} \setminus \Delta$ est fini
2) $\mathrm{card}(\mathbb{N} \setminus \Delta) = c/2$, où $c-1$ est le plus grand élément de $\mathbb{N} \setminus \Delta$.

Remarque: si $\Delta = \sum_{i=1}^{n} \mathbb{N}d_i$, la finitude de $\mathbb{N} \setminus \Delta$ équivaut à $\mathrm{pgcd}(d_1,\ldots,d_n) = 1$.

<u>Théorème</u> (Bresinsky, $[B_2]$)

Soit $\Gamma \subset \mathbb{A}^4$ une courbe monomiale, paramétrée par $x_i = t^{d_i}$ $(i = 1,\ldots,4)$, t.q. le semi-groupe $\Delta = \sum_{i=1}^{4} \mathbb{N}d_i$ est symétrique. Alors Γ est I.C.E.

<u>Preuve</u>. Selon $[B_2]$, si Δ est symétrique, l'idéal P de Γ est engendré par 3 ou 5 éléments. Dans le premier cas, il n'y a rien à faire; dans le second, le système de générateurs est le suivant:

$$f_1 = X_1^{a_1} - X_3^{a_{13}}X_4^{a_{14}}$$

$$f_2 = X_2^{a_2} - X_1^{a_{21}}X_4^{a_{24}}$$

$$f_3 = X_3^{a_3} - X_1^{a_{31}}X_2^{a_{32}}$$

$$f_4 = X_4^{a_4} - X_2^{a_{42}}X_3^{a_{43}}$$

$$f_5 = X_3^{a_{43}}X_1^{a_{21}} - X_2^{a_{32}}X_4^{a_{14}},$$

avec $a_{ij} > 0 \ \forall i,j$, et les relations arithmétiques:

$a_1 = a_{21} + a_{31}$, $a_2 = a_{32} + a_{42}$, $a_3 = a_{13} + a_{43}$,
$a_4 = a_{14} + a_{24}$.

Assertion: les binômes $f_1,\ldots,f_4$ satisfont les conditions de la proposition 6. En effet:
c) est banalement satisfaite; b) est conséquence directe des relations arithmétiques. Pour a): on a $\langle\hat{f}_1,\ldots,\hat{f}_4\rangle = \langle\hat{f}_1,\ldots,\hat{f}_5\rangle = M_0$, la première égalité découlant de $\hat{f}_5 = (\hat{f}_2\hat{f}_4)^{-1}$, et la seconde de $\Gamma = V(f_1,\ldots,f_5)$. $\square$

II. La courbe (t^4,t^b,t^c,t^d) est I.C.E., si on a simultanément

i) $b,c,d \geq 4$

ii) $b \equiv 1$, $c \equiv 2$, $d \equiv 3 \bmod 4$

iii) $2c \geq b+d$.

Par exemple: (t^4,t^5,t^6,t^7).

Remarque: On ne perd rien à supposer i), ni à supposer que les classes mod 4 de $4,b,c,d$ sont 2-à-2 distinctes. Sinon, l'un des quatre entiers serait dans le semi-groupe engendré par les trois autres, et l'assertion découlerait banalement du résultat pour les courbes monomiales dans $\mathbb{A}^3$.

preuve: par le point ii), il existe $\beta,\gamma,\delta \in \mathbb{Z}$ t.q. $2b-c = 4\beta$, $2c-(b+d) = 4\gamma$, $2d-c = 4\delta$. D'une part, $\gamma \in \mathbb{N}$ à cause de iii). D'autre part, on peut supposer

$\beta,\delta \in \mathbb{N}$, sinon c serait dans le semi-groupe engendré par $4,b,d$ et l'assertion serait banale.

Posons donc
$$g_2 = X_2^2 - X_1^\beta X_3$$
$$g_3 = X_3^2 - X_1^\gamma X_2 X_4$$
$$g_4 = X_4^2 - X_1^\delta X_3.$$

On vérifie facilement que $\hat{g}_2,\hat{g}_3,\hat{g}_4$ engendrent M_0 (en éliminant X_4 avec $\hat{g}_3$, puis X_3 avec $\hat{g}_2$). D'autre part, $\hat{g}_2\hat{g}_3\hat{g}_4 = X_2X_4/X_1^{\beta+\gamma+\delta}$, donc en posant $g_1 = X_1^{\beta+\gamma+\delta} - X_2X_4$, on obtient $\hat{g}_1\hat{g}_2\hat{g}_3\hat{g}_4 = 1$.

Il est maintenant facile de vérifier les conditions de la proposition 6 pour $g_1,\ldots,g_4$. □

Remarques: Le semi-groupe du cas II. n'est pas symétrique, car on peut vérifier que le nombre minimal de générateurs de P est 6 (dans le cas symétrique, ce nombre est 3 ou 5 [B_2]). J'ignore s'il existe un cas satisfaisant la proposition 6, et t.q. P nécessite plus de 6 générateurs.

Questions:

1) Etant donné N, existe-t'il une courbe (monomiale) dans $\mathbb{A}^4$, qui soit I.C.E., et t.q. P nécessite au moins N générateurs?
2) La courbe (t^4,t^6,t^7,t^9) est-elle I.C.E.? Dans l'ordre lexicographique sur $\mathbb{N}^4$, c'est à ma connaissance le plus petit cas ouvert.

Références

[B_1] BRESINSKY. Monomial space curves in $\mathbb{A}^3$ as set-theoretic complete intersections. Proceedings of the Amer. Math. Soc., vol 75, no. 1 (1979) 23-24.

[B_2] BRESINSKY. Monomial Gorenstein curves in $\mathbb{A}^4$ as set-theoretic complete intersections. Manuscripta Math. 45 (1978) 111-114.

[E-E] EISENBUD-EVANS. Every algebraic set in n-space is the intersection of n hypersurfaces. Inv. Math. 19 (1973) 107-112.

[H] HERZOG. Note on complete intersections. In KUNZ, Einführung in die kommutative Algebra und algebraische Geometrie, Vieweg 1980 (142-144).

[V] VALLA. On determinantal ideals which are set-theoretic complete intersections. Comp. Math. 42 (1981) 3-11.

CURVES ON RATIONAL AND ELLIPTIC NORMAL CONES WHICH ARE SET THEORETICALLY COMPLETE INTERSECTION (†)

*Andrea Del Centina** *Alessandro Gimigliano**

INTRODUCTION

The aim of this paper is the study of set-theoretically complete intersection properties for curves lying on rational and elliptic normal cones.

In section 1 we will denote by R^n the rational normal curve in $\mathbf{P}^n(k) = \mathbf{P}^n$, where the ground field k is always supposed to be algebraically closed of any characteristic, and by Γ^n a cone in $\mathbf{P}^{n+1}$ over R^n.

The main result of the first section is: Any curve on Γ^n is set-theoretically complete intersection (s.t.c.i.).

From this theorem follows a very simple geometric proof of the following known fact: The R^n's are s.t.c.i. [V].

In the later section we will denote by E^{n+1} the elliptic normal curve in $\mathbf{P}^n(k) = \mathbf{P}^n$, where k is supposed to be an algebraically closed field of characteristic zero, and by Λ^{n+1} a cone in $\mathbf{P}^{n+1}$ over E^{n+1}.

(†) *This is a draft of a paper that will be published elsewhere.*

* *As a member of G.N.S.A.G.A. of C.N.R.*

The main result of this section is: Let C be a curve of degree r on Λ^{n+1} and let $P_1, \ldots, P_r$ be the points where C meets E^{n+1}. Then, fixed a n-osculating point 0 as zero element of the group structure on E^{n+1}, C is s.t.c.i. on Λ^{n+1} if and only if $h(P_1 \oplus \ldots \oplus P_r)=0$ (where $\oplus$ is the group operation) for a suitable $h>0$.

This theorem shows that the condition to be s.t.c.i. for a curve on Λ^{n+1} is an arithmetical one, and also it gives a sufficient condition for a curve lying on Λ^{n+1} to be s.t.c.i. in $\mathbf{P}^{n+1}$.

As a corollary in a very simple geometric way we get the following known fact: The E^{n+1}'s are s.t.c.i. [G]. Moreover we prove that: E^{n+1} is s.t.c.i. on Λ^n if and only if Λ^n has the vertex in a principal point of E^{n+1}.

Linkage methods give some more information about ideals of smooth irreducible curves lying on the cubic or quadric cone in $\mathbf{P}^3$.

NOTATIONS

k	An algebraically closed field of any characteristic up to section 2, where it is supposed to be of characteristic zero.
$\mathbf{P}^n(k)=\mathbf{P}^n$	the projective n-dimensional space on k.
R^n	the rational (projectively) normal curve in $\mathbf{P}^n$.
E^{n+1}	" elliptic " " " " " "
Γ^n, Λ^{n+1}	cones in $\mathbf{P}^{n+1}$ over R^n and E^{n+1}, respectively.

H	the hyperplane section either of Γ^n or Λ^{n+1}.
$\sim$	linear equivalence of divisors.
$\equiv$	numerical equivalence of divisors.
Cl(X)	the class group of linear equivalence on the variety X.
$\mathcal{O}_X$	the structure sheaf of the variety X.
C(A)	the divisor class group of a ring A.
Pic°(E)	the Picard group of invertible sheaves of degree zero on an elliptic curve E.

Also we will write $H^i(\mathcal{F})$ instead of $H^i(X,\mathcal{F})$ for a variety X and a sheaf $\mathcal{F}$, moreover we will denote by $h^i(\mathcal{F})$ the dimension of $H^i(\mathcal{F})$.

1. CURVES ON RATIONAL NORMAL CONES.

Through this section we will denote by Γ^n the rational normal cone of degree n, i.e. the cone in $\mathbf{P}^{n+1}$ over the rational normal curve $R^n \subset \mathbf{P}^n$. Also we will denote by V_o the vertex of Γ^n.

Let us start with the following:

THEOREM 1.1: *Every curve (even singular or reducible) on* Γ^n, *with multiplicity at most* n , *is c.i. on* Γ^n .

Proof.

A ring A is said to be an *almost factorial domain* or A.F.D. for short, if it is a Krull domain and its divisor class group C(A) is torsion.

Any subvariety of codimension one of a projective variety V is s.t.c.i. on V if and only if the homogeneus coordinate ring A_V of V is A.F.D. (see [F] Proposition 6.7).

Now A_{R^n} is A.F.D. (from a geometric point of view $C(A_{R^n})$ is torsion since $\forall$ $P \in R^n$ there exists a hyperplane H_P s.t. $H_P . R^n = nP$), then A_{Γ^n} is A.F.D.. since $A_{\Gamma^n} \simeq A_{R^n}[x]$ and (see [F] Theorem 8.1) a ring A is A.F.D. if and only if A[x] is .

To end the proof we have just to observe that $C(A_{\Gamma^n}) \simeq C(A_{R^n})$.

COROLLARY 1.2: *Every curve on a rational normal cone* $\Gamma^n \subset \mathbf{P}^{n+1}$ *is s.t.c.i. in* $\mathbf{P}^{n+1}$. *In particular any rational normal curve* R^n *in* $\mathbf{P}^n$ *is s.t.c.i.*

Proof.

From Theorem 1.1 it is enough to show that Γ^n itself is s.t.c.i. in $\mathbf{P}^{n+1}$. Indeed: the quadric cone Γ^2 is ideal-theoretically a complete intersection (so s.t.c.i.) in $\mathbf{P}^3$ so R^3 is s.t.c.i. by Theorem 1.1 and so $\Gamma^3 \subset \mathbf{P}^4$ is s.t.c.i. and so on.

Any $R^n \subset \Gamma^{n-1}$ is given as s.t.c.i. with multeplicity n-1 on Γ^{n-1} by an homogeneus polynomial f_n of degree n. So one can easily see from Corollary 1.2 that R^n, as s.t.c.i. in $\mathbf{P}^n$ with multiplicity (n-1)!, is given by an ideal $J = (f_2, f_3, \ldots, f_n)$, where the f_i's are homogeneus polynomials of degree i, in particular $f_2=0$ represent in $\mathbf{P}^3$ the cone Γ^2.

When one consider smooth curves on quadric cones something more can be said (see also [Ga]). If a curve C on Γ^2 does not contain the vertex then degC = 2a and it is well known that C is a ideally complete intersection of Γ^2 and a surface of degree a (e.g. [H] page 384). If $V_o \in$ C then deg C = = 2a+1 and C has genus a^2-a.

In this case more explicit calculations can be done and (using linkage) we can find a rather explicit way to write down ideals which give our curve as s.t.c.i. Namely we have:

THEOREM 1.3: *Let* C *be a smooth curve on* Γ^2, *passing through the vertex of* Γ^2, *of degree* 2a+1 . *Let* $\{x_0; x_1; x_2; x_3\}$ *be homogeneus coordinate in* $\mathbf{P}^3$ *s.t.* $x_0x_2 - x_1^2 = 0$ *is the equation of* Γ^2, *then* C *is s.t.c.i. of* Γ^2 *and a surface of equation:*

$$(1) \qquad \begin{vmatrix} x_0 & x_1 & F_1 \\ x_1 & x_2 & F_2 \\ F_1 & F_2 & 0 \end{vmatrix} = 0$$

where F_i, $i=1,2$, *are homogeneus polynomials of degree* a .

One example

Let C be a hyperelliptic curve of genus g, and let denote by K_o a divisor of the g_2^1 on C. We have that $|(g+1)K_o|$ is a g_{2g+2}^{g+2} . The embedding $\tilde{C}$ of C in $\mathbf{P}^{g+2}$ associated with the $|(g+1)K_o|$ is called *hypercanonical embedding*. This series contains the *minimal sum* $(g+1)g_2^1$ and this means that all the lines joining two corresponding points on $\tilde{C}$ under the hyperelliptic involution, pass through a same point. Then $\tilde{C}$ lies in $\mathbf{P}^{g+2}$ on a cone of degree g+1 (as one can easily see) so on a rational normal (being $\tilde{C}$ normal) cone Γ^{g+1}. Applying our results it is immediate to see $\tilde{C}$ as s.t.c.i. of Γ^{g+1} and a quadric hypersurface. The converse holds. For more details see [E].

2. CURVES WHICH ARE S.T.C.I. ON ELLIPTIC CONES.

In this section the ground field k is supposed to be algebraically closed of char.zero.
Let us recall a few facts about elliptic curves which we need in the following (see [H] for instance).
Let E be an elliptic curve. There exists in E a group structure given by the bijection $E \to Pic^{\circ}(E)$ which, fixed a point $P_o \in E$ (the zero element), maps $P \in E$ into the equivalence class of $P-P_o$. $\oplus$ will denote the group operation.
For every divisor D on E of degree $n \geq 3$, $\mathcal{O}_E(D)$ is very ample and this defines an embedding of E in $\mathbf{P}^{n-1}$ as a curve of degree n which is (projectively) normal. Let denote these

curves by E^n.

A point $P_o \in E^n$ is said to be *principal* if it is s.t.c.i on E^n i.e. (being E^n projectively normal) if there exist two positive integer h and l such that $hP_o \sim lH'$, where H' denote a hyperplane section of E^n. The points P_o for which $nP_o \sim H'$ are said *n-osculation points*; E^n posseses n^2 n-osculation points and then infinite principal points [Ma].

Λ^n will denote an elliptic normal cone, i.e. a cone in $\mathbb{P}^n$ over an elliptic normal curve $E^n \subset \mathbb{P}^{n-1}$.

It is well known that $Cl(\Lambda^n) = Cl(E^n)$ and that for every curve C of degree r on Λ^n we have $C \sim L_1+...+L_r$, where the L_i's are the lines of Λ^n through the points where C and E^n meet ([H] 6.3 page 146).

The following theorem shows that the condition for a curve C to be s.t.c.i. is an arithmetical one.

THEOREM 2.1. *Let* C *be a curve of degree* r *on* Λ^n *and let* $P_1,...,P_r$ *be the* r *points where* C *meet* E^n. *Then, fixing a* n-*osculation point* 0 *as zero element in the group structure on* E^n, C *is s.t.c.i. on* Λ^n *if and only if*

$$h(P_1 \oplus ... \oplus P_r) = 0 \tag{1}$$

for a suitable integer h>0.

Proof.

If C is s.t.c.i. on Λ^n, then for a suitable integers h, m>0 (must be hr = mn) we have $hC \sim mH$. So $h(P_1+...+P_r) \sim mH'$, where H' denotes a hyperplane section of E^n; since $H' \sim n0$, we have $h(P_1 \oplus ... \oplus P_r) = mn0$ hence $h(P_1 \oplus ... \oplus P_r) = 0$.

Suppose $C \subset \Lambda^n$ satisfies (1), it is straightforward to prove that $nhC \sim rhH$ and we conclude just observing that

Λ^n is projectively normal.

Will be useful the following:

PROPOSITION 2.2: *Let* 0 *be an* n*-osculation point of* E^n *. If we take* 0 *as zero element for the group structure of* E^n *, then a set of* r *points* $P_1,\ldots,P_r$ *on* E^n *is s.t.c.i. on* E^n *if and only if*

$$(2) \qquad n_1P_1 \oplus \ldots \oplus n_rP_r = 0$$

for suitable integers $n_i > 0$, $i = 1,\ldots,r$.

Proof.

The proof goes as the one of Theorem 2.1.

Remark 1. Proposition 2.2.holds also if 0 is only a principal point and extends a known result on plane cubics [Fu. p.128]. As a corollary of the above theorem we get:

<u>THEOREM 2.3</u>: *Every elliptic normal curve* E^{n+1} *is s.t.c.i. in* $\mathbf{P}^n$.

Sketch of proof.

Let 0 be an (n+1)-osculation point of E^{n+1}; in the projection from 0 into a $\mathbf{P}^{n-1}$ E^{n+1} projects onto an elliptic normal curve E^n, and the image of 0 is a point $0' \in E^n$ which is an n-osculation point for E^n. Let denote by Λ^n the cone on E^n with vertex 0. We can see that does exist a quadric hypersurface of $\mathbf{P}^n$ containg E^{n+1} but not 0' and such that its intersection with Λ^n is: $E^{n+1} \cup L_1 \cup \ldots \cup L_{n-1}$, where the L_i's are lines.
Then if $E^{n+1} \cap E^n = P_1 + \ldots + P_{n+1}$ and $L_i \cap E^n = P_i'$ $(i=1,\ldots,n-1)$, one can see that $P_1 \oplus \ldots \oplus P_{n+1} \oplus P_1' \oplus \ldots P_{n-1}' = 0'$ and $P_1' \oplus \ldots \oplus P_{n-1}' = 0$, so that $P_1 \oplus \ldots \oplus P_{n+1} = 0'$ too, and then, by Theorem 2.1, E^{n+1}

is the intersection, with multiplicity n, of Λ^n and a hypersurface of degree n+1.
To conclude, one can see that E^{n+1} is s.t.c.i. in $\mathbb{P}^n$ in a similar way than the one used for the R^n's.

Remark 2. An ideal which gives E^{n+1} as s.t.c.i. with multiplicity $\frac{n!}{2}$ is of type: $(f_3, f_4, \ldots, f_{n+1})$, where the f_i's are homogeneus polynomials in $k[x_0, \ldots, x_n]$ of degrees i, and $f_3 = 0$ represents, in $\mathbb{P}^3$, the cone Λ^3, and $f_4=0$ cuts on Λ^3_3, in $\mathbb{P}^3$, three times E^4, etc. etc.
Let us observe that by representing Λ^4 as ideally complete intersection of two quadric hypersurfaces in $\mathbb{P}^4$ of equations $g_2=0$, $g'_2=0$ we can obtain E^{n+1} as s.t.c.i. also by the ideal $(g_2, g'_2, f_5, \ldots, f_{n+1})$ which gives E^{n+1} with multeplicity $\frac{n!}{2}$ (see [G]).

We can see then, with a similar construction as the one of Theorem 2.3, that for $E^{n+1} \subset \Lambda^n$ to be a s.t.c.i. strictly depends on the point we choose as the vertex of Λ^n, as the following proposition points out:

PROPOSITION 2.4: *An elliptic normal curve* E^{n+1} *on an elliptic normal cone* Λ^n *is s.t.c.i. on* Λ^n *if and only if the vertex of* Λ^n *is a principal point of* E^{n+1}.

Let us make some remarks on curves on Λ^3.
Via linkage (in a similar way to that used for quadric cones) we can prove the following:

THEOREM 2.5 *Let* C *be a smooth irreducible curve on* $\Lambda^3 \subset \mathbb{P}^3$. *Then* C *is s.t.c.i. in* $\mathbb{P}^3$.

Proof.

Let V_o be the vertex of Λ^3. If $V_o \notin C$ then $\deg C = 3a$ and the theorem follows from Riemann-Roch's theorem. If $V_o \in C$ and the base curve E^3 of Λ^3 is represented by the equation:

$$x_2 x_o^2 - x_1^3 + (\lambda-1) x_2 x_1^2 - \lambda x_1 x_2^2 = 0$$

(it is always possible, see [H] page 319), the ideal of C is generated by the 2×2 minors of a matrix:

$$\begin{pmatrix} x_1^2 & x_o^2 + (\lambda+1)x_1 - \lambda x_1 x_2 & f \\ x_2 & x_1 & g \end{pmatrix}$$

so the theorem descends from a result of Robbiano and Valla [Ro-Va] about this kind of ideals.

Remark 3. In positive characteristic $\neq 2$, the above results still hold if E^n is supposed to have one (and then infinite [Ma]) principal points, for instance the (projectively) normal model of elliptic curves E associated with sheaves of type $\mathcal{O}_E(nP_o)$.

This is the case of elliptic (projectively) normal curves given by a parametrization of type:

$$\{1; \wp; \wp^2; \ldots; \wp^m; \wp'; \wp'\wp; \ldots; \wp'\wp^{m-2}\} \text{ for } E^{2m} \subset P^{2m-1}$$
$$\{1; \wp; \wp^2; \ldots; \wp^m; \wp'; \wp'\wp; \ldots; \wp'\wp^{m-1}\} \text{ " } E^{2m+1} \subset P^{2m}$$

where $\wp$ and $\wp'$ are the extensions of the usual Weierstrass function for $k = \mathbb{C}$ and of its derivative (see [G]).
If $k = \bar{\mathbb{F}}_p$ and $p \neq 2,3$ by [Ma] Corollary 1.6, any point of

E^n is principal and then we have, by our results, that every curve on Λ^n is s.t.c.i.

REFERENCES

[B] Burch L. *On ideals of finite homological dimension in local rings.* Proc.Cambr.Phil.Soc. 64, (1968), 941-948.

[E] Eisenbud D. *Transcanonical embeddings of hyperelliptic curves.* Journal of Pure and Appl. Algebra 19, (1980), 77-83.

[F] Fossum R.M. *The divisor class of a Krull domain.* Erg. der Math. und ihrer Grenz.Bd. 74, Springer-Verlag, Berlin-Heidelberg-New York, (1973).

[Fu] Fulton W. *Algebraic curves.* Benjamin, New York-Amsterdam, 1979.

[Ga] Gallarati D. *Sul contatto di superficie algebriche lungo curve,* Ann. di Mat. Pura e Applicata (4) 38 (1955) 225-251.

[G] Gattazzo R. *La curva ellittica normale* $C_n \subset P^{n-1}$, $n \geq 4$ *è intersezione completa insiemistica di n-2 ipersuperficie.* Boll. U.M.I. 18-B, (1981), 977-988.

[H] Hartshorne R. *Algebraic Geometry.* Springer-Verlag, New York-Heidelberg-Berlin, 1977.

[Ma] Maroscia P. *Alcune osservazioni sulle varietà intersezioni complete.* Rendiconti Acc. dei Lincei LXVI (1973), 365-371.

[P-S] Peskine C. - Szpiro L. *Liason de varietes algebriques.* Inv.Math. 26 (1974) 271-302.

[Ro-Va] Robbiano L. - Valla G. *Some curves in $\mathbb{P}^3$ are set-theoretic complete intersections.* Preprint.

[Va] Valla G. *On determinantal ideals which are set-theoretic complete intersections.* Comp.Math. 42 (1981) 3-11.

[V] Verdi L. *Le curve razionali normali come intersezioni complete insiemistiche.* Boll. U.M.I. 16-A (1979) 385-390.

A.Del Centina, A.Gimigliano
Istituto Matematico "U.Dini"
Firenze, Italy

COMPLETE INTERSECTIONS IN $\mathbb{P}^2$: CAYLEY-BACHARACH CHARACTERIZATIONS

E. Davis[1)] and P. Maroscia[1)]

Introduction: The Euler-Cramer Problem.

Many have contributed to our knowledge of the special properties of a group of points in the plane which occurs as the complete intersection of two curves: in the 18th century, MacLaurin, Euler, Cramer; in the last century, Jacobi, Cayley, Max Noether; and in our own time, Beniamino Segre, Griffiths, Harris. These last three have explicitly addressed the natural problem -- and generalizations thereof -- of characterizing complete intersections in the plane. Nevertheless, our reading of the historical record suggests linking that problem to the names of Euler and Cramer: for it is that problem, especially for the famous case of a group of 9 points, that underlies the entire Euler-Cramer correspondence on the "Cramer Paradox" [E, pp. XI-XIII].

We must state the problem precisely. In the sequel: *Z denotes a nonempty finite set of closed points in $\mathbb{P} = \mathbb{P}^2(k)$, k an algebraically closed field (so Z is a reduced 0-dimensional subscheme of $\mathbb{P}$); "curve" means "1-dimensional closed subscheme of $\mathbb{P}$, without isolated or embedded points" (so globally defined by one equation); "intersection" is to be understood scheme-theoretically (so "Z is a complete intersection" means that Z is the intersection of two curves, and hence that Z is, as a scheme, globally defined by two equations)*. Thus:

ECP (Euler-Cramer Problem): Find "good" necessary and sufficient conditions on Z for the existence of curves A and B with $Z = A \cap B$.

[1)] Thanks are due to: the Consiglio Nazionale delle Ricerche for support; the Dipartimento di Matematica, Politecnico di Torino, for including a preliminary version of this work in its preprint series (On the Cayley-Bacharach theorem, Preprint No. 1983-34); the mathematical communities of Torino, Genova, Roma, Napoli, Palermo and Catania for most gracious hospitality extended to the first named author.

Quite early on were discovered primitive forms of two important properties of complete intersections:

Bezout Theorem (MacLaurin, 1720). Two curves, of degrees m and n respectively, having no component in common, can have at most mn points in common.

Theorem of the 9 Points (Euler, 1744). If a group of 9 points is the complete intersection of two cubics, then any cubic passing through 8 of the 9 passes through all 9.[2)]

These two "almost" characterize the complete intersection of cubics:

Elementary Exercise. If a group of 9 points satisfies the conclusion of the Theorem of the 9 Points and lies on no conic, then it is the complete intersection of two cubics.

The generalization of this last fact is the main theorem of this paper; its corollaries include (we claim) solutions to both ECP and a quite natural restricted version of it.

To support the claim just made, we appeal to authority. Both [S] and [GH] view ECP as the problem of "inverting the Bezout Theorem"; and the latter goes on to suggest (implicitly) that a good solution might well be formulated in terms of the "Cayley-Bacharach Property" (the conclusion of Cayley's generalization of the Theorem of the 9 Points -- see [SR, p. 97] or (2.3) below). [M] transforms that suggestion into a formal conjecture, which [DGM] affirms. This note, drawing on elements of [GH], [M] and [DGM], develops a strikingly sharper treatment of ECP than is contained in the union of those three works. In fact, if a Cayley-Bacharach treatment is indeed admitted as "good", we offer in this note a definitive resolution of the problem.

---

2) This can be deduced easily from the Bezout Theorem [EC, p. 237]. (Most of the modern textbooks ignore this fact.) The question of whether Euler knew that proof is moot: we know only that he asserted the result in a letter to Cramer, and that his reply to the letter in which Cramer asks for justification of that assertion avoids the issue.

§1. Statement of principal results.

Throughout this paper X denotes a 0-dimensional subscheme of $\mathbb{P}$ -- not necessarily reduced. (Although our principal results can be formulated within the confines of "reduced", our proofs cannot. So in the interest of economy, we maintain the level of notational generality required for the proofs.) As usual, $\mathcal{I}_Y$ denotes the ideal sheaf on $\mathbb{P}$ of the closed subscheme Y, and $h^0(\mathcal{I}_Y(t))$ denotes the k-length of $H^0(\mathcal{I}_Y(t))$. So $h^0(\mathcal{I}_X(t))-1$ is the dimension of the linear system of curves of degree t passing through X.

Special Notation (t, α, δ, $CI(a,b)$, $h(a,b,t)$). t always denotes an integer variable, and (a,b) an ordered pair of positive integers with $a \leq b$. δ denotes the degree function on closed subschemes of $\mathbb{P}$. (So: $\delta(X) \geq \text{card} X$, equality $\Leftrightarrow X = X_{red}$; δ(curve) is the degree of a polynomial globally defining the curve.) $\alpha(X) = \min\{t: H^0(\mathcal{I}_X(t)) \neq 0\}$. X is a complete intersection provided that there are curves A and B with $X = A \cap B$; X is a $CI(a,b)$ provided that A and B can be chosen such that $\delta(A) = a$ and $\delta(B) = b$. (Note: X is a $CI(a,b) \Rightarrow \alpha(X) = a$.) In case X is a $CI(a,b)$, the integer $h^0(\mathcal{I}_X(t))$ depends only on (a,b) and t (e.g., see §2); we denote this integer by $h(a,b,t)$.

Definition (Cayley-Bacharach Property). Given $d \in \mathbb{Z}$, we say that X has $CB(d)$ provided that: for every subscheme Y of X with $\delta(Y) = \delta(X)-1$, $H^0(\mathcal{I}_Y(t)) = H^0(\mathcal{I}_X(t))$ $(t \leq d)$. We say that X is $CB(a,b)$ provided that X has $CB(a+b-3)$ and $\delta(X) = ab$.

Now $CI(a,b) \Rightarrow CB(a,b)$ (see (2.3) below), but $CB(a,b) \not\Rightarrow$ complete intersection: e.g., consider a group of 9 points on an irreducible conic. Question [GH]: $CB(a,b) + (?) = CI(a,b)$? In §4 we give an answer valid for arbitrary X, but not so satisfying an answer as that furnished by the following theorem in the reduced case. We give the proof in §3, after a bit of algebraic reformulation.

(1.1) THEOREM. Suppose that Z is $CB(a,b)$, and let $\alpha = \alpha(Z)$. Then:

(a) $h^0(\mathcal{I}_Z(t)) \geq h(a,b,t)$ $(t \leq b)$. (Hence: $\alpha \leq a$.)

(b) If there exists $a-1 \leq t \leq b$ such that equality holds in (a), then Z is a $CI(a,b)$. (Hence: $\alpha = a \Rightarrow Z$ is a $CI(a,b)$.)

(c) If Z is not a $CI(a,b)$, then $h^0(\mathcal{I}_Z(\alpha)) = 1$, and for $\alpha \leq t \leq a+b-\alpha$, the canonical maps $H^0(\mathcal{I}_Z(\alpha)) \otimes H^0(\mathcal{O}_{\mathbb{P}}(t-\alpha)) \to H^0(\mathcal{I}_Z(t))$ are surjective. (Hence: If the linear system of curves of degree $b+1$ passing through Z has no fixed curve, then Z is a $CI(a,b)$.)

Remarks. (1) [GH] and [GHP] both prove (1.1a) for $a = b$, in which case [GH] shows that the $CI(a,b)$'s are Zariski open in the $CB(a,b)$'s; (1.1a,b) shows that this is so in any case. (2) The interval in (1.1b) is as large as possible: let Z be the 9 points on an irreducible conic. (3) One can somewhat more weakly take the degree b instead of $b+1$ for the criterion in (1.1c) -- as is done in [GH]; but $b+1$ is the largest degree possible: again the 9 points on an irreducible conic.

(1.2) COROLLARY (solution to ECP). Let $\alpha = \alpha(Z)$, $\delta = \delta(Z)$. Then: Z is a complete intersection $\Longleftrightarrow \alpha^{-1}\delta \in \mathbb{Z}$ and Z has $CB(\alpha+\alpha^{-1}\delta-3)$.

Proof. Enough to prove $\Leftarrow$. Let $a = \min\{\alpha, \alpha^{-1}\delta\}$, $b = \max\{\alpha, \alpha^{-1}\delta\}$. Then Z is $CB(a,b)$; whence by (1.1a), $\alpha \leq a$. So $\alpha = a$; whence by (1.1b), Z is a $CI(a,b)$.

Restricted ECP. Suppose now $\delta(Z) = mn$, Z on a reduced curve M with $\delta(M) = m$. Problem: Find necessary and sufficient conditions on Z for the existence of a curve N with $Z = N \cap M$. By (2.3) below (i.e., by the Bezout and Cayley-Bacharach theorems), two conditions are necessary: (i) Z has $CB(m+n-3)$; (ii) for every irreducible component D of M, $n\delta(D) = \delta(Z \cap D)$. Trivial examples show that neither of the two implies the other, and it is merely an exercise in the use of the Bezout theorem to deduce the following corollary from (1.1a,b): treat separately $m \leq n$ and $n < m$.

(1.3) COROLLARY (solution to Restricted ECP). In the notation of the discussion immediately preceding: (i) + (ii) $\Rightarrow$ N exists.

Segre's Theorem [S, p. 3]. Consider now the Restricted ECP with M irreducible, in which case (ii) becomes irrelevant. Segre proves (1.3) with (i) weakened to: Z does not impose independent conditions on the curves of degree m+n-3 (i.e., $H^1(\mathcal{I}_Z(m+n-3) \neq 0)$. But his proof employs the unstated hypothesis: $Z \cap \mathrm{Sing} M = \emptyset$. (Of course this must be so if the theorem is valid.) In fact this unstated hypothesis is implied by those that are stated; however, the best way we know to prove that fact is to prove Segre's theorem directly, in fact for arbitrary X over any ground field [D].

Remarks on the proof of (1.1). First note that the theorem is trivially valid if a = 1 or b = 2. In fact one can give a special proof of (1.1) for $a \leq 2$ that is valid without "reduced" (see §4); however, that argument gives no insight into a proof of (1.1). Such an insight is best gained by examining CB(a,b) for small values of b. For example, using only the Bezout theorem, supplemented by basic elementary facts about systems of linear equations, and proceeding quite straightforwardly, one can prove (1.1) for $b \leq 5$. We recommend that the reader do this, and then afterward take note of the obstructions presented to those methods by the cases (3,7), (5,6), (a,a) for large a. The proof we eventually give in §3 is basically elementary, but not straightforward. If the reader does as we have suggested, we think that he will thereby grasp the general outlines of our proof and gain insight into the motivation behind its several steps. In any case, something more than the bare-handed methods suggested above seems to be required. For us that something more is "Hilbert function".

Definitions. $H(X,t) = h^0(\mathcal{O}_{\mathbb{P}}(t)) - h^0(\mathcal{I}_X(t))$. H(a,b,-) denotes the Hilbert function of a CI(a,b): $H(a,b,t) = h^0(\mathcal{O}_{\mathbb{P}}(t)) - h(a,b,t)$. We say that X is a HCI(a,b) provided that H(X,-) = H(a,b,-).

We sketch a development of the basic properties of H(X,-) needed for our proof of (1.1) in §2. Obviously CI(a,b)⇒HCI(a,b); but except for the trivial case a = 1, HCI(a,b)⇏CI(a,b). [DGM, §4] analyzes the algebraic and geometric implications of HCI(a,b); and we use that analysis in §4 to extend partially the results stated above to the nonreduced case. Translating (1.1a,b) into Hilbert function notation gives:

(1.4) COROLLARY. Suppose Z is CB(a,b). Then: $H(X,t) \leq H(a,b,t)$ $(t \leq b)$; if equality holds for some $a-1 \leq t \leq b$, then Z is a CI(a,b).

Remarks. [DGM, (4.21)], is a weak form of (1.4) which affirms the conjecture of [M] for Z: HCI(a,b) + CB(a+b-3) ⇒ CI(a,b). (We prove this in §4 without "reduced".) But even the weak form of (1.4), with CI and CB appropriately redefined for $\mathbb{P}^n$ $(n>2)$, is false for $n > 2$ -- however, only because "complete intersection" ≠ "arithmetically Gorenstein" for $n > 2$. Indeed, the reduced, 0-dimensional, arithmetically Gorenstein subschemes of $\mathbb{P}^n$ are characterized by symmetry of Hilbert function and validity of the Cayley-Bacharach theorem. (See [DGO] for the definitions of the underlined terms and a proof.)

§2. Algebraic preparations for the proof of (1.1).

We retain in force all notation previously introduced. However, the standard cohomological notation used to describe our results is not so convenient for our proofs. Therefore, in the course of this section, we introduce further notation to remain in force from then on.

Further Notation. Fix a homogeneous coordinate algebra K for $\mathbb{P}$. As usual, we regard K as $\mathbb{Z}$-graded with $K_t = 0$ $(t<0)$. Hereafter, this property is subsumed under $\mathbb{Z}$-graded algebra. Let L denote a general linear form in K, and let $x \mapsto \bar{x}$ denote the canonical map $K \to K/LK = \bar{K}$ (the homogeneous coordinate algebra for a general line in $\mathbb{P}$). Let $F \in K_a$ and $G \in K_b$, (a,b) as above, denote forms with trivial GCD (greatest common divisor). Because L is general, the same is true for $\bar{F}$ and $\bar{G}$; let $R = \bar{K}/(\bar{F},\bar{G})$. We denote the length function on k-modules by /-/; i.e., if M is a k-module, then /M/ is the k-length of M.

One can prove the following proposition directly by elementary computation, or more sophisticatedly by appealing to the fact that R is an artinian Gorenstein k-algebra. The corollary stated immediately thereafter is an immediate consequence of the proposition; and as we shall soon see, the corollary in effect contains both the Bezout and Cayley-Bacharach theorems. (This almost trivial observation is further developed in [DGO].)

(2.1) PROPOSITION. $\{x \in R : xR_1 = 0\} = R_{a+b-2}$. $R_t \neq 0 \Leftrightarrow 0 \leq t \leq a+b-2$. Moreover: $/R_t/ = a$ $(a-1 \leq t \leq b-1)$; $/R_t/ = /R_{a+b-2-t}/ = t+1$ $(0 \leq t \leq a-1)$.

(2.2) COROLLARY. Let $J \neq R$ be a homogeneous ideal of R. Then:

(a) $/(R/J)/ = ab \Leftrightarrow J = 0$.

(b) $/(R/J)/ = ab-1 \Leftrightarrow J_t = 0$ $(t \neq a+b-2)$, $J_{a+b-2} \neq 0$.

Further Notation. Let I(X) denote the homogeneous ideal of X in K. So $h^0(\mathcal{J}_X(t)) = /I(X)_t/$. In particular, if X is a CI(a,b), then I(X) is of the form (F,G); consequently, if X is a HCI(a,b), then $/I(X)_t/ = /(F,G)_t/ = h(a,b,t)$. For $0 \neq J$ a homogeneous ideal of a $\mathbb{Z}$-graded ring, define $\alpha(J) = \min\{t : J_t \neq 0\}$. So $\alpha(I(X)) = \alpha(X)$. Let $A = K/I(X)$ (the homogeneous coordinate algebra of X). We denote the multiplicity of the algebra A by $\delta(A)$; so $\delta(A) = \delta(X)$.

Observations. (1) A is 1-dimensional and Cohen-Macaulay. (2) So L is A-regular by the generality of L; whence $\delta(A) = /(A/LA)/$. (3) If X is a subscheme of a CI(a,b), i.e., if $I(X) \supset (F,G)$, then: $A/LA = R/J$, where $J = \overline{I(X)}/(\bar{F},\bar{G})$; whence $\delta(X) = /(R/J)/$.

(2.3) COROLLARY. Suppose X is a CI(a,b). Then:

(a) Bezout Theorem: $\delta(X) = ab$.

(b) Cayley-Bacharach Theorem: X has CB(a+b-3).

Proof. We let $I(X) = (F,G)$, and use the foregoing observations. Since $A/LA = R$, (a) follows immediately from observation (2) and (2.2a). To prove (b), we may assume $ab \neq 1$. Let Y be a subscheme of X such that $\delta(Y) = \delta(X)-1$. We must prove: $I(Y)_t = (F,G)_t$ ($t<a+b-2$). So it suffices to prove: $\alpha(I(Y)/(F,G)) = a+b-2$. Let $J = \overline{I(Y)}/(\bar{F},\bar{G})$. Now because L is general, it is $K/I(Y)$-regular; whence $\alpha(I(Y)/(F,G)) = \alpha(J)$. (This is well known and easy to prove -- e.g., see [DGM, §3].) So applying observation (3) and (2.2b) gives: $\alpha(I(Y)/(F,G)) = a+b-2$. QED.

Further Notation. Given a $\mathbb{Z}$-graded k-algebra S, let H(S,-) denote its Hilbert function: $H(S,t) = /S_t/$. So: $H(K,t) = /K_t/ = h^0(\mathcal{O}_{\mathbb{P}}(t))$; $H(X,t) = /K_t/-/I(X)_t/ = H(A,t)$; $h^0(\mathcal{I}_X(t)) = H(K,t)-H(A,t)$.

Observations on H(K/(F,G),-). Because $H(a,b,t) = H(K/(F,G),t)$, $h(a,b,t) = H(K,t)-H(K/(F,G),t)$. Now since F is K-regular, and G is K/FK-regular, it follows that $H(K/FK,t) = H(K,t)-H(K,t-a)$, and that $H(K/(F,G),t) = H(K/FK,t)-H(K/FK,t-b)$. Moreover, $H(K,t) = /K_t/$, and for $t \geq 0$, $/K_t/$ is precisely the number of distinct monomials of degree t in three indeterminates. Summarizing then:

$H(K/FK,t) = /K_t/-/K_{t-a}/$.

$H(a,b,t) = /K_t/-/K_{t-a}/-/K_{t-b}/+/K_{t-a-b}/$.

$h(a,b,t) = /K_{t-a}/+/K_{t-b}/-/K_{t-a-b}/$.

$/K_t/ = (t+2)(t+1)/2$ $(0 \leq t)$.

Hereafter, whenever these formulas are needed for an explicit computation, we use them without comment.

The next proposition collects the properties of H(X,-) needed for the proof of (1.1) and for use in §4. But first a useful definition, which is justified by the proposition.

Definition. $\tau(X) = \min\{t : H(X,t)=\delta(X)\}$.

(2.4) PROPOSITION. Let $\alpha = \alpha(X)$, $\beta = \inf\{t: I(X)_t \neq I(X)_\alpha K_{t-\alpha}\}$, and $\tau = \sup\{t: H(X,t) > H(X,t-1)\}$. Then:

(a) $H(X,t) = /K_t/$ $(t<\alpha)$.

(b) $/I(X)_\alpha/ = 1 \Rightarrow H(X,t) = /K_t/-/K_{t-\alpha}/$ $(t<\beta)$.

(c) $\tau(X) = \tau \in \mathbb{N}$; $H(X,t) = \delta(X)$ $(\tau \leq t)$; $H(X,t) > H(X,t-1)$ $(0 \leq t \leq \tau)$.

(d) For $Y \neq X$ a subscheme of X, $I(Y)_t \neq I(X)_t$ $(\tau \leq t)$; hence: X has $CB(d) \Rightarrow d < \tau$.

(e) X is a $HCI(a,b) \Rightarrow \delta(X) = ab$; $\tau = a+b-2$; $H(X,a+b-3) = ab-1$.

(f) X is a $HCI(a,b) \Longleftrightarrow H(X,t) = H(a,b,t)$ $(t \leq a+b-3)$ and $\delta(X) = ab$.

Proof. $A_t = K_t$ $(t<\alpha)$; hence (a). Let $0 \neq P \in I(X)_\alpha$. Under the hypothesis of (b), $A_t = (K/PK)_t$ $(t<\beta)$; hence (b). Because L is A-regular, $H(A/LA,t) = H(X,t)-H(X,t-1)$. That observation immediately yields (c): $\tau = \max\{t: (A/LA)_t \neq 0\}$. (c) implies (d) because $\delta(Y) < \delta(X)$. If X is a $HCI(a,b)$, then $H(A/LA,-) = H(R,-)$; so (e) follows from (c) and (2.1). (f) follows immediately from (c) and (e).

Our proof of (1.1) involves a certain amount of manipulation of inequalities, among them the following elementary fact:

(&) $0<x \leq y \leq z \leq w$ and $yz \leq xw \Rightarrow y+z \leq x+w$, equality $\Longleftrightarrow x=y$ and $z=w$.

We use freely and without comment the following standard facts:

X is $HCI(a,b) \Rightarrow \delta(X) = ab$.

$Y \neq X$ a subscheme of $X \Rightarrow \delta(Y) < \delta(X)$.

$\delta(X) \geq \mathrm{card}(\mathrm{Reg}X) + 2\mathrm{card}(\mathrm{Sing}X) \geq \mathrm{card}X = \delta(X_{red})$.

§3. Proof of (1.1).

We retain in force all the notation previously introduced. Moreover: throughout this section we assume that our reduced scheme Z is $CB(a,b)$, and we denote its ideal $I(X)$ by I. We first reformulate (1.1) as (3.1), immediately thereafter deducing (1.1) from (3.1). The rest of this section is devoted to the proof of (3.1).

(3.1) THEOREM. Let $\alpha = \alpha(Z)$, $\beta = \inf\{t: I_t \neq I_\alpha K_{t-\alpha}\}$. Then:

(a) $\alpha \leq a$.

(b) $\alpha = a \Rightarrow Z$ is a CI(a,b).

(c) $\alpha < a \Rightarrow /I_\alpha/ = 1$, and $a+b < \alpha+\beta$. (Recall: $/I_t/ = h^0(\mathcal{I}_Z(t))$.)

Proof of (3.1)⇒(1.1). Given (3.1a,b), (1.1c) merely translates (3.1c) into the notation of §1. Since $h(a,b,a-1) = 0 < /I_{a-1}/$ if $I_{a-1} \neq 0$, it suffices to prove: $I_{a-1} \neq 0 \Rightarrow h(a,b,t) < /I_t/$ $(a \leq t \leq b)$. Now for $0 \neq P \in I_{a-1}$: $/I_t/ \geq /PK_{t-a+1}/ = /K_{t-a+1}/ \geq /K_{t-a}/ + 2$ $(a \leq t)$. And on the other hand: $h(a,b,t) = /K_{t-a}/ + /K_{t-b}/ - /K_{t-a-b}/ \leq /K_{t-a}/ + 1$ $(t \leq b)$. Done.

Now for the proof of (3.1). Again we remark that (3.1) is trivially valid if $a = 1$ or $b = 2$. Therefore, in order to avoid the annoyance of special consideration of the trivial cases, all further arguments in this section assume $a \geq 2$ and $b \geq 3$. The proof rests on three lemmas, the hypotheses of which all subsume the hypothesis and notation of (3.1). The first two are abstractions of the proof of a weak form of (1.1b): $/I_t/ = h(a,b,t)$ $(t \leq b) \Rightarrow Z$ is a CI(a,b). Two lemmas for the two cases $a = b$ and $a \neq b$. It may help the reader to bear this in mind.

(3.2) LEMMA. Z is not a CI(a,b) and $\alpha^2 \leq ab \Rightarrow /I_\alpha/ = 1$.

(3.3) LEMMA. Z is not a CI(a,b), $/I_\alpha/ = 1$ and $\alpha\beta \leq ab \Rightarrow a+b < \alpha+\beta$.

Before the proofs observe that (3.1c) is a formal consequence of these two lemmas and the inequality (&). The third lemma comes after the proofs of these two; and all three are used to prove (3.1a,b), which comprise, in effect, the main theorem of this paper.

Proof of (3.2). Suppose (A): $/I_\alpha/ \neq 1$. So under the hypothesis of (3.2), it suffices to prove: (A) ⇒ Z does not have CB(a+b-3). Let D denote a GCD of I_α, and let $d = \deg D$.

CLAIM: $0<d<\alpha$. Proof. If $d = \alpha$, then contrary to (A), $/I_\alpha/ = 1$. So suppose $d=0$. Then I_α contains two forms with trivial GCD, i.e., Z is a subscheme of a $CI(\alpha,\alpha)$; and since $\alpha^2 \leq ab = \delta(Z)$, it follows that Z is that $CI(\alpha,\alpha)$ and $\alpha^2 = ab$. Therefore $a<\alpha<b$: for otherwise, $\alpha=a=b$, whence contrary to hypothesis, Z is a $CI(a,b)$. By (&) then, $2\alpha < a+b$. On the other hand, by (2.4d,e), $a+b-3 < 2\alpha-2$. This contradiction proves CLAIM.

By CLAIM, $Y = \{p\in Z:D(p)\neq 0\} \neq \emptyset$. Observe: $DI(Y)_{t-d} = I_t$ $(t\leq\alpha)$. So: $\alpha(Y) = \alpha-d$; and $I(Y)_{\alpha-d}$ contains two forms with trivial GCD, i.e., Y is a subscheme of a $CI(\alpha-d,\alpha-d)$, say X. Sublemma: $Y\not\subset \mathrm{Sing}X$. Proof. Otherwise, $(\alpha-d)^2 \geq 2\delta(Y)$; so it suffices to prove $2\delta(Y) > (\alpha-d)^2$. By (2.4a,c) and CLAIM: $2\delta(Y) \geq 2H(Y,\alpha-d-1) = (\alpha-d+1)(\alpha-d) > (\alpha-d)^2$. This proves the sublemma. So Y contains a simple point of X, say p; and it makes sense to write $X-\{p\}$ to denote the unique subscheme of X having degree $\delta(X)-1$, and not having $\{p\}$ as an irreducible component.

By (2.4d,e), there exists $P \in I(X-\{p\})_{2\alpha-2d-2}\backslash I(X)$. (By the usual convention, I(null scheme) = K; actually one can prove $X\neq\{p\}$.) Hence $DP \in I(Z-\{p\})_{2\alpha-d-2}\backslash I$; so Z does not have $CB(2\alpha-d-2)$. Now by CLAIM, $2\alpha-d-2 \leq 2\alpha-3$; and by "$\alpha^2 \leq ab$" and (&), $2\alpha \leq a+b$. So: (A) $\Rightarrow$ Z does not have $CB(a+b-3)$. QED.

Proof of (3.3). We shall skip lightly over those details of this proof which are in principle almost identical with the corresponding points in the proof of (3.2). Suppose (B): $\alpha+\beta \leq a+b$. So under the hypothesis of (3.3), it suffices to prove: (B) $\Rightarrow$ Z does not have $CB(a+b-3)$. Let D denote a GCD of I_β, and let $d = \deg D$.

CLAIM: $0<d<\alpha$. Proof. Since $I_\alpha \subset DK_{\alpha-d}$ and $I_\beta \not\subset I_\alpha K_{\beta-\alpha}$, it follows that $d < \alpha$. So suppose $d=0$. Then I contains two forms, one in I_α, the other in I_β, with trivial GCD, i.e., Z is a subscheme of a $CI(\alpha,\beta)$; and since $\alpha\beta \leq ab = \delta(Z)$, it follows that Z is that $CI(\alpha,\beta)$ and $\alpha\beta = ab$. Consequently, $\alpha+\beta < a+b$: for otherwise, by (B), $\alpha+\beta = a+b$, in which case $(\alpha,\beta) = (a,b)$, whence contrary to hypothesis, Z is a $CI(a,b)$. But we have, by (2.4d,e), $a+b-3 < \alpha+\beta-2$. This contradiction proves CLAIM.

For $Y = \{p\in Z:D(p)\neq 0\}$: $DI(Y)_{t-d} = I_t$ $(t\leq\beta)$. It follows that Y is a subscheme of a $CI(\alpha-d,\beta-d)$, say X. Then arguing as in the proof of (3.2), but using (2.4b) in place of (2.4a), one shows that Y contains a simple point p of X. Finally, using X, $X-\{p\}$ and (2.4d,e), one shows that Z does not have $CB(\alpha+\beta-d-2)$. So: (B) $\Rightarrow$ Z does not have $CB(a+b-3)$.

The third lemma is an abstraction of the old trick employed by [SR, p. 98] to deduce an "extended" Cayley-Bacharach theorem from the one stated above, and by [GHP, p. 674] to prove (1.1a) for the case $a=b$. Those arguments consist primarily of translating into geometric form the most basic facts in the theory of systems of homogeneous linear equations: a set of m such equations in n unknowns has nontrivial solutions if $m<n$, in which case the number of independent solutions is at least $n-m$; if there are no nontrivial solutions, then at least one subset consisting of n of the equations has no nontrivial solutions. Our translation is:

(3.4) LEMMA. Let u and v be nonnegative integers such that $u+v \le a+b-3$ and $u<\alpha$. Then $H(Z,v) \le ab-/K_u/$. If also $/K_u/+/K_v/ > ab$, then $\alpha \le v$.

Proof. Note that the first assertion implies the second. (Proof: $H(Z,v) \le ab-/K_u/$ and $/K_u/+/K_v/ > ab \Rightarrow H(Z,v) < /K_v/ \Rightarrow I_v \ne 0$.) So it suffices to prove: $I_u = 0 \Rightarrow H(Z,v) \le ab-/K_u/$. Now we do "linear equations":

Since $I_u = 0$, there is a subset Y of Z such that $\text{card}Y = /K_u/$ and $I(Y)_u = 0$. CLAIM: $I_v = I(Z-Y)_v$. Proof. Since $I \subset I(Z-Y)$, it suffices to prove: $p \in Y$ and $V \in I(Z-Y)_v \Rightarrow V(p) = 0$. Because $\text{card}(Y-\{p\}) < /K_u/$, there exists $0 \ne U \in I(Y-\{p\})$; whence because Z has $CB(u+v)$, $UV \in I_{u+v}$. Therefore $V(p) = 0$. This proves CLAIM. By CLAIM and (2.4c), $H(Z,v) = H(Z-Y,v) \le \delta(Z-Y) = ab-/K_u/$. QED.

Remark. Observe that one can eliminate the use of (2.4) in the proof of (3.4) by using instead the elementary facts about systems of linear equations mentioned above. The proofs of (3.2) and (3.3) really need (2.4) for the possibly nonreduced scheme X.

Proof of (3.1a). Suppose $a < \alpha$. Suppose also for a moment $b-a \leq 3$. Then since $/K_a/+/K_{b-3}/ > ab$, by (3.4), $\alpha \leq b-3 \leq a < \alpha$ -- absurd. So for the rest of this proof we assume $b-a > 3$.

Let r be any nonnegative integer such that $a+r \leq b-r-3$, equivalently, $2r+3 \leq b-a$. With the latter inequality one shows: $/K_{a+r}/+/K_{b-r-3}/ > ab$. So (3.4) gives: $a+r < \alpha \Rightarrow \alpha \leq b-r-3$ and $H(Z,b-r-3) \leq ab-/K_{a+r}/$. Now let r be the largest integer such that $a+r \leq b-r-3$; so $a+r = b-r-3$, or $a+r+1 = b-r-3$. By the observation just made: in the first case, $\alpha \leq a+r$; in the second case, $\alpha \leq a+r+1$. Now fix r to be the largest integer such that $a+r < \alpha$. In view of the foregoing observations:

$$(\dagger) \quad \alpha = a+r+1 \leq b-r-3; \; H(Z,b-r-3) \leq ab-/K_{\alpha-1}/ = ab-/K_{a+r}/.$$

Now by (2.4a,c) and ($\dagger$): $/K_{\alpha-1}/ = H(Z,\alpha-1) \leq H(Z,b-r-3) \leq ab-/K_{\alpha-1}/$. Thus $2/K_{\alpha-1}/ \leq ab$; whence $\alpha^2 < ab$. By (3.2) then, $/I_\alpha/ = 1$. CLAIM: $\beta \leq b-r-3$. Otherwise: by ($\dagger$) and (2.4b), $ab-/K_{a+r}/ \geq H(Z,b-r-3) = /K_{b-r-3}/-/K_{(b-r-3)-(a+r+1)}/$; and this inequality implies $b \leq a+r+1$ -- a contradiction of ($\dagger$). This proves CLAIM. Now by ($\dagger$), CLAIM and (2.4b,c): $ab-/K_{\alpha-1}/ \geq H(Z,b-r-3) \geq H(Z,\beta-1) = /K_{\beta-1}/-/K_{\beta-1-\alpha}/$. This inequality implies $\alpha\beta < ab$. By (3.3) then, $a+b < \alpha+\beta$. On the other hand, CLAIM and ($\dagger$) give: $\alpha+\beta \leq (a+r+1)+(b-r-3) = a+b-2$. So: $a < \alpha \Rightarrow$ absurdity. This proves (3.1a).

Proof of (3.1b). By (3.2), we may assume $/I_a/ = 1$, in which case, by (3.3), it suffices to prove $\beta \leq b$. Suppose not. Then by (2.4b), $H(Z,b) = /K_b/-/K_{b-a}/$. Explicit computation then gives:

$$\text{(i)} \quad H(Z,b) = ab-(a(a-3)/2).$$

If $a = 2$, (i) gives: $H(Z,b) = 2b+1 > 2b = \delta(Z)$ -- impossible by (2.4c). So assume $a \geq 3$. By (3.4) then, $H(Z,b) \leq ab-/K_{a-3}/$. Computation gives:

$$\text{(ii)} \quad H(Z,b) \leq ab-((a-1)(a-2)/2).$$

(i) and (ii) give: $(a-1)(a-2) \leq a(a-3)$. So: $b < \beta \Rightarrow$ absurdity. Therefore $\beta \leq b$. QED.

§4. Complements on the nonreduced case.

(1.4) above strengthens [DGM, (4.21)] by weakening its hypothesis on the Hilbert function while retaining the hypothesis "reduced". (4.1) below is another strengthening, which does the reverse with those two hypotheses; its corollary (4.2) is a solution to the nonreduced ECP -- but not so satisfying a solution as is (1.2) in the reduced case.

This section uses the notation and results of §2, <u>and very heavily the results of [DGM, §4]</u>. The latter are formulated in terms of an ideal P and an integer $\sigma(P)$. <u>Translation</u>: $P = I(X)$; $\sigma(P) = \tau(X)+1$. (To recall the definition and properties of $\tau(X)$ see (2.4) above.)

(4.1) THEOREM. The following are equivalent.

(a) X is a CI(a,b).

(b) X is CB(a,b) and $H(X,t) = H(a,b,t)$ $(t \le a+b-3)$.

(c) X is a HCI(a,b) and has CB(a+b-3).

(4.2) COROLLARY. X is a complete intersection if, and only if, the following two conditions are satisfied. (Let $\alpha = \alpha(X)$, $\delta = \delta(X)$.)

(a) $\alpha^{-1}\delta \in \mathbb{Z}$ and X has $CB(\alpha+\alpha^{-1}\delta-3)$.

(b) $\alpha \le \alpha^{-1}\delta$ and $H(X,t) = H(\alpha,\alpha^{-1}\delta,t)$ $(t \le \alpha+\alpha^{-1}\delta-3)$.

<u>Proof</u>. (4.1) obviously implies (4.2); and in view of (2.3) and (2.4f), to prove (4.1) it suffices to prove (c) $\Rightarrow$ (a). We do so by showing: (c) and X is not a CI(a,b) $\Rightarrow$ absurdity.

Under the hypothesis "HCI(a,b), but not CI(a,b)", [DGM, (4.4)] asserts that there is a subscheme Y of X, and a positive integer $d < a$, such that: Y is a HCI(a-d,b-d); $I = (DJ,E)$, where $I = I(X)$, $J = I(Y)$, $D \in K_d$ and $E \in J_{a+b-d}$; $\delta(X) = \delta(Y)+\delta(W)$, where $I(W) = (D,E)$. Observe: $d \le a+b-3$. (Proof: $0<d<a$; so $2 \le a \le b$, and $d<a+b-2$.) Consequently, $\delta(Y) \neq 1$: otherwise, the fact that X has CB(a+b-3) implies the absurdity $X = W$.

Now take $\hat{Y}$ any subscheme of Y with $\delta(\hat{Y}) = \delta(Y)-1$, and let $\hat{J} = I(\hat{Y})$.

By [DGM, (4.3)], $\hat{I} = (D\hat{J}, E)$ is an unmixed ideal of height 2; so $\hat{I} = I(\hat{X})$ for a subscheme $\hat{X}$ of X. Moreover -- see [DGM, (4.5)] -- the relationship that necessarily exists between $H(\hat{X},-)$ and $H(\hat{Y},-)$ implies that $\delta(\hat{X}) = \delta(\hat{Y})+\delta(W)$. Hence $\delta(\hat{X}) = \delta(X)-1$. So because X has CB(a+b-3), $\hat{I}_{a+b-d-2} = I_{a+b-d-2}$. Therefore, since $\deg E > a+b-d-2$, $D\hat{J}_{a+b-2d-2} = DJ_{a+b-2d-2}$. So: $\hat{J}_{a+b-2d-2} = J_{a+b-2d-2}$. This last assertion is absurd by (2.4d,e). QED.

We conclude with several observations on the relationship between the Cayley-Bacharach property and the growth of Hilbert function. Now [DGM, §4] in fact analyzes the concept "tail of a complete intersection", a weaker approximation of CI than is HCI:

Definition. X has TCI(a,b) if $\Delta H(X,t) = \Delta H(a,b,t)$ $(b-2 \le t)$. Here Δ is the difference operator: $\Delta f(t) = f(t)-f(t-1)$. Since by (2.4c,e), $\Delta H(a,b,t) = 0$ $(a+b-1 \le t)$, we can equally well take the finite interval $b-2 \le t \le a+b-1$ in the definition of TCI(a,b).

A careful examination of the role of [DGM, (4.3)–(4.5)] in the proof of (4.1) reveals the following fact. (ASIDE: the integer m that appears in the following statement is, in the language of [DGM], $a+\tau(Y)$, where Y is the subscheme of X "residual to the initial tail of X".)

(4.3) LEMMA. Suppose that X has TCI(a,b) and is not a HCI(a,b). Let $m = \max\{t : \Delta H(X,t) \ne \Delta H(a,b,t)\}$. Then X does not have CB(m).

Now by (2.4a), $\Delta H(X,t) = t+1$ $(0 \le t \le \alpha-1)$, $\alpha = \alpha(X)$; and also well known is the fact that $\Delta H(X,-)$ is a nonincreasing function for $t \ge \alpha-1$. (Simple proof in the spirit of §2: (1) $\Delta H(X,t) = H(A/LA,t) = H(S/J,t)$, where $S = \bar{K}/\bar{P}\bar{K}$, $0 \ne P \in I(X)_\alpha$, $J = \overline{I(X)}/\bar{P}\bar{K}$; (2) $/S_t/ = \alpha$ $(\alpha-1 \le t)$; (3) Because the general linear form in $\bar{K}$ is S-regular, for any ideal Q of S, $/Q_t/$ is a nondecreasing function of t.) Now with (4.3), and (2.4) supplemented by these standard facts about $\Delta H(X,-)$, one easily deduces the following two corollaries of (4.1). The first strengthens (4.1) and contains new information even if X is reduced.

(4.4) COROLLARY. Suppose $\Delta H(X,t) = \Delta H(a,b,t)$ $(b-2 \leq t \leq a+b-2)$ and X has CB(a+b-3). Then X is a complete intersection: CI(a,b) if $a \neq 1$; otherwise CI(1,δ(X)).

(4.5) COROLLARY. In case $a \leq 2$ or $b = 3$, (3.1) is valid as stated, but with Z replaced by the possibly nonreduced scheme X. Consequently:

(a) If X is CB(a,b), then X is: CI(1,b) if $a = 1$; CI(1,2b) or CI(2,b) if $a = 2$ (according to whether $\alpha(X) = 1$ or 2).

(b) Mutatis mutandis, the "converse" to the Theorem of the 9 Points (see Introduction) is valid for X if $\delta(X) = 9$.

Obviously, given (4.4), one can strengthen (4.2) by weakening the hypothesis (b). But the significant question remains: Is (4.2) valid, as in the reduced case, without (b)? More generally: To what extent does (3.1) survive absent "reduced"? The methods of this section can nibble at the edges of this question, e.g., (4.5); but to prove (3.1) in nonreduced form will certainly require the infusion of a fresh idea.

References

[DGM] E. DAVIS, A. GERAMITA and P. MAROSCIA, Perfect homogeneous ideals: Dubreil's theorems revisited, Bull. Sc. Math. (in press).

[D] E. DAVIS, On a theorem of Beniamino Segre, The Curves Seminar at Queen's, vol. III (A. Geramita, editor) (in press).

[DGO] E. DAVIS, A. GERAMITA and F. ORECCHIA, Gorenstein algebras and the Cayley-Bacharach theorem, preprint (1983).

[EC] F. ENRIQUES and O. CHISINI, Lezioni sulla Teoria geometrica delle Equazioni algebriche e delle Funzioni algebriche, Bologna: Zanichelli, 1915.

[E] L. EULER, Commentationes Geometricae, Opera omnia, ser. 1 vol. 26, (A. Speiser, editor), Zürich: Orell Füssli, 1953.

[GH] P. GRIFFITHS and J. HARRIS, Residues and zero-cycles on algebraic varieties, Ann. Math. 108 (1978) 461-505.

[GHP] ———————————, Principles of Algebraic Geometry, New York: Wiley, 1978.

[M] P. MAROSCIA, Some problems and results on finite sets of points in $\mathbb{P}^n$, Proceedings of the Ravello Conference (C. Ciliberto, F. Ghione and F. Orecchia, editors), Springer Lecture Notes in Mathematics 997 (1983) 290-314.

[S] B. SEGRE, Sui teoremi di Bézout, Jacobi e Reiss, Ann. Mat. Pura Appl. (4) 26 (1947) 1-26.

[SR] J. SEMPLE and L. ROTH, Introduction to Algebraic Geometry, Oxford: Oxford University Press, 1949.

Authors' Addresses

DAVIS: MATH DEPT SUNYA, ALBANY NY 12222 (USA)

MAROSCIA: DIPARTIMENTO DI MATEMATICA, ISTITUTO MATEMATICO "G. CASTELNUOVO", CITTA' UNIVERSITARIA, 00100 ROMA (ITALIA)

POINCARÉ FORMS, GORENSTEIN ALGEBRAS AND SET THEORETIC COMPLETE INTERSECTIONS

by

Maksymilian Boratyński

Introduction: Let I and J be the ideals of $A = k[X_1, X_2, \ldots, X_n]$ such that $J \subset I$ and $I^s \subset J$ for some s. Then the natural epimorphism $f : A/J \to A/I$ is an infinitesimal extension of A/I in the sense that $\ker f$ is a nilpotent ideal of A/J. Moreover one can make the obvious remark that the natural epimorphism $A \to A/I$ factors through an epimorphism $A \to A/J$. An infinitesimal extension with this property will be called an embedded one.

Let now $S \to R = A/I$ be an embedded infinitesimal extension of A/I and let $h : A \to S$ be a factorization of $A \to A/I$. Then $\ker h \subset I$ and some power of I is contained in $\ker h$.

So if one contemplates the problem of whether an ideal $I \subset A$ is a set-theoretic complete intersection, one should investigate the existence of $S \to R = A/I$ - an embedded infinitesimal extension of R with S a complete intersection.

The purpose of this paper is to study the existence problem of Gorenstein, embedded, infinitesimal extensions $S \to R = A/I$ i.e. such that S is Gorenstein in the sense that S is CM and a canonical module of S is trivial in case $R = A/I$ is smooth over k. It turns out that the existence of Gorenstein embedded, infinitesimal extensions of $R = A/I$ of a certain kind (which will be specified later) is equivalent with the existence of some extra structure (non-trivial Poincaré forms) on I/I^2. We could prove that its existence in case $\mathrm{ht} I = 2$ implies that $c_1^2(I/I^2)$ is torsion. This condition is not always satisfied if $R = \mathbb{C}[X_1, X_2, X_3, X_4]/I$ is an affine ring of a smooth surface in $\mathbb{C}^4$. Nevertheles we could point out the case when the non-trivial Poincare forms exist and using the fact that "Gorenstein in codimension 2" implies "complete intersection" one obtains the following.

Corollary: Let I be an ideal of $A = k[X_1, X_2, \ldots, X_n]$, $\mathrm{ht}\, I = 2$ such that $R = A/I$ is smooth over k. Suppose that $P = I/I^2 \simeq P_1 \oplus P_2$, P_i-projective of rank 1 with the property that there exist the positive integers n_1, n_2 such that $P_1^{\otimes n_1} \otimes P_2^{\otimes n_2} \simeq R$. Then I is a set theoretic complete intersection ideal.

Note that if $\dim R = 1$ (a case of a curve) then $P \simeq \omega^{\otimes -2} \oplus \omega$ where $\omega \simeq \Lambda^2 P^*$ since in the one-dimensional case the projective modules of the same rank are isomorphic if their determinants are isomorphic. The assumption of the above Corollary will be satisfied for $n_1 = 1$ and $n_2 = 2$. So one obtains the result of Ferrand--Szpiro ([4]) which asserts that a smooth curve in $\mathbb{A}_k^3$ is a set theoretic complete intersection.

Note also that the assumption of the above Corollary will be satisfied in case $\omega^{\otimes r} \simeq R$ for some $r > 0$. So one concludes that every smooth codimension 2 variety in the affine space with a decomposable conormal boundle and torsion canonical bundle is a set theoretic complete intersection. This result has been obtained by Murthy ([4]) in case of surfaces over an algebraically closed field.

I would like to thank R.Rubinsztein, P.Pragacz and J.Weyman for the helpful conversations in the course of this work.

I also with to thank S. Greco for allowing this paper to appear in these proceedings even though I did not participate in the conference itself.

1°. Canonical modules.

In this section we shall recall some basic facts about the canonical modules which are more or less well known at least in the projective case. For the details we refer the readers to [1], [7] and [8].

Let R be an affine Cohen Macaulay k-algebra with $\dim R = r$ and let $f : A = k[X_1, X_2, \dots, X_n] \to R$ be an epimorphism. One defines a canonical module of R - ω_R as $\operatorname{Ext}_A^{n-r}(R,A)$ (the R-module structure of ω_R is induced by that of R). It turns out that this definition does not depend on the choice of n and f. In case R is smooth over k we have $\omega_R \simeq \wedge^r \Omega_{R/k}$. In particuliar ω_R is projective of rank 1. In the sequel we shall need the following.

Proposition 1.1. Let $f : R \to S$ be a homomorphism of affine CM k-algebras which makes S into a finitely generated R-module. If $\dim R = \dim S$ then $\omega_s = \operatorname{Hom}_R(S, \omega_R)$.

Proposition 1.2. Let $f : R \to S$ be a homomorphism of affine CM k-algebras which is an epimorphism, Then $\omega_S \simeq \operatorname{Ext}_R^{n-r}(S, \omega_R)$ where $n = \dim R$ and $r = \dim S$.

Definition: Let R be an affine CM k-algebra. R is called Gorenstein if $\omega_R \simeq R$.

Caution: Note that unlike in the local case R regular does not imply R is Gorenstein. Any smooth affine variety with a non-trivial canonical bundle gives a counterexample.

Proposition 1.3. (Serre). Let I be an ideal of $A = k[X_1,X_2,\dots,X_n]$ with $\operatorname{ht} I = 2$. If $R = A/I$ is Gorenstein then I is a complete intersection.

Remark: Let $A = k[X_{ij}]_{\substack{1\leq i\leq t\\ 1\leq j\leq t+1}}$, $t \geq 2$, char $k = 0$. By [9] I-the ideal of maximal minors of (X_{ij}) is not a set theoretic complete intersection. It follows from Proposition 1.3 that $R = A/I$ has no Gorenstein, embedded, infinitesimal extensions since $\operatorname{ht} I = 2$. This example explains partly the fact, that we restrict ourselves to the case when $R = A/I$ is smooth over k.

2°. Infinitesimal extensions of smooth algebras.

Let I be an ideal of $A = k[X_1,X_2,\dots,X_n]$. We denote by $Gr(I,A)$ the associated graded ring $\bigoplus_{i\geq 0} I^i/I^{i+1}$ and by $Gr(I,A)_{\leq t}$ its t-th truncation $Gr(I,A)/\bigoplus_{i>t} I^i/I^{i+1}$.

Theorem 2.1. Let I be an ideal of $A = k[X_1,X_2,\dots,X_n]$ such that $R = A/I$ is smooth over k. Then $A/I^{t+1} \simeq Gr(I,A)_{\leq t}$ for $t \geq 0$.

Proof: We denote by d_o the natural homomorphism $A \longrightarrow A/I$. Since R is smooth we have the following splitting exact sequence

$$0 \longrightarrow I/I^2 \longrightarrow \Omega_{A/k} \otimes A/I \longrightarrow \Omega_{R/k} \longrightarrow 0$$

Let $\gamma : \Omega_{A/k} \otimes A/I \longrightarrow I/I^2$ be a splitting of $I/I^2 \longrightarrow \Omega_{A/k} \otimes A/I$. For $f \in A$ we put $d_1 f = \gamma(\overline{df})$ where $\overline{df}$

denotes the image of df in $\Omega_{A/k} \otimes A/I$. It follows from the definition of $d_1 : A \to I/I^2$ that d_1 is an extension of the natural homomorphism $I \to I/I^2$.

The association $X_i \to (d_o X_i, d_1 X_i, 0, 0, \ldots, 0) \in Gr(I,A)_{\leq t}$ defines $d : A \to Gr(I,A)_{\leq t}$ - the homomorphism of k-algebras. Let $d_i : A \to I^i/I^{i+1}$ denote the i-th component of d. It is easy to check that d_o and d_1 coincide with the previously defined d_o and d_1.

We have $d_i(fg) = \sum_{j+k=i} d_j(f) d_k(g)$ for $f, g \in A$ since d is a homomorpism.

We claim that d_i restricted to I^i is the natural homomorphism $I^i \to I^i/I^{i+1}$ for $1 \leq i \leq t$. This is true for $i = 1$. Let $f \in I^{i-1}$ and $g \in I$. By induction $d_i(fg) = d_{i-1}(f) \cdot d_1(g) =$ (image of f in I^{i-1}/I^i)(image of g in I/I^2) = image of fg in I^i/I^{i+1}. This proves the claim since I^i is additively generated by the elements of the form $f \cdot g$ with $f \in I^{i-1}$ and $g \in I$. It follows from the claim that $\ker d = I^{t+1}$.

We want to prove now that d is onto. It suffices to show that the homogenous elements of $Gr(I,A)_{\leq t}$ are in the image of d. Let $x \in Gr(I,A)_{\leq t}$ with $\deg x = i$. Then there exists $a \in I^i$ such that $d_i(a) = x$. If $i = t$ then $d(a) = x$ and we are done. Suppose that $i < t$. We shall define inductively a sequence of elements $\{a_j\}_{i \leq j \leq t}$ with $a_j \in I^j$. Put $a_i = a$. Suppose the sequence $a_i, a_{i+1}, \ldots, a_k$ is defined with $a_j \in I^j$ for $i \leq j \leq k$. Let a_{k+1} be such an element of I^{k+1} that $d_{k+1}(a_i + a_{i+1} + \ldots + a_k) = - d_{k+1}a_{k+1}$. Then it is easy to check that $d(\sum_{j=i}^{t} a_j) = x$.

3°. Poincaré forms.

Let P be a projective R-module (R-commutative, noetherian) and let $\varphi \in \mathrm{Hom}(S^tP,\omega)$ $(t \geq 0)$ where ω is a projective R-module of rank 1. Then for any $i,j \geq 0$ with $i+j = t$ φ induces the map $\varphi'_{ij} : S^i(P) \to \mathrm{Hom}(S^j(P),\omega)$. Let K_i denote its kernel. It is easily seen that the image of φ'_{ij} is contained in $\mathrm{Hom}(S^j(P)/K_j,\omega)$. So we have an induced homomorphism $\varphi_{ij} : S^i(P)/K_i \to \mathrm{Hom}(S^j(P)/K_j,\omega)$ which is always a monomorphism.

Definition: $\varphi \in \mathrm{Hom}(S^tP,\omega)$ $(t \geq 0)$ is called a t-linear Poincaré ω-valued form on P if the following two conditions hold:

1° K_i is a direct summand of $S^i(P)$ for $0 \leq i \leq t$

2° φ_{ij} is an isomorphism for $i,j \geq 0$ such that $i+j=t$.

Theorem 3.1. Let R be a connected ring which has a minimal prime ideal p such that R_p is Gorenstein. Then $\varphi \in \mathrm{Hom}(S^tP,\omega)$ is a Poincaré form if and only if coker φ'_{ij} is projective for $i,j \geq 0$ such that $i+j = t$.

Remark: Let $f : P_1 \to P_2$ with P_1, P_2 projective over a reduced (connected) ring. It is easy to prove (and well known) that coker f is projective if and only if $\mathrm{rank}(f \otimes k(x))$ does not depend on $x \in \mathrm{Spec}\, R$ where $k(x)$ denotes the residue field of x and $f \otimes k(x)$ denotes the induced map $P_1 \otimes k(x) \to P_2 \otimes k(x)$.

Hence coker φ'_{ij} is projective if and only if $rk(\varphi'_{ij} \otimes k(x))$ does not depend on $x \in \operatorname{Spec} R$.

Proof: Let $\varphi \in \operatorname{Hom}(S^t P, \omega)$ be a t-linear Poincaré ω-valued form on P. Then the splitting exact sequence

$$0 \to K_j \to S^j(P) \to S^j(P)/K_j \to 0$$

induces the splitting exact sequence

$$0 \to \operatorname{Hom}(S^j(P)/K_j,\omega) \to \operatorname{Hom}(S^j(P),\omega) \to \operatorname{Hom}(K_j,\omega) \to 0$$

coker $\varphi'_{ij} = \operatorname{Hom}(K_j,\omega)$ since the φ_{ij}'s are the isomorphisms. Therefore coker φ'_{ij} is projective for $i,j \geq 0$ such that $i+j=t$.

Suppose now that coker φ'_{ij} is projective for $i,j \geq 0$ such that $i+j = t$. Then Im $\varphi'_{ij} = S^i(P)/K_i$ is projective since Im φ'_{ij} is a direct summand of $\operatorname{Hom}(S^j(P),\omega)$. Therefore K_i is a direct summand of $S^i(P)$ for $0 \leq i \leq t$.

Let p be a minimal prime ideal of R such that R_p is Gorenstein. We claim that φ_{ij} localized at p is an isomorphism. We can assume that R is O-dimensional, local, Gorenstein since everything localizes nicely.

Let $l(M)$ denote the length of an R-module M. Then $l(S^i(P)/K_i) \leq l(\operatorname{Hom}(S^j(P)/K_j,\omega) = l(S^j(P)/K_j)$ since φ_{ij} is a monomorphism and R is Gorenstein. (Note that $\omega = R$). By symmetry $l(S^j(P)/K_j) \leq l(S^i(P)/K_i)$ and $l(S^i(P)/K_i) = l(S^j(P)/K_j)$. So $l(S^i(P)/K_i) = l(\operatorname{Hom}(S^j(P)/K_j),\omega))$ and φ_{ij} is an isomorphism which proves the claim.

It is easily seen that coker φ_{ij} is projective. We have

$(\text{coker } \varphi_{ij})_p = 0$. So $\text{coker } \varphi_{ij} = 0$ and φ_{ij} is an isomorphism.

4°. The basic construction.

Let $\varphi \in \text{Hom}(S^t P, \omega)$. It is easily seen that $\bigoplus_{i=0}^{t} K_i$ is a graded ideal of $S(P)_{\leq t} = \bigoplus_{i=0}^{t} S^i(P)$ the truncated symmetric algebra of P.

Put $S(\varphi) = \bigoplus_{i=0}^{t} S_i(\varphi) = \bigoplus_{i=0}^{t} S^i(P)/K_i$.

Theorem 4.1. Suppose R is a smooth (connected) finitely generated k-algebra and let $\varphi \in \text{Hom}(S^t P, \omega)$ be a non-zero t-linear Poincaré ω-valued from on P where P is a projective R-module and ω is a canonical module of R. Then $S(\varphi)$ is Gorenstein.

And conversely, if $K = \bigoplus_{i=0}^{t} K_i$ is a graded ideal of $S(P)_{\leq t}$ with $K_o = 0$ such that $S = S(P)_{\leq t}/K$ is Gorenstein with $S_t = S^t(P)/K_t \neq 0$ then $S \simeq S(\varphi)$ where φ is a non-zero t-linear Poincaré ω-valued form on P.

Proof: Let φ be a non-zero t-linear Poincaré ω-valued form on P. It follows from our assumptions on R and φ that $K_o = 0$. So every prime ideal p of $S(\varphi)$ is of the form $q \oplus \bigoplus_{i=1}^{t} S_i(\varphi)$ where q is a prime ideal of R. $(S(\varphi))_p = \bigoplus_{i=0}^{t} (S_i(\varphi))_q$ is a finite extension of $(S_o(\varphi))_q = R_q$. $(S(\varphi))_p$ is CM since $(S_i(\varphi))_q$ is free over R_q for $0 \leq i \leq t$ and R_q is regular. So $S(\varphi)$ is CM.

We have to prove that $\omega_S \simeq S$ where $S = S(\varphi)$. By Proposition 1.1, $\omega_{S(P)_{\leq t}} = \text{Hom}_R(S(P)_{\leq t}, \omega) = \bigoplus_{i=0}^{t} \text{Hom}_R(S^i P, \omega)$. The $S(P)_{\leq t}$-module structure on $\omega_{S(P)_{\leq t}}$ looks as follows:

Let $x_i \in S^i(P)$, $f_j \in \mathrm{Hom}(S^j(P),\omega)$ with $j \geq i$. Then $x_i f_j = g_{j-i} : S^{j-i}(P) \longrightarrow \omega$ where $g_{j-i}(\alpha) = f_j(\alpha x_i)$. If $j < i$ then $x_i f_j = 0$.

We get again by Proposition 1.1 that $\omega_S \simeq \mathrm{Hom}_{S(P)_{\leq t}}(S,\omega_{S(P)_{\leq t}}) = \mathrm{Hom}_{S(P)_{\leq t}}(S(P)_{\leq t}/\bigoplus_{i=0}^{t} K_i, \omega_{S(P)_{\leq t}}) = \mathrm{Anih}_K \omega_{S(P)_{\leq t}}$ where $K = \bigoplus_{i=0}^{t} K_i$.

It follows from the definition of K_i's that $f = (f_j) \in \mathrm{Anih}_K \omega_{S(P)_{\leq t}}$ where $f_j = 0$ for $0 \leq j < t$ and $f_t = \varphi$.

We want to prove that $\mathrm{Anih}_K \omega_{S(P)_{\leq t}}$ is generated by f. Let $g = (g_j)_{0 \leq j \leq t} \in \mathrm{Anih}_K \omega_{S(P)_{\leq t}}$. Then $\sum_{j-i=s} x_i g_j = 0$ for $x_i \in K_i$ and $0 \leq s \leq t$. We get in particular that $K_j g_j = 0$ for $0 \leq j \leq t$. This means that $g_j \in \mathrm{Hom}(S^j(P)/K_j,\omega)$. So $g_j = a_{t-j}\varphi = a_{t-j} f_t$ with $a_{t-j} \in S^{t-j}(P)$ since $\varphi_{t-j,j}$ is an isomorphism. So we obtain that $g = af$ where $a = (a_{t-j})_{j=t,t-1,0}$. It follows from the definition of K_i's that $K = \{x \in S(P)_{\leq t} \mid xf = 0\}$. So $\omega_S \simeq S(P)_{\leq t}/K = S$ which was to be proved.

Let now $S = S(P)_{\leq t}/\bigoplus K_i$ be Gorenstein with $K_0 = 0$ and $S_t \neq 0$.

We have $\omega_S \simeq \mathrm{Hom}_{S(P)_{\leq t}}(S(P)_{\leq t}/K, \omega_{S(P)_{\leq t}}) \simeq S(P)_{\leq t}/K$. So there exists $f = (f_j)_{0 \leq j \leq t} \in \omega_{S(P)_{\leq t}} = \bigoplus_{i=0}^{t} \mathrm{Hom}(S^i(P),\omega)$ such that $\{x \in S(P)_{\leq t} \mid xf = 0\} = K$ and $\mathrm{Anih}_K \omega_{S(P)_{\leq t}} = S(P)_{\leq t} \cdot f$. We have $f_t \neq 0$ since $S_t \neq 0$. Moreover $K_i f_t = 0$ for $0 \leq i \leq t$. Hence $K f_t = 0$. So there exists $a = (a_i)_{0 \leq i \leq t} \in S(P)_{\leq t}$ such that

$a_o f_t = f_t$ and $\sum_{j-i=s} a_i f_j = 0$ where $0 \leq s < t$. We have $a_o = 1$ and $0 = f_{t-1} + a_1 f_t$. Suppose that $f_j = b_{t-j} f_t$ for $j_o \leq j \leq t-1$ with $b_{t-j} \in S^{t-j}(P)$. We have $0 = f_{j_o-1} + \sum_{\substack{j-i=j_o-1 \\ j \geq j_o}} a_i f_j$.

So we get by induction that $f_j = b_{t-j} f_t$ for $0 \leq j \leq t-1$ with $b_{t-j} \in S^{t-j}(P)$. It follows that $bf' = f$ where $b = (b_{t-j})_{j=t,t-1,\ldots,0} \in S(P)_{\leq t}$ with $b_o = 1$ and $f' = (f_i')$ where $f_i' = 0$ for $0 \leq i < t$ and $f_t' = f_t$. $K = \{x \in S(P)_{\leq t} | xf' = 0\}$ and $\mathrm{Anih}_{K^\omega S(P)_{\leq t}} = S(P)_{\leq t} f'$ since b is invertible in $S(P)_{\leq t}$. We claim that f_t is a Poincaré form. K_i is a direct summand of $S^i(P)$ for $0 \leq i \leq t$ since the arguments used in the first part of the proof can be reversed. Let $g_{j_o} \in \mathrm{Hom}(S^{j_o}(P)/K_{j_o}, \omega)$, It follows easily from the definition of K_i's that $Kg = 0$ where $g = (g_j)_{0 \leq j \leq t}$ with $g_j = 0$ for $j \neq j_o$. So $g = (a_i)f'$ and $g_{j_o} = a_{t-j_o} f_t$ with $a_{t-j_o} \in S^{t-j_o}(P)$. Therefore f_t is a Poincaré form and $S = S(\varphi)$ where $\varphi = f_t$ which proves the theorem.

Corollary 4.2. Let R be a smooth (connected) finitely generated k-algebra and let P be a projective R-module. Then $S(P)_{\leq t}$ has a graded Gorenstein factor ring $S = \bigoplus_{i=0}^{t} S_i$ with $S_o = R$ and $S_t \neq 0$ if and only if there exists a non-zero t-linear Poincaré ω-valued form on P.

5. Set theoretic Gorenstein ideals.

Let I be an ideal of $A = k[X_1,\ldots,X_n]$ such that

$k[X_1,X_2,\ldots,X_n]/I$ is a smooth (connected) k-algebra. Let $J \subset I$ such that $I^{t+1} \subset J$. Then in view of Theorem 2.1 there is an epimorphism $\gamma : Gr(I,A)_{\leq t} \to A/J$.

Definition: I is called a homogenous , set theoretic Gorenstein ideal if there exists an ideal $J \subset I$ such that $I^{t+1} \subset J$ for some t, A/J is Gorenstein and $\ker \gamma$ is a homogenous ideal of $Gr(I,A)_{\leq t}$.

Remark: Note that for dimension reasons $K_o = 0$ where

$$\ker \gamma = \bigoplus_{i=0}^{t} K_i .$$

If we replace in the above definition "Gorenstein" by "complete intersection" we get a notion of homogenous set theoretic complete intersection ideal.

We can now state one of the main results of this paper.

Theorem 5.1. Let I be an ideal of $A = k[X_1,X_2,\ldots,X_n]$ such that $R = A/I$ is smooth over k. Then I is a homogenous set theoretic Gorenstein ideal if an only if there exists a non-zero t-linear Pointcaré ω-valued symmetric form on P for some $t \geq 0$ where $P = I/I^2$ and ω is a canonical module of R.

Proof: Let $\varphi \in Hom(S^t(P),\omega)$ be a non-zero t-linear Poincaré form. Let f denote the composition $A \to A/I^{t+1} \simeq Gr(I,A)_{\leq t} \simeq S(P)_{\leq t} \to S(\varphi)$ and let J denote its kernel. Then $A/J \simeq S(\varphi)$ since f is an epimorphism. I is a homogenous set theoretic Gorenstein ideal since $I^{t+1} \subset J \subset I$ and A/J is Gorenstein by Theorem 4.1

Suppose now that I is a homogenous set theoretic Gorenstein ideal. Then there exists $J \subset I$ such that $I^{t+1} \subset J$, $I^t \not\subset J$ for some t and A/J-Gorenstein. Let $K = \bigoplus_{i=1}^{t} K_i$ denote the kernel of $Gr(I,A)_{\leq t} \to A/J$.

By Theorem 4.1 $A/J = S(\varphi)$ where φ is a non-zero t-linear Poincaré ω-valued form on P since $Gr(I.A)_{\leq t} \simeq S(P)_{\leq t}$ and $S_t = S^t(P)/K_t \neq 0$.

6°. Examples of Poincaré forms

Proposition 6.1: Let R be a domain and let $0 \neq \varphi \in Hom(S^1 P,\omega) = Hom_R(P,\omega)$ with P-projective and ω projective of rank one. Then φ is Poincaré if and only if φ is an epimorphism.

Proof: Follows easily from the definition of Poincare forms.

Corollary: Let I be an ideal of $A = k[X_1,\ldots,X_n]$ such that $R = A/I$ is smooth (connected) over k. There exists an ideal $J \subset I$ such that $I^2 \subset J$ and A/J is Gorenstein if and only if R is either Gorenstein or a canonical module of R is a homomorphic image of I/I^2,

Proof: One has to note only that $\ker \gamma$ where $\gamma : S(P)_{\leq 1} = Gr(I,A)_{\leq 1} \to A/J$ $(P = I/I^2)$ is a homogenous ideal.

Corollary: (Ferrand, Szpiro). Let I satisfy the same hypotheses as above and suppose moreover that $ht\, I > \dim R$. Then there exists an ideal $J \subset I$ such that $I^2 \subset J$ and A/J is Gorenstein.

Proof: $rk(I/I^2 \otimes \omega^{-1}) = htI > \dim R$. So by the Bass unimodular theorem there exists an exact sequence

$I/I^2 \otimes \omega^{-1} \to R \to O$. After tensoring it with ω one gets the exact sequence $I/I^2 \to \omega \to O$ which was to be proved. In the sequel R will denote an arbitrary domain and P a projective R module of rank k.

Proposition 6.2: Let $f_1, f_2, \ldots, f_k \in S(P)$ be a homogenous system of parameters of $A = S(P) = \underset{i \geq 0}{\oplus} S^i(P)$ i.e. $(A^+)^s \subset (f_1, f_2, \ldots, f_k)$ for some s with $f_1, f_2, \ldots, f_k$ homogenous of degree ≥ 1 and $A^+ = \underset{i>0}{\oplus} S^i(P)$. Then there exists a non-zero t-linear Poincaré ω-valued form on P with $t = \sum_{i=1}^{k} \deg f_i - k$ and $\omega = \Lambda^k P^*$.

Since we are not going to use the above Proposition in the sequel we shall indicate only the construction of the Poincaré form on P without going into the details.

Let $S = A/(f_1, \ldots, f_k) = \bigoplus_{i=0}^{t} S_i$ with $S_t \neq O$. It turns out that $S_t \simeq \Lambda^k P^*$ and $t = \sum_{i=1}^{k} \deg f_i - k$. Let $\varphi \in \mathrm{Hom}(S^t P, \Lambda^k P^*)$ denote the composition of $S^t P \to S_t$ and $S_t \xrightarrow{\simeq} \Lambda^k P^*$. It can be proved that φ is a Poincaré t-linear symmetric $\Lambda^k P^*$-valued form on P.

Remarks: It was proved in [3] that the existence of a homogenous system of parameters of $S(P)$ implies (in the nonsingular, geometric case) that all the Chern classes are torsion.
On the other hand if P is a direct sum of rank 1 projectives which define the torsion elements of $\mathrm{Pic}\, R$ then $S(P)$ has a system of homogenous parameters ([2]).

Proposition 6.3: Let $P = \bigoplus_{i=1}^{k} I_i$ with I_i's projective of rank 1
Let n_i be the integers for $1 \leq i \leq k$ such that $n_i \geq O$. Then

there exists a non-zero t-linear Poincaré ω-valued form on P with $t = \sum_{i=1}^{k} n_i$ and $\omega = \bigotimes_{i=1}^{k} I_i^{\otimes n_i}$.

Proof: We have an obvious surjection $\varphi : S^t P \to \omega$. We claim that φ is Poincaré. Let $\varphi'_{ij} : S^i(P) \to \mathrm{Hom}(S^j(P), \omega)$ and let $k(x)$ denote the residue field of a point $x \in \mathrm{Spec}\, R$.

The map $\varphi'_{ij} \otimes k(x)$ is induced by $\varphi \otimes k(x) : S^t(P) \otimes k(x) = S^t(P \otimes k(x)) \to \omega \otimes k(x)$. Put $\varphi_x = \varphi \otimes k(x)$. Let e_i denote a generator of $I_i \otimes k(x)$ for $1 \leq i \leq k$. Then

$\varphi_x(e_1^{n_1} \otimes e_2^{n_2} \otimes \ldots \otimes e_k^{n_k}) = a_x$ where $0 \neq a_x \in k(x)$ and

$\varphi_x(e_1^{i_1} \otimes e_2^{i_2} \otimes \ldots \otimes e_k^{i_k}) = 0$ if $i_1 + i_2 + \ldots + i_k = n_1 + n_2 + \ldots + n_k$

and $(i_1, i_2, \ldots, i_k) \neq (n_1, n_2, \ldots, n_k)$. Now it is clear that $\mathrm{rk}(\varphi'_{ij} \otimes k(x))$ does not depend on a_x. This finishes the proof of our claim in view of the remark following Theorem 3.1.

The next result guarantees the existence of non-zero Poincaré forms in the (possibly) non-indecomposable case.

Proposition 6.4: Suppose $2 = 0$ in R. Then there exists a non-zero t-linear Poincaré ω-valued form on P with $t = k = \mathrm{rank}\, P$ and $\omega = \Lambda^k P$.

Proof: We have non-trivial canonical $\varphi : S^k P \to \Lambda^k P$ since each alternating form is symmetric. Using the same kind of argument as above one can prove that φ is Poincaré.

We can now state the following

Corollary: Let I be an ideal of $A = k[X_1, X_2, \ldots, X_n]$ such that $R = A/I$ is smooth over k. Then I is a homogenous set-theoretic Gorenstein ideal in each of the following cases.

1°. $P = I/I^2 \simeq \bigoplus_{i=1}^{k} P_i$ with $k = \mathrm{ht}\, I$ and P_i-projective of rank 1 with the property that there exist the integers $n_i \geq 1$ for $1 \leq i \leq k$ such that $\bigotimes_{i=1}^{k} P_i^{\otimes n_i} \simeq R$.

2°. char $k = 2$ and $\omega^{\otimes 2} \simeq R$ where ω is a canonical module of R.

Proof: Let $P = I/I^2 = \bigoplus_{i=1}^{k} P_i$ with P_i-projective of rank 1. Then the canonical module of $R - \omega \simeq \Lambda^k P^* \simeq \bigotimes_{i=1}^{k} P_i^{\otimes(-1)} \simeq \bigotimes_{i=1}^{k} P_i^{\otimes(n_i-1)}$

So there exists a non-zero t-linear Poincare ω-valued form on P with $t = \sum_{i=1}^{k} (n_i - 1)$ (Proposition 6.3) and by Theorem 5.1 I is a homogenous set theoretic Gorenstein ideal. In case "2°" note only that our assumption implies that $\omega \simeq \Lambda^k P^* \simeq \Lambda^k P$. So the conclusion follows from Proposition 6.4 and Theorem 5.1.

7°. Basic properties of Chern classes.

We shall recall some basic results about Chern classes we are going to use in the sequel, For the details we refer to [5].

Let V be a smooth variety over an algebraically closed field k and let $A(V) = \bigoplus_{i=0}^{\dim V} A^i(V)$ denote the Chow ring of V. $A^i(V)$ consists of rational equivalence classes of codimension i cycles of V.

For any locally free sheaf $\mathfrak{F}$ over V one defines for $i \geq 0$ the elements $c_i(\mathfrak{F}) \in A^i(V)$ such that $c_o(\mathfrak{F}) = 1$ and $c_i(\mathfrak{F}) = 0$ for $i > \mathrm{rk}\,\mathfrak{F}$. If $\mathrm{rk}\,\mathfrak{F} = 1$ then $c_1(\mathfrak{F})$ is equal to the image of $\mathcal{F}$ in $A^1(V) = \text{Pic } V$. In general $c_1(\mathcal{F}) = c_1(\bigwedge^{\text{rank}\,\mathcal{F}} \mathcal{F})$. Let $0 \to \mathcal{F}' \to \mathcal{F} \to \mathcal{F}'' \to 0$ be an exact sequence of locally free sheaves over V. Then

$$c_k(\mathcal{F}) = \sum_{i+j=k} c_i(\mathcal{F}')c_j(\mathcal{F}'').$$

In particular we have $c_i(\mathcal{F}) = 0$ for $i > 0$ if $\mathcal{F}$ is free.

We shall need later a certain fact about $c_k(\mathcal{F})$ where $k = \text{rank}\,\mathcal{F}$. Namely suppose that there exists a section of $\mathcal{F}$ whose scheme of zeros is of codimension k. Then $c_k(\mathcal{F})$ is represented by the corresponding cycle of zeros ([6]).

Let I be an ideal of $A = k[X_1, X_2, \ldots, X_n]$ (k-algebraically closed) such that $R = R/I$ is smooth over k and let V be the corresponding affine (smooth) variety. Then for any projective R-module P one defines for $i \geq 0$ $c_i(P) = c_i(\tilde{P})$ where $\tilde{P}$ is the corresponding locally free sheaf over V.

Let π denote the structural map $X = \text{Proj } S(P) \to V$. Then π^* makes $A(X)$ into a free $A(V)$-module generated by $1, t, \ldots, t^{r-1}$ where $r = \text{rank } P$ and $t = c_1(\mathcal{O}_X(1))$. Moreover we have $\sum_{i=0}^{r} (-1)^i \pi^* c_i(P) t^{r-1} = 0$ in $A(X)$.

8°. Codimension two case.

The last Corollary of Section 6 specialized to the height 2 case gives in view of Proposition 1.3 the following

Corollary: Let I be an ideal of $A = k[X_1, X_2, \ldots X_n]$ such that $R = A/I$ is smooth over k and $\text{ht} I = 2$. Then I is a homogenous set-theoretic complete intersection ideal in each of the following cases.

1°. $I/I^2 = P_1 \oplus P_2$, P_1, P_2 projective of rank 1 with the property that there exist positive integers n_1, n_2 such that $P_1^{\otimes n_1} \otimes P_2^{\otimes n_2} \simeq R$

2°. char k = 2 and $\omega^{\otimes 2} \simeq R$ where ω is a canonical module of R.

The following result shows that the conditions of the above Corollary on I/I^2 can not be relaxed too much.

Theorem 8.1: Let I be an ideal of $A=k[X_1,X_2,\ldots,X_n]$ (k-algebraically closed) such that $R = A/I$ is smooth over k and ht I = 2. If I is a homogenous set-theoretic complete intersection ideal then $c_1^2(I/I^2)$ is a torsion element in $A^2(\text{Spec } R)$.

Remark: $c_1^2(I/I^2)$ is an invariant of Spec R which does not depend on its imbedding into Spec A since $c_1^2(I/I^2) = c_1^2(\omega_R)$.

For the proof of Theorem 8.1 we shall need two following lemmas.

Lemma 8.2: Let R be a smooth affine algebra over k and let P be a projective R-module such that $\overset{\text{rank } P}{\Lambda} P \simeq \omega_R^{\otimes -1}$. Then $S = S(P)$ is a Gorenstein ring

Proof: S is obviously a Cohen-Macaulay ring since S is smooth over k. Moreover since S is smooth over R we have an exact sequence of projective S-modules

$$0 \to \Omega_{R/k} \underset{R}{\otimes} S \to \Omega_{S/k} \to \Omega_{S/R} \to 0$$

We get $\omega_S \simeq \bigwedge^{\dim S} \Omega_{S/k} \simeq \bigwedge^{\dim R} \Omega_{R/k} \otimes_R S \otimes_S \bigwedge^{\operatorname{rank} P} \Omega_{S/R} \simeq$

$\bigwedge^{\dim R} \Omega_{R/k} \otimes_R S \otimes_S \bigwedge^{\operatorname{rank} P} P \otimes_R S \simeq \omega_R \otimes_R \omega_R^{\otimes -1} \otimes_R S \simeq S$

Lemma 8.3: Let I be an ideal of $A = k[X_1, X_2, \ldots, X_n]$ (k-algebraically closed) such that $R = A/I$ is smooth over k and $\operatorname{ht} I = 2$. Then $c_2(I/I^2) = 0$.

Proof: I/I^2 is stably isomorphic to $A/I \oplus \wedge^2(I/I^2)$ by [4]. We have $c_2(I/I^2) = 0$ since the Chern classes are stably invariant.

Proof of Theorem 8.1. Let $I \subset A$ satisfy the assumptions of Theorem 8.1. Then there exists an ideal $J \subset I$ such that $I^{t+1} \subset J$, for some t, A/J is Gorenstein (a complete intersection) and $\ker \gamma$ is a homogenous ideal of $Gr(I,A)_{\leq t}$ where γ is an epimorphism $Gr(I,A)_{\leq t} \to A/J$. We have $Gr(I,A)_{\leq t} \simeq S(P)/ \bigoplus_{i \geq t+1} S^i(P)$ where $P = I/I^2$. So there exists a homogenous ideal K such that $\bigoplus_{i \geq t+1} S^i(P) \subset K \subset \bigoplus_{i>0} S^i(P) = S^+$ and S/K is Gorenstein. $(S=S(P))$. We claim that K is a locally complete intersection ideal. By Proposition 1.2 $\operatorname{Ext}^2_S(S/K,S) \simeq S/K$ since S and S/K are Gorenstein. It follows from the lemma of Serre that there exists an exact sequence $0 \to S \to Q \to K \to 0$ with Q S-projective of rank 2. So K is a locally complete intersection ideal which proves the claim. K/S^+K is a projective R-module of rank 2 since $K/S^+K \simeq K/K^2 \otimes S/S^+$. Put $K = \oplus K_i$ $(K_o = 0)$. Then

$K/S^+K = \bigoplus_{i\geq 1} K_i/S_1K_{i-1}$ $(S_1 = S^1(P))$. S_1K_{i-1} is an R-direct summand of K_i for $i \geq 1$, It is clear that $L = \bigoplus_{i\geq 1} L_i$ where L_i denotes a complement of S_1K_{i-1} in K_i is a graded projective R-submodule of K of rank 2 which generates K as an ideal of S.

We claim that $c_1(L) = c_2(L) = 0$. We have $\mathrm{Hom}(\Lambda^2(K/K^2), S/K) \simeq \mathrm{Ext}^2_S(S/K, S) \simeq S/K$. The first isomorphism holds since K is a locally complete intersection ideal ([1]): It follows that $\Lambda^2(K/K^2) \simeq S/K$ and $\Lambda^2(K/S^+K) \simeq \Lambda^2(K/K^2) \otimes S/S^+ \simeq S/K \otimes S/S^+ \simeq R$. So $\Lambda^2 L \simeq R$ and $c_1(L) = 0$ since $L \simeq K/S^+K$. $\mathrm{Spec}(S/S^+)$ is a scheme of zeros of the canonical section of $(P \otimes S)^*$. So $[\mathrm{Spec}\ S/S^+] = 0$ in $A^2(\mathrm{Spec}\ S)$ since $c_2(P \otimes S)^* = c_2(P \otimes S) =$ The imbedding $L \subset K$ induces the exact sequence $L \otimes S \to K \to 0$ which shows that $\mathrm{Spec}\ S/K$ is a scheme of zeros of a section of $(L \otimes S)^*$. It follows that $c_2(L \otimes S) = c_2(L \otimes S)^* = 0$ and $c_2(L) = 0$ since $[\mathrm{Spec}\ S/K] = 0$ in $A^2(\mathrm{Spec}\ S)$ being an integer multiple of $[\mathrm{Spec}\ S/S^+]$ $(\bigoplus_{i\geq t+1} S^i(P) \subset K \subset S^+)$. This finishes the proof of the claim. We shall distinguish now two cases

1° L is decomposable. Then $L = L_1 \oplus L_2$ with L_1, L_2 projective of rank 1 and $L_1 \subset S^m P$, $L_2 \subset S^n P$ for some $m, n > 0$. The exact sequence $L \otimes S \to K \to 0$ induces the following exact sequence of locally free sheaves over $X = \mathrm{Proj}\ S(P)$

$$\widetilde{L_1} \otimes S(-m) \oplus \widetilde{L_2} \otimes S(-n) \to O_X \to 0$$

We infer that $c_2(\widetilde{L_1 \otimes S(-m)} \oplus \widetilde{L_2 \otimes S(-n)}) = 0$

Let $a = c_1(L_1)$ and $b = c_1(L_2)$. Then we obtain

$(a-mt)(b-nt) = 0$ where $t = c_1(O_X(1))$. (We should have written π^*a and π^*b. This abuse of the notation is not serious since $\pi^* : A(\mathrm{Spec}\ R) \to A(X)$ is injective). We

have $a+b = 0$ and $ab = 0$. So we get $mnt^2 + (m-n)at = 0$. Taking into account the relation $t^2 - c_1(P)t = 0$ we obtain $(m-n)at + mnc_1(P)t = 0$. It follows that $mnc_1(P) = (n-m)a$ and $m^2n^2c_1^2(P) = (n-m)^2a^2 = 0$ which was to be proved.

2° L is indecomposable. Then $L \subset S^mP$ for some $m > 0$. We have the exact sequence

$$\widetilde{L \otimes S(-m)} \to O_X \to 0$$

It follows that $c_2(L \otimes S(-m)) = 0$. By "the splitting principle" for the purpose of Chern classes calculation we can assume that L is decomposable. So the arguments used in case 1° apply with $n = m$. This finishes the proof of Theorem 8.1.

Remark: There exists ([10]) a smooth affine surface in $\mathbb{C}^4$ with $c_1^2(I/I^2)$ non-torsion where I denotes the ideal of its imbedding. It follows that this surface is not a homogenous set-theoretic complete intersection.

References

[1] A.Altman, S.Kleiman; Introduction to the Grothendieck duality theory, Lecture Notes in Mathematics 146, Springer Verlag 1970.

[2] M.Boratyński; Generating ideals up to radical and systems of parameters of graded rings, J.Alg. 78(1982), 20-24.

[3] M.Boratyński; Generating ideals up to radical and systems of parameters of graded rings II-submitted.

[4] M.P.Murthy; Complete intersections; Conference on commutative algebra, 196-211, Kingston 1975.

[5] A.Grothendieck; La theorie des classes de Chern; Bull.Soc. Math. France 86(1958), 137-154.

[6] R.Hartshorne; Algebraic geometry, Springer Verlag, 1977.

[7] G.Scheja, U.Storch; Quasi-Frobenius Algebra und local vollstandige Durchschnitte, Manuscripta Math. 19(1976), 75-104.

[8] L.Szpiro; Lectures on equations defining space curves, Berlin 1979, TIFR.

[9] G.Valla; On determinantal ideals which are set-theoretic complete intersections, Comp.Math. 42(1981), 3-11

[10] M.G.Murthy; Affine varieties as complete intersections; Intl. Symp. on Algebraic Geometry; Kyoto 1977, pp.231-236.

FONDAZIONE C.I.M.E.
CENTRO INTERNAZIONALE MATEMATICO ESTIVO
INTERNATIONAL MATHEMATICAL SUMMER CENTER

"Harmonic Mappings and Minimal Immersions"

is the subject of the First 1984 C.I.M.E. Session.

The Session, sponsored by the Consiglio Nazionale delle Ricerche and the Ministero della Pubblica Istruzione, will take place under the scientific direction of Prof. ENRICO GIUSTI (Università di Firenze, Italy) at Villa «La Querceta», Montecatini Terme (Pistoia), Italy, ***from June 24 to July 3, 1984.***

Courses

a) ***Harmonic Mapping of Riemannian Manifolds.*** (8 lectures in English).
Prof. Stefan HILDEBRANDT (University of Bonn).

Lecture 1 : Dirichlet's problem for harmonic mappings.
Lecture 2 : Harmonic maps into spheres and into Grassman manifolds.
Lecture 3 : Liouville theorems for harmonic mappings and removable singularities. Application to minimal submanifolds of Euclidean space.
Lecture 4 : Interior regularity of weakly harmonic maps. Differential geometric tools derived from Rauch's estimates.
Lecture 5 : Boundary regularity and boundary estimate for weakly harmonic mappings.
Lecture 6 : A uniqueness theorem for harmonic mappings.
Lecture 7 : Harmonic diffeomorphisms between two-dimensional manifolds. The basic estimates of the determinant due to E. Heinz.
Lecture 8 : The Plateau problem for two-dimensional minimal surfaces in Riemannian manifolds. Regularity of stationary surfaces.

References

1. F. EELLS and L. LEMAIRE, A report on harmonic maps, Bull. London Math. Soc. 10 (1978), 1-68.
2. F. EELLS and L. LEMAIRE, Selected topics in harmonic maps, Regional Conference Series in Math. 50 (1983), ed. by Conf. Board of the Math. Scienc., Amer. Math. Soc..
3. M. GIAQUINTA, Multiple integrals in the calculus of variations and nonlinear elliptic systems, Princeton University Press, Princeton 1983; also: SFB 72, Vorlesungsreihe No. 6, Bonn 1981.
4. M. GIAQUINTA and E. GIUSTI, On the regularity of the minima of variational integrals, Acta, Math. 148 (1982), 31-46.
5. M. GIAQUINTA and S. HILDEBRANDT, A priori estimates for harmonic mappings, J. reine und angew. Math. 336 (1982), 124-164.
6. S. HILDEBRANDT, J. JOST and K.-O. WIDMAN, Harmonic mappings and minimal submanifolds, Invent. Math. 62 (1980), 269-298.
7. S. HILDEBRANDT, H. KAUL and K.-O WIDMAN, An existence theorem for harmonic mappings of Riemannian manifolds, Acta Math. 138 (1977), 1-16.
8. S. HILDEBRANDT, Nonlinear elliptic systems and harmonic mappings, Proc. Beijing Sympos. Diff. Geometry and Diff. Equ. 1980, Science Press, Beijing 1982, Vol. I, 481-615; also: SFB 72, Vorlesungsreihe No. 3, Bonn 1980.
9. S. HILDEBRANDT, Quasilinear elliptic systems in diagonal form, in: «System of nonlinear partial differential equations» edited by F.M. Ball, Proc. of the NATO Advanced Study Institute July 25-August 7, 1982, 173-217, NATO ASI Series C, No. 111, D. Reidel Publ.
10. W. JÄGER and H. KAUL, Uniqueness and stability of harmonic maps and their Jacobi fields, manuscripta math. 28 (1979), 269-291.
11. J. JOST, Harmonic maps between surfaces, Vorlesungsreihe SFB 72 No. 15, Bonn 1983. To appear in Springer Lecture Notes in Math..
12. R. SCHOEN and K. UHLENBECK, A regularity theory for harmonic maps, J. Diff. Geom. 17 (1982), 307-335.

b) ***Harmonic and conformal maps between surfaces.*** (8 lectures in English).
Prof. Jürgen JOST (University of Bonn).

Lecture 1 : Existence theorems for harmonic maps between surfaces.

Lecture 2 : A variational method that produces conformal diffeomorphisms.

Lecture 3 : The Plateau-Douglas problem for minimal surfaces of higher topological structure in Riemannian manifolds.

Lecture 4 : Harmonic diffeomorphisms between surfaces.

Lecture 5 : Harmonic maps and Teichmüller theory.

Lecture 6 : Approximate fundamental solutions and representation formulae for functions on Riemannian manifolds. Almost linear functions.

Lecture 7 : Existence and uniqueness of harmonic maps for a nonpositional curved image.

Lecture 8 : Harmonic coordinates. $C^{2,\alpha}$-estimates for harmonic maps depending only on curvature bounds, injéctivity radii, and dimensions.

References

1. EELLS, J., and L. LEMAIRE, A report on harmonic maps, Bull. London Math. Soc. 10, (1978), 1-68.
2. EELLS, J., and J.H. SAMPSON, Harmonic mappings of Riemannian manifolds, Am. J. Math. 86 (1964), 109-160.
3. JOST, J., Univalency of harmonic mappings between surfaces, J. reine angew. Math 324 (1981), 141-153.
4. JOST, J., Harmonic maps between surfaces, Springer Lecture Notes, to appear, also Preprint SFB 72, Bonn, 1983.
5. JOST, J., Conformal mappings and the Plateau-Douglas problem for minimal surfaces in Riemannian manifolds, Preprint.
6. JOST, J., and H. KARCHER, Geometrische Methoden zur Gewinnung von a-priori-Schranken fur harmonische Abbildungen, man. math. 40 (1982), 27-77.
7. LEMAIRE, L., Applications harmoniques de surfaces Riemanniènnes, J. Diff. Geom. 13 (1978), 51-78.

c) ***Partial Differential Equations Aspects of the Study of Minimal Surfaces.*** (6 lectures in English).
Prof. Leon SIMON (Australian National University, Canberra).

Outline.

Results concerning asymptotic behaviour of minimal surfaces on approach to an isolated singular points, and asymptotic behaviour of entire minimal graphs near.

Most of the results discussed depend on recent new work on approximation of solutions on non-linear equations by solutions of the linearized equation. Some of the results carry over to other geometric problems - for example to the study of the behaviour of harmonic maps near an isolated singular point.

Basic literature

1. D. GILBARG - N.S. TRUDINGER, Second order elliptic equations 2nd Ed.), Springer 1984.
2. E. GIUSTI, Birkhauser book on minimal surfaces (to be published).

Other more specific references will be mentioned during the talks.

Seminars

A number of seminars and special lectures will be offered during the Session.

FONDAZIONE C.I.M.E.
CENTRO INTERNAZIONALE MATEMATICO ESTIVO
INTERNATIONAL MATHEMATICAL SUMMER CENTER

"Schrödinger Operators"

is the subject of the Second 1984 C.I.M.E. Session.

The Session, sponsored by the Consiglio Nazionale delle Ricerche and the Ministero della Pubblica Istruzione, will take place under the scientific direction of Prof. SANDRO GRAFFI (Università di Bologna, Italy) at «Villa Olmo», Como, Italy, ***from August 26 to September 4, 1984.***

Courses

a) ***Bounds on Exponential Decay of Eigenfunctions of Schrödinger Operators.*** (8 lectures in English).
Prof. Shmuel AGMON (Hebrew University, Jerusalem).

1. The Schrödinger operator $P = -\Delta + V$ on R^n and its spectrum. Persson's formula for the bottom of the essential spectrum.
2. On the existence of positive solutions of $(P - \lambda)u = 0$ at a neighborhood of infinity. Spectral implications.
3. Positive super-solutions of $(P - \lambda)u = 0$ at a neighborhood of infinity as majorants of L^2 solutions at infinity.
4. In search of positive solutions. Existence of exponentially decaying positive solutions for a class of Schrödinger equations. Measuring non-isotropic exponential decay by a metric.
5. Applications to exponential decay of eigenfunctions of Schrödinger operators.
6. The N-body Schrödinger operator.
7. The HVZ theorem.
8. Exponential decay of eigenfunctions of the N-body Schrödinger operators.
9. The Carmona-Simon theorem.
10. Remarks on Schrödinger operators with periodic coefficients.

References

a) Special topic references.

1. S. AGMON, Lectures on exponential decay of solutions of second-order elliptic equations: bounds on eigenfunctions of N-body Schrödinger operators, Mathematical Notes No. 29, Princeton University Press, 1982.
2. R. CARMONA and B. SIMON, Pointwise bounds on eigenfunctions and wave packets in N-body quantum systems V: lower bounds and path integrals, Comm. Math. Phys. 80 (1981), 59-98.

b) General background references.

1. S. AGMON, Lectures on elliptic boundary value problems, Van Nostrand, Princeton, 1965.
2. R. REED and B. SIMON, Methods of modern mathematical physics, Volumes II and IV, Academic Press, New York 1975 and 1978.
3. G. STAMPACCHIA, Le problème de Dirichlet pour les équations elliptiques du seconde ordre à coéfficients discontinus, Ann. Inst. Fourier 15 (1965), 189-258.

b) ***Scattering Theory of Multiparticle Schrödinger Operators.*** (8 lectures in English).
Prof. Volker ENSS (Freie Universität Berlin).

The model, expectations from physics. Geometric methods in scattering theory: asymptotic observables and phase space localization, localization properties of free and interacting time evolutions. Asymptotic stability of channel decompositions. Existence and completeness of wave operators. We will treat systems of three (and possibly more) particles interacting with short- and long-range forces.

References

For the (simpler) two-body case the methods are given in

— P.A. PERRY, Scattering theory by the Enss Method, Mathematical Reports Nr. 1 Harwood, New York, to appear.

— V. ENSS, Geometric methods in spectral and scattering theory of Schrödinger operators, in «Rigorous Atomic and Molecular Physics», G. Velo and A.S. Wightman eds., Plenum, New York, 1981.
— V. ENSS, Asymptotic observables on scattering states, Commun. Math. Phys. 89 (1983), 245-268.

Three-body systems are treated in

— V. ENSS, Completeness of three-body quantum scattering, in «Dynamics, Algebras, Processes», P. Blanchard and L. Streit eds., Springer Lecture Notes in Mathematics 1031.
— V. ENSS, Scattering and spectral theory for three particle systems, in «Proceed. Int. Conf. on Differential Equations», Birmingham AL, 1983, I. Knowles and R. Lewis eds., North-Holland, to appear.
— V. ENSS, Three-body Coulomb scattering theory, in «Proceed. Int. Conf. on Operator Algebras, Ideals, and their Applications in Theoretical Physics», Leipzig 1983, Teubner, Leipzig to appear.

c) ***Some Aspects of the Theory of Schrödinger Operators.*** (8 lectures in English).
Prof. Barry SIMON (California Institute of Technology).

1. Self-adjointness, properties of eigenfunctions, and all that.
2. Bound state problems.
3. The basic notions of scattering theory.
4. The N-body Mourre estimates.
5. An introduction to the theory of stochastic Jacobi matrices.

References

1. M. REED and B. SIMON, Methods of Modern Mathematical Physics, Volumes 1-4, Academic Press.
2. Schrödinger semigroups, Bull. AMS 7 (1982), 447-526.
3. R. FROESE and I. HERBST, Duke Math. J. 49 (1982), 1075-1085.
4. I. SIGAL, Comm. Math. Phys. 85 (1982), 309-324.
5. J. AVRON and B. SIMON, Duke Math. J. 50 (1983), 369-391.
6. F. DELYON, H. KUNZ and B. SOUILLARD, to appear in J. Phys. A.

Seminars

The following seminars will be offered:

— G. JONA-LASINIO (Università di Roma I, «La Sapienza»)
«Stochastic Processes and Quantum Mechanics».
— J. BELLISSARD (Université de Provence et C.N.R.S., GPT II, Marseille-Luminy)
«Stability and Instability in Quantum Mechanics».
— K. YAJIMA (The University of Tokyo)
«The Semiclassical Limit by the Methods of the Fourier Integral Operators».
— E. MOURRE (C.N.R.S., CPT II, Marseille-Luminy)
«Propagation estimates in Scattering Theory».

Applications

Those who wish to attend the Session should fill in an application form and mail it to the Director of the Fondazione C.I.M.E. at the address below, ***not later than June 15, 1984***.

An important consideration in the acceptance of applications is the scientific relevance of the Session to the field of interest of the applicant.

Applicants are requested, therefore, to submit, along with their application, a scientific curriculum and a letter of recommendation.

Participation will only be allowed to persons who have applied in due time and have had their application accepted.

FONDAZIONE C.I.M.E.
CENTRO INTERNAZIONALE MATEMATICO ESTIVO
INTERNATIONAL MATHEMATICAL SUMMER CENTER

"Buildings and the Geometry of Diagrams"

is the subject of the Third 1984 C.I.M.E. Session.

The Session, sponsored by the Consiglio Nazionale delle Ricerche and the Ministero della Pubblica Istruzione, will take place under the scientific direction of Prof. LUIGI A. ROSATI (Università di Firenze, Italy) at «Villa Olmo», Como, Italy, ***from August 26 to September 4, 1984.***

Courses

a) ***The Geometry of the Finite Simple Groups.*** (8 lectures in English).
Prof. Francis BUEKENHOUT (Université Libre de Bruxelles).

1. Transitive permutation groups, their graphs and permutation characters. The lattice of subgroups of a group.
2. Maximal subgroups and their orbit structure.
3. Geometries for and from groups, using graphs, orbits and lattices.
4. Representations in projective spaces inspired by quadrics.
5. Classifications of geometries for groups.
6. Classifications of geometries for groups emphasizing a prime characteristic.
7. Classifications of diagram geometries.
8. Classifications of point-line geometries.

References

1. F. BUEKENHOUT, Diagram for geometries and groups. J. Comb. Th. (A) 27 (1979), 121-151.
2. F. BUEKENHOUT, Diagram geometries of sporadic groups. (Preprint).
3. J. TITS, Buildings and Buekenhout Geometries. Finite Simple Groups II, ed. M. Collins, Academic Press, New York 1981, 309-320.
4. D. GORENSTEIN, Finite Simple Groups, Plenum Press, New York 1982.
5. S.A. SYSKIN, Abstract properties of the simple sporadic groups. Russian Math. Surveys 35:5 (1980), 209-246.
6. J.H. CONWAY, R.T. CURTIS, S.P. NORTON, R.A. PARKER, Atlas of Finite groups (unpublished).
7. J. McKAY, The simple groups, G, $|G| < 10^6$. Character tables. Comm. Algebra 7 (1979), 1407-1445.
8. J. FISHER, J. McKAY, The nonabelian simple groups G, $|G| < 10^6$. Maximal subgroups. Math. Comp. 32 (1978), 1293-1302.
9. A.A. IVANOV, M.H. KLIN, I.A. FARADJEV, The primitive representations of the nonabelian simple groups of order less than 10^6. Part I. The Institute for System Studies, Moscow 1982 (Preprint).

b) ***General Polygons and Building-Like Geometries.*** (8 lectures in English).
Prof. William M. KANTOR (University of Oregon, USA).

There will be two lectures on generalized polygons. The remaining six lectures will be split between (1) Tits' local characterization of buildings, and (2) examples of finite morphic images of buildings and their group theoretic uses.

Partial Bibliography

1. M. ASCHBACHER, Flag structures on Tits geometries (to appear).
2. J.L. BELL and A.B. SLOMSON, Models and ultraproducts, North-Holland 1969.
3. W.M. KANTOR, Some geometries that are almost building, Europ. J. Comb. 2 (1981), 239-247.
4. W.M. KANTOR, Some exceptional 2-adic building (to appear in J. Alg.).
5. W.M. KANTOR, Some locally finite flag-transitive buildings (to appear in Europ. J. Comb.).
6. P. KOHLER, T. MEIXNER and M. WESTER, Triangle groups (to appear).
7. P. KOHLER, The 2-adic affine building of type A_2 and its finite projections (to appear in JCT (A)).
8. R. NILES, BN-pairs and finite groups with parabolic-type subgroups, J. Alg. 75 (1982), 484-494.
9. M. RONAN, Triangle geometries (to appear).
10. F. TIMMESFELD, Tits geometries and parabolic systems in finite groups (to appear in Math. Z.).
11. F. TIMMESFELD, Tits geometries and parabolic systems of rank 3 (to appear).

12. J. TITS, Buildings of spherical type and finite BN-pairs, Springer Lecture Notes 386, 1974.
13. J. TITS, A local approach to buildings, pp. 519-574 in «The Geometric Vein. The Coxeter Festschrift», Springer 1981.
14. D.G. HIGMAN, Invariant relations, coherent configurations and generalized polygons, pp. 347-363 in Combinatorics, Reidel 1975.
15. W.M. KANTOR, Generalized quadrangles associated with $G_2(q)$. JCT (A) 29 (1980), 212-219.
16. J.A. THAS, Combinatorics of generalized quadrangles: a survey. Ann. of Disc. Math. 14 (1982), 57-76.
17. J. TITS, Classification of buildings of spherical type and Moufang polygons: a survey, pp. 229-246 in Atti Coll. Comb., Rome 1976.

c) ***Geometries with Coxeter Diagrams and Buildings.*** (8 lectures in English).
Prof. Jacques TITS (Collège de France).

1. ***Basic notions.*** Definitions and examples of geometries with Coxeter diagrams and building (ref. T1, T2, T4). Relations between the two notions: statement of known results (ref. T5).

2. ***Classification of buildings and geometries of spherical type and rank ≥ 3.*** Statement of known results (ref. T1, T5, BC). The case of weak buildings. Remark on the rank 2 case: the Moufang condition (ref. T1, T2).

3. ***Classification of buildings and geometries of affine type and rank ≥ 4.*** Classification, with proof (possibly only in typical cases), of the buildings in question in the title. They turn out to be just those which arise in the study of classical and algebraic simple groups over fields (or skew fields) with discrete valuations (ref. BT1, BT2, T3). No reference can be given for the classification itself, which is unpublished as yet. Application to geometries - in particular finite geometries - by means of results of A. Borel and G.A. Margulis on arithmetic groups (ref. B, M, T6).

4. ***Free geometries with Coxeter diagrams.*** For «most» Coxeter diagrams, the corresponding category of geometries admits «free constructions» analogous to the construction of free projective planes starting from partial planes. Necessary and sufficient conditions for the existence of such constructions (i.e. for the non-existence of obstructions to them) will be given, with proofs. Applications to buildings and BN-pairs. (These results are unpublished).

References

B A. BOREL, Some properties of adele groups attached to algebraic groups, Bull. A.M.S. 67 (1961), 583-585.

BC A.E. BROUWER and A. COHEN, Some remarks on Tits' geometries, Indagationes Math., to appear.

BT1 F. BRUHAT et J. TITS, Groupes algébriques simples sur un corps local, Proc. Conf. on Local Fields (Driebergen 1966), Springer 1967, 23-26.

BT2 F. BRUHAT et J. TITS, Groupes réductifs sur un corps local, I. Données radicielles valuées, Publ. Math. I.H.E.S. 41 (1972), 5-252.

M G.A. MARGULIS, Diskretnye gruppy dvizenii mnogoobrazii nepolozitel'noi krivizny (= Discrete groups of motions of varieties with nonpositive curvature), Proc. Intern. Congress Math. Vancouver 1974, vol. 2, 21-34, (1975).

T1 J. TITS, Buildings of spherical type and finite BN-pairs, Springer Lecture Notes No. 386, 1974.

T2 J. TITS, Classification of buildings of spherical type and Moufang polygons: A survey, Atti Coll. Intern. Teorie Combinatorie, Accad. Naz. dei Lincei, Roma, 1973, 229-246 (1976).

T3 J. TITS, Reductive groups over local fields, Proc. Symp. Pure Math. 33 (1979), Part 1, 29-69.

T4 J. TITS, Buildings and Buekenhout geometries, *in* «Finite Simple Groups II», ed. M. Collins, Academic Press, New York 1981, 309-320.

T5 J. TITS, A local approach to buildings, *in* «The geometric vein, the Coxeter Festschrift», Springer 1981, 519-547.

T6 J. TITS, Travaux de Margulis sur les sous-groupes discrets de groupes de Lie, Sém. Bourbaki, février 1976, exposé n. 482, Springer Lecture Notes no. 567, 1977, 174-190.

Seminars

A number of seminars and special lectures will be offered during the Session.

Applications

Those who wish to attend the Session should fill in an application form and mail it to the Director of the Fondazione C.I.M.E. at the address below, ***not later than June 15, 1984***.

An important consideration in the acceptance of applications is the scientific relevance of the Session to the field of interest of the applicant.

Applicants are requested, therefore, to submit, along with their application, a scientific curriculum and a letter of recommendation.

Participation will only be allowed to persons who have applied in due time and have had their application accepted.

LIST OF C.I.M.E. SEMINARS

		Publisher
1954 -	1. Analisi funzionale	C.I.M.E.
	2. Quadratura delle superficie e questioni connesse	"
	3. Equazioni differenziali non lineari	"
1955 -	4. Teorema di Riemann-Roch e questioni connesse	"
	5. Teoria dei numeri	"
	6. Topologia	"
	7. Teorie non linearizzate in elasticità, idrodinamica, aerodinamica	"
	8. Geometria proiettivo-differenziale	"
1956 -	9. Equazioni alle derivate parziali a caratteristiche reali	"
	10. Propagazione delle onde elettromagnetiche	"
	11. Teoria della funzioni di più variabili complesse e delle funzioni automorfe	"
1957 -	12. Geometria aritmetica e algebrica (2 vol.)	"
	13. Integrali singolari e questioni connesse	"
	14. Teoria della turbolenza (2 vol.)	"
1958 -	15. Vedute e problemi attuali in relatività generale	"
	16. Problemi di geometria differenziale in grande	"
	17. Il principio di minimo e le sue applicazioni alle equazioni funzionali	"
1959 -	18. Induzione e statistica	"
	19. Teoria algebrica dei meccanismi automatici (2 vol.)	"
	20. Gruppi, anelli di Lie e teoria della coomologia	"
1960 -	21. Sistemi dinamici e teoremi ergodici	"
	22. Forme differenziali e loro integrali	"
1961 -	23. Geometria del calcolo delle variazioni (2 vol.)	"
	24. Teoria delle distribuzioni	"
	25. Onde superficiali	"
1962 -	26. Topologia differenziale	"
	27. Autovalori e autosoluzioni	"
	28. Magnetofluidodinamica	"
1963 -	29. Equazioni differenziali astratte	"
	30. Funzioni e varietà complesse	"
	31. Proprietà di media e teoremi di confronto in Fisica Matematica	"

1964 - 32. Relatività generale — C.I.M.E.
33. Dinamica dei gas rarefatti — "
34. Alcune questioni di analisi numerica — "
35. Equazioni differenziali non lineari — "

1965 - 36. Non-linear continuum theories — "
37. Some aspects of ring theory — "
38. Mathematical optimization in economics — "

1966 - 39. Calculus of variations — Ed. Cremonese, Firenze
40. Economia matematica — "
41. Classi caratteristiche e questioni connesse — "
42. Some aspects of diffusion theory — "

1967 - 43. Modern questions of celestial mechanics — "
44. Numerical analysis of partial differential equations — "
45. Geometry of homogeneous bounded domains — "

1968 - 46. Controllability and observability — "
47. Pseudo-differential operators — "
48. Aspects of mathematical logic — "

1969 - 49. Potential theory — "
50. Non-linear continuum theories in mechanics and physics and their applications — "
51. Questions of algebraic varieties — "

1970 - 52. Relativistic fluid dynamics — "
53. Theory of group representations and Fourier analysis — "
54. Functional equations and inequalities — "
55. Problems in non-linear analysis — "

1971 - 56. Stereodynamics — "
57. Constructive aspects of functional analysis (2 vol.) — "
58. Categories and commutative algebra — "

1972 - 59. Non-linear mechanics — "
60. Finite geometric structures and their applications — "
61. Geometric measure theory and minimal surfaces — "

1973 - 62. Complex analysis — "
63. New variational techniques in mathematical physics — "
64. Spectral analysis — "

1974 - 65. Stability problems		Ed. Cremonese, Firenze
66. Singularities of analytic spaces		"
67. Eigenvalues of non linear problems		"
1975 - 68. Theoretical computer sciences		"
69. Model theory and applications		"
70. Differential operators and manifolds		"
1976 - 71. Statistical Mechanics		Ed. Liguori, Napoli
72. Hyperbolicity		"
73. Differential topology		"
1977 - 74. Materials with memory		"
75. Pseudodifferential operators with applications		"
76. Algebraic surfaces		"
1978 - 77. Stochastic differential equations		"
78. Dynamical systems		Ed. Liguori, Napoli and Birkhäuser Verlag
1979 - 79. Recursion theory and computational complexity		Ed. Liguori, Napoli
80. Mathematics of biology		"
1980 - 81. Wave propagation		"
82. Harmonic analysis and group representations		"
83. Matroid theory and its applications		"
1981 - 84. Kinetic Theories and the Boltzmann Equation	(LNM 1048)	Springer-Verlag
85. Algebraic Threefolds	(LNM 947)	"
86. Nonlinear Filtering and Stochastic Control	(LNM 972)	"
1982 - 87. Invariant Theory	(LNM 996)	"
88. Thermodynamics and Constitutive Equations	to appear	"
89. Fluid Dynamics	(LNM 1047)	"
1983 - 90. Complete Intersections	(LNM 1092)	"
91. Bifurcation Theory and Applications	(LNM 1057)	"
92. Numerical Methods in Fluid Dynamics	to appear	"

Note: Volumes 1 to 38 are out of print. A few copies of volumes 23,28,31,32,33,34,36,38 are available on request from C.I.M.E.